Lecture Notes in Mathematics 1568

Ferenc Weisz

Martingale Hardy Spaces and their Applications in Fourier Analysis

Springer-Verlag
Berlin Heidelberg New York
London Paris Tokyo
Hong Kong Barcelona
Budapest

Author

Ferenc Weisz
Department of Numerical Analysis
Eötvös L. University
Bogdánfy u.10/B
H-1117 Budapest, Hungary

Mathematics Subject Classification (1991): 60G42, 60G46, 42B30, 42C10, 60G48, 46E30, 42A20, 42A50

ISBN 3-540-57623-1 Springer-Verlag Berlin Heidelberg New York
ISBN 0-387-57623-1 Springer-Verlag New York Berlin Heidelberg

Library of Congress Cataloging-in-Publication Data. Weisz, Ferenc, 1964- Martingale Hardy spaces and their applications in Fourier analysis / Ferenc Weisz. p. cm. – (Lecture notes in mathematics; 1568) Includes bibliographical references and index. ISBN 0-387-57623-1
1. Martingales (Mathematics) 2. Hardy spaces. 3. Fourier analysis. I. Title II. Series: Lecture notes in mathematics (Springer-Verlag); 1568. QA3.L28 no. 1568 [QA274.5] 510 s–dc20 [519.2'87] 93-49415

Printed in Germany

SPIN: 10078770 2146/3140-543210 - Printed on acid-free paper

TABLE OF CONTENTS

PREFACE

The history of martingale theory goes back to the early fifties when Doob [61] pointed out the connection between martingales and analytic functions. On the basis of Burkholder's scientific achievements ([17], [20], [24]) the theory of continuous parameter martingales can perfectly well be applied in complex analysis and in the theory of classical Hardy spaces. This connection is the main point of Durrett's book [64]. Attention was also drawn to martingale Hardy spaces when Burkholder and Gundy [27] proved the inequality named after them ever since, which states that the L_p norms of the maximal function and the quadratic variation of a one-parameter martingale are equivalent for $1 < p < \infty$. Some years later this result was extended to $p = 1$ by Davis [55]. In 1973 the dual of the one-parameter martingale Hardy space generated by the maximal function was characterized by Garsia [82] and Herz [94] as the space of the functions of bounded mean oscillation (BMO). The beginning of the study of two-parameter martingales can be dated to 1970 when Doob's inequality was shown by Cairoli [29]. The Burkholder-Gundy inequality was also proved later for two-parameter martingales by Metraux [125].

While the theory of the martingale Hardy spaces is discussed briefly, their application in Fourier analysis has not yet been presented at all in books on martingale theory (e.g. Neveu [142], Garsia [82], Dellacherie, Meyer [58]). In the only book on two-parameter martingales (see Imkeller [105]) the theory of the martingale Hardy spaces is not studied in great detail. Just a few papers have been published to date on tree martingales. This book is intended to fill this gap, giving, on the one hand, an exhaustive study of one- and two- (discrete-) parameter martingale Hardy spaces, and on the other hand, demonstrating some of their applications in Fourier analysis. Moreover, several new and still unpublished results are presented as well.

The methods of proof for one and two parameters are entirely different; in most cases the theorems stated for two parameters are much more difficult to verify. A method that can be applied both in the one- and in the two-parameter cases, the so-called atomic decomposition method was improved by the author. Nevertheless, in a simpler form, it has already appeared in the writings on this field (see e.g. Herz [94], Bernard, Maisonneuve [10]). With the help of this method a new, common construction of the theory of the one- and two-parameter martingale Hardy spaces is presented. Most of the one- and two-parameter theorems as well as quite a few of the theorems about tree martingales and of the applications can be traced back to this method.

The book is structured as follows. In Chapter 1 the rudiments are summarized. In Chapter 2 and 3 one- and two-parameter martingales are studied in a similar structural way. The method of the atomic decomposition is described in Sections 2.1 and 3.1. Then some martingale inequalities follow; the Burkholder-Gundy inequality is proved in both cases while Davis's inequality is shown for arbitrary one-parameter and for special (regular and strong) two-parameter martingales. The latter is still unknown for general two-parameter martingales. In Sections 2.3 and 3.3 some duality theorems are verified. The equivalence between the BMO spaces is examined and the John-Nirenberg theorem is proved. Furthermore, in these two chapters martingale transforms are considered. The first three sections of Chapter

2 most resemble Garsia's book ([82]), however, their structure is much simpler and several new theorems are given here.

In Chapter 4 tree martingales are dealt with. In Section 4.1, amongst other things, the Burkholder-Gundy inequality together with one relative to martingale transforms are shown. As an important application, one of the most difficult theorems in Fourier analysis, Carleson's theorem is obtained; it asserts that the one-parameter Walsh-Fourier series and its generalization, the so-called Vilenkin-Fourier series of a function in L_p $(1 < p < \infty)$, converge almost everywhere to the function itself. In addition to this, the foregoing convergence in L_p norm is verified as well.

In Chapter 5 the still young theory of interpolation is applied and the interpolation spaces of the martingale Hardy spaces are characterized. These results, which are very interesting in themselves, will be used in the last chapter.

In Chapter 6 Hardy type inequalities are presented for Walsh-Fourier and Vilenkin-Fourier coefficients. Applying these, one obtains Carleson's theorem for two-parameter systems and for the functions of L_p $(1 < p)$ with monotone Vilenkin-Fourier coefficients.

Since this book deals with the relation between analysis and probability theory, I have tried to write it in such a way that it is accessible to everyone familiar with the fundamentals of any of the two fields.

First of all, I would like to express my thanks to my dear colleague, Prof. Ferenc Schipp who supported me in the very first steps of my career, drew my attention to this field; in addition, I am thankful for his valuable ideas to improve this work. I am very grateful to Prof. Peter Imkeller who read through the manuscript and made many suggestions. I would like to thank the "Deutscher Akademischer Austauschdienst" with whose financial support I had the opportunity to bring this book to its final version at the Ludwig-Maximilians-Universität, München. My thanks are due to the Hungarian Scientific Research Fund, as well as to the Hungarian Scientific Foundation for supporting my research. Above all, I am particularly indebted to my wife for reading through the manuscript and even more for her patience and love.

I would like to ask the readers to overlook the quality of the English in my book.

CHAPTER 1

PRELIMINARIES AND NOTATIONS

1.1. MARTINGALES

Let us denote the set of integers, the set of non-negative integers, the set of positive integers, the set of real numbers and the set of complex numbers by $\mathbf{Z}$, $\mathbf{N}$, $\mathbf{P}$, $\mathbf{R}$ and $\mathbf{C}$, respectively. For a set $\mathbf{X} \neq \emptyset$ let $\mathbf{X}^1 := \mathbf{X}$ and $\mathbf{X}^2$ be its Descartes product $\mathbf{X} \times \mathbf{X}$ taken with itself. To denote the sequences the notation $z = (z_\alpha, \alpha \in A)$ will be used where A is an arbitrary index set. Thus the one- and two-parameter sequences will be denoted by $z = (z_n, n \in \mathbf{N})$ and $z = (z_n, n \in \mathbf{N}^2)$, respectively.

A pair of non-negative integers from $\mathbf{N}^2$ is denoted by (n, m), or, simply by n depending on the text environment. We write n_1 and n_2 for the first and the second coordinate of a pair $n \in \mathbf{N}^2$, respectively. Let us introduce the following partial ordering on $\mathbf{N}^2$: for $n = (n_1, n_2)$, $m = (m_1, m_2) \in \mathbf{N}^2$ set $n \leq m$ if $n_1 \leq m_1$ and $n_2 \leq m_2$. We say that $n < m$ if $n \leq m$ and $n \neq m$ $(n, m \in \mathbf{N}^2)$. Of course, $n < \infty$ for all $n \in \mathbf{N}^2$. Moreover, $n \ll m$ means that both the inequalities $n_1 < m_1$ and $n_2 < m_2$ hold. For $n = (n_1, n_2) \in \mathbf{N}^2$ we set $n - 1 := (n_1 - 1, n_2 - 1)$.

For two arbitrary sets $H, G \subset \mathbf{N}^2$ consisting of *incomparable number-pairs* (i.e. if $n, m \in H$ or G then neither of the inequalities $n \leq m$ and $m \leq n$ hold) we write $H \ll G$ resp. $H \leq G$ if for all $n \in G$ there exists $m \in H$, such that $m \ll n$ resp. $m \leq n$. An elementary computation shows that the relation $\leq$ is reflexive and transitive. If the elements of H and G are incomparable then $H \leq G$ and $G \leq H$ imply $G = H$. Denote by $\inf H$ the set of the number-pairs $m \in H$ for which there does not exist any $n \in H$ such that $n < m$ $(H \subset \mathbf{N}^2)$. We shall use the convention $\inf \emptyset = \infty$. For two arbitrary sets $H, G \subset \mathbf{N}^2$ we write $H \ll G$ resp. $H \leq G$ if and only if $\inf H \ll \inf G$ resp. $\inf H \leq \inf G$. It is easy to see that for every $H, G \subset \mathbf{N}^2$, $\inf H \leq H$ and that $H \subset G$ implies $G \leq H$.

Let $(\Omega, \mathcal{A}, P)$ be a probability space and let $\mathcal{F} = (\mathcal{F}_n, n \in \mathbf{N}^j)$ $(j = 1, 2)$ be a non-decreasing sequence of σ-algebras with respect to the complete ordering on $\mathbf{N}$ or to the partial ordering on $\mathbf{N}^2$. The σ-algebra generated by an arbitrary set system $\mathcal{H}$ will be denoted by $\sigma(\mathcal{H})$. Introduce the following σ-algebras:

$$\mathcal{F}_\infty := \sigma\Big(\bigcup_{n \in \mathbf{N}^j} \mathcal{F}_n\Big) \qquad (j = 1, 2),$$

$$\mathcal{F}_{n_1, \infty} := \sigma\Big(\bigcup_{k=0}^{\infty} \mathcal{F}_{n_1, k}\Big), \quad \mathcal{F}_{\infty, n_2} := \sigma\Big(\bigcup_{k=0}^{\infty} \mathcal{F}_{k, n_2}\Big) \qquad (n \in \mathbf{N}^2).$$

For the sake of simplicity, suppose that $\mathcal{F}_\infty = \mathcal{A}$ and let $\mathcal{F}_{-1} := \mathcal{F}_0$, $\mathcal{F}_{-1,-1} := \mathcal{F}_{0,0}$, $\mathcal{F}_{-1,i} := \mathcal{F}_{0,i}$ and $\mathcal{F}_{i,-1} := \mathcal{F}_{i,0}$ $(i \in \mathbf{N})$.

The expectation operator and the conditional expectation operators relative to $\mathcal{F}_n$ $(n \in \mathbf{N}^j \cup \{\infty\}, j = 1, 2)$, $\mathcal{F}_{n_1, \infty}$ and $\mathcal{F}_{\infty, n_2}$ $(n \in \mathbf{N}^2)$ are denoted by E, E_n, $E_{n_1, \infty}$ and E_{∞, n_2}, respectively. We briefly write L_p instead of the real or complex $L_p(\Omega, \mathcal{A}, P)$ space while the norm (or quasinorm) of this space is defined by $\|f\|_p := (E|f|^p)^{1/p}$ $(0 < p \leq \infty)$. The space l_p consists of those sequences $b = (b_n, n \in \mathbf{Z}^j$ [or $\mathbf{N}^j$]) of real or complex numbers for which

$$\|b\|_{l_p} := (\sum_{n \in \mathbf{Z}^j} |b_n|^p)^{1/p} < \infty.$$

For simplicity, we assume that for a function $f \in L_1$ we have in the one-parameter case $E_0 f = 0$ and in the two-parameter case $E_n f = 0$ if $n_1 = 0$ or $n_2 = 0$. The space $\mathcal{M}(\mathcal{B})$ is the set of $\mathcal{B}$-measurable functions for an arbitrary σ-algebra $\mathcal{B}$.

In the two-parameter case we also suppose the *condition* F_4 introduced by Cairoli and Walsh [31] for the stochastic basis $\mathcal{F}$: the σ-algebras $\mathcal{F}_{n_1,\infty}$ and $\mathcal{F}_{\infty,n_2}$ are *conditionally independent* with respect to the σ-algebra $\mathcal{F}_n$, i.e. every bounded function $f \in \mathcal{M}(\mathcal{F}_{n_1,\infty})$ and $g \in \mathcal{M}(\mathcal{F}_{\infty,n_2})$ satisfy the equation

$$(F_4) \qquad E_n(fg) = E_n f E_n g \qquad (n \in \mathbf{N}^2).$$

An equivalent condition to this (see Cairoli, Walsh [31] p.114) is the following: for all $n \in \mathbf{N}^2$ and for all bounded function f

$$(F_4) \qquad E_n f = E_{n_1,\infty}(E_{\infty,n_2} f) = E_{\infty,n_2}(E_{n_1,\infty} f).$$

Of course, the equalities between random variables mean *P-almost everywhere* equalities in the whole book.

An integrable sequence $f = (f_n, n \in \mathbf{N}^j)$ $(j = 1, 2)$ is said to be a *martingale* if

(i) it is *adapted*, i.e. f_n is $\mathcal{F}_n$ measurable for all $n \in \mathbf{N}^j$

(ii) $E_n f_m = f_n$ for all $n \leq m$.

If $E_n f_m \geq (\leq) f_n$ for every $n \leq m$ then f is called *sub- (super-) martingale*. For simplicity, we always suppose that for a martingale f we have in the one-parameter case $f_0 = 0$ and in the two-parameter case $f_n = 0$ if $n_1 = 0$ or $n_2 = 0$. Of course, the theorems that are to be proved later are true with a slightly modification without this condition, too.

The stochastic basis $\mathcal{F}$ is said to be *regular* if there exists a number $R > 0$ such that

$$f_n \leq R f_{n-1} \qquad (n \in \mathbf{N})$$

(for one index) and

$$f_{n_1,n_2} \leq R f_{n_1-1,n_2}, \quad f_{n_1,n_2} \leq R f_{n_1,n_2-1} \qquad (n \in \mathbf{N}^2)$$

(for two indices) hold for all non-negative martingales $(f_n, n \in \mathbf{N}^j)$ $(j = 1, 2)$. Some examples for a regular sequence of σ-algebras are given in Section 1.2.

The martingale $f = (f_n, n \in \mathbf{N}^j)$ $(j = 1, 2)$ is said to be *L_p-bounded* $(0 < p \leq \infty)$ if $f_n \in L_p$ $(n \in \mathbf{N}^j)$ and

$$\|f\|_p := \sup_{n \in \mathbf{N}^j} \|f_n\|_p < \infty.$$

If $f \in L_1$ then it is easy to show that the sequence $\tilde{f} = (E_n f, n \in \mathbb{N}^j)$ $(j = 1, 2)$ is a martingale. Moreover, if $1 \le p < \infty$ and $f \in L_p$ then $\tilde{f}$ is L_p-bounded and

$$\lim_{n \to \infty} \|E_n f - f\|_p = 0, \tag{1.1}$$

consequently, $\|\tilde{f}\|_p = \|f\|_p$ (see Neveu [142]). In this book both the a.e. and the L_p limit of a two-parameter sequence $(z_n, n \in \mathbb{N}^2)$ are taken in *Pringsheim sense* (see e.g. Móricz [137]), namely, the a.e. resp. L_p limit of the sequence (z_n) is z if for all $\epsilon > 0$ there exists an index $N \in \mathbb{N}^2$ such that $|z_n - z| < \epsilon$ resp. $\|z_n - z\|_p < \epsilon$ whenever $n \ge N$ $(0 < p \le \infty)$. The converse of the lattest proposition holds also if $1 < p < \infty$ (see Neveu [142]): for an arbitrary martingale $f = (f_n, n \in \mathbb{N}^j)$ $(j = 1, 2)$ there exists a function $g \in L_p$ for which $f_n = E_n g$ if and only if f is L_p-bounded. If $p = 1$ then there exists a function $g \in L_1$ of the preceding type if and only if f is *uniformly integrable* (Neveu [142]), namely, if

$$\lim_{y \to \infty} \sup_{n \in \mathbb{N}^j} \int_{\{|f_n| > y\}} |f_n| \, dP = 0.$$

This representation of uniformly integrable martingales is valid for martingales indexed with a directed index set, too (see Neveu [142]).

Note that in case $f \in L_p$ $(1 < p < \infty)$ besides the L_p convergence in (1.1) the conditional expectation $E_n f$ converges also a.e. to f $(n \in \mathbb{N}^j, j = 1, 2)$. If $p = 1$ then in the one-parameter case every L_1-bounded martingale converges a.e. (but, of course, not necessarily in L_1 norm), and in the two-parameter case every L_1-bounded martingale converges in probability (Neveu [142]).

Thus the map $f \mapsto \tilde{f} := (E_n f, n \in \mathbb{N}^j)$ $(j = 1, 2)$ is isometric from L_p onto the space of L_p-bounded martingales when $1 < p < \infty$. Consequently, these two spaces can be identified with each other. Similarly, the L_1 space can be identified with the space of uniformly integrable martingales. For this reason a function $f \in L_1$ and the corresponding martingale $(E_n f, n \in \mathbb{N}^j)$ $(j = 1, 2)$ will be denoted by the same symbol f.

The concept of a *stopping time* will be of primary importance in the book. In the one-parameter case a map $\nu : \Omega \longrightarrow \mathbb{N} \cup \{\infty\}$ is called a stopping time relative to $(\mathcal{F}_n, n \in \mathbb{N})$ if

$$\{\omega \in \Omega : \nu(\omega) = n\} =: \{\nu = n\} \in \mathcal{F}_n.$$

It is well known that the last condition is equivalent to the conditions

$$\{\nu \le n\} \in \mathcal{F}_n \qquad (n \in \mathbb{N})$$

and

$$\{\nu \ge n\} \in \mathcal{F}_{n-1} \qquad (n \in \mathbb{N}).$$

Keeping these properties we generalize the concept of a stopping time for two parameters (see Weisz [199]). A function ν which maps Ω into the set of subspaces of $\mathbb{N}^2 \cup \{\infty\}$ is said to be a stopping time relative to $(\mathcal{F}_n, n \in \mathbb{N}^2)$ if the elements of $\nu(\omega)$ are incomparable for all $\omega \in \Omega$, furthermore, if for an arbitrary $n \in \mathbb{N}^2$

$$\{\omega \in \Omega : n \in \nu(\omega)\} =: \{n \in \nu\} \in \mathcal{F}_n.$$

The set of stopping times will be denoted by $\mathcal{T}_1$ in the one-parameter case and by $\mathcal{T}_2$ in the two-parameter case, or – if it does not lead to misunderstanding – by $\mathcal{T}$ in both cases. We get immediately that if the elements of $\nu(\omega)$ $(\omega \in \Omega)$ are incomparable then $\nu \in \mathcal{T}_2$ if and only if

$$\{\nu \leq n\} \in \mathcal{F}_n \qquad (n \in \mathbf{N}^2),$$

since

$$\{\nu \leq n\} = \bigcup_{m \leq n} \{m \in \nu\} \qquad (n \in \mathbf{N}^2).$$

Similarly, if $\nu \in \mathcal{T}_2$ is a stopping time then

$$\{\nu \not\ll n\} \in \mathcal{F}_{n-1} \qquad (n \in \mathbf{N}^2). \tag{1.2}$$

The converse of this last statement follows from the equality

$$\begin{aligned}&\{(n_1,n_2) \in \nu\}\\ &\quad = \{\nu \ll (n_1+1, n_2+1)\} \cap \{\nu \not\ll (n_1+1, n_2)\} \cap \{\nu \not\ll (n_1, n_2+1)\}\end{aligned}$$

where $n \in \mathbf{N}^2$ is arbitrary. If $\nu, \mu \in \mathcal{T}_2$ then $\nu \wedge \mu, \nu \vee \mu \in \mathcal{T}_2$, too, because for all $n \in \mathbf{N}^2$

$$\{\nu \wedge \mu \leq n\} = \{\nu \leq n\} \cap \{\mu \leq n\}, \quad \{\nu \vee \mu \leq n\} = \{\nu \leq n\} \cup \{\mu \leq n\}.$$

Our definition of stopping times is similar to the one introduced by Schipp [162] for a tree stochastic basis. Moreover, we can define a probability domain D with $n \in D$ if and only if $\nu \not\ll n$. These domains are called stopping domains and were introduced by Walsh [187]. Other types of stopping times (called stopping lines) are investigated in Wong, Zakai [202], Merzbach [123] and Imkeller [105] for continuous parameter.

The *hitting time* ν_H of a Borel set H relative to a sequence $(f_n, n \in \mathbf{N}^j)$ of adapted functions is defined by

$$\nu_H(\omega) := \inf\{n \in \mathbf{N}^j : f_n(\omega) \in H\} \qquad (\omega \in \Omega).$$

Since the elements of $\nu_H(\omega)$ are incomparable and

$$\{n \in \nu_H\} = \Big(\bigcap_{m<n} \{f_m \notin H\}\Big) \cap \{f_n \in H\} \in \mathcal{F}_n,$$

ν_H is a stopping time relative to $(\mathcal{F}_n, n \in \mathbf{N}^j)$. Note that if $m \not\geq \nu_H(\omega)$ then $f_m(\omega) \notin H$.

The *maximal function* of a martingale $f = (f_n, n \in \mathbf{N}^j)$ is denoted by

$$f^* := \sup_{m \in \mathbf{N}^j} |f_m|$$

and let us agree on the notation

$$f_n^* := \sup_{m \leq n} |f_m| \qquad (n \in \mathbf{N}^j).$$

We define the *martingale differences* as follows. For one parameter let

$$d_0 f := 0, \quad d_n f := f_n - f_{n-1} \qquad (n \in \mathbf{P})$$

and, for two parameters, let

$$d_m f := 0 \quad \text{if} \quad m_1 = 0 \quad \text{or} \quad m_2 = 0$$

or else

$$d_m f := f_{m_1,m_2} - f_{m_1-1,m_2} - f_{m_1,m_2-1} + f_{m_1-1,m_2-1}.$$

It is easy to show that $(d_n f, n \in \mathbf{N}^j)$ is an integrable and adapted sequence, moreover, one can conclude that

$$E_n d_m f = 0 \qquad (n \not\geq m). \tag{1.3}$$

Reversely, if an integrable and adapted function sequence $(d_n, n \in \mathbf{N}^j)$ has the property (1.3) then $(f_n, n \in \mathbf{N}^j)$ $(j = 1, 2)$ is a martingale where

$$f_n := \sum_{m \leq n} d_m.$$

The concept of a *stopped martingale* is well known in the one-parameter martingale theory: if ν is a stopping time and f is a martingale then the stopped martingale $f^\nu = (f_n^\nu, n \in \mathbf{N})$ is defined by

$$f_n^\nu := \sum_{m=0}^{n} \chi(\nu \geq m) d_m f$$

where $\chi(A)$ is the characteristic function of a set A.

Let us generalize this concept for two parameters. For an arbitrary stopping time ν and a martingale f the sequence $f^\nu = (f_n^\nu, n \in \mathbf{N}^2)$ is said to be a stopped martingale if

$$f_n^\nu := \sum_{m \leq n} \chi(\nu \not< m) d_m f.$$

In consequence of (1.2) and (1.3) it is clear that f^ν is really a martingale. It is easy to check that in the one-parameter case $f_n^\nu = f_m$ on the set $\{\nu = m\}$ whenever $n \geq m$. The stopped martingale does not have this property in the two-parameter case. This is one of the fundamental differences between the one- and two-parameter techniques.

$S(f)$ and $s(f)$ are called the *quadratic variation* and the *conditional quadratic variation* of a martingale f:

$$S_m(f) := \Big(\sum_{n \leq m} |d_n f|^2\Big)^{1/2}, \qquad S(f) := \Big(\sum_{n \in \mathbf{N}^j} |d_n f|^2\Big)^{1/2},$$

$$s_m(f) := \Big(\sum_{n \leq m} E_{n-1} |d_n f|^2\Big)^{1/2}, \qquad s(f) := \Big(\sum_{n \in \mathbf{N}^j} E_{n-1} |d_n f|^2\Big)^{1/2}.$$

Let us introduce the *martingale Hardy spaces* for $0 < p \leq \infty$; denote by H_p^s, H_p^S and H_p^* the spaces of martingales for which

$$\|f\|_{H_p^s} := \|s(f)\|_p < \infty,$$

$$\|f\|_{H_p^S} := \|S(f)\|_p < \infty$$

and

$$\|f\|_{H_p^*} := \|f^*\|_p < \infty,$$

respectively.

We shall say that a martingale $f = (f_n, n \in \mathbf{N}^j)$ $(j = 1,2)$ is *predictable in* L_p $(0 < p \leq \infty)$ if there exists a sequence $(\lambda_n, n \in \mathbf{N}^j)$ of non-decreasing, non-negative and adapted functions such that

$$|f_n| \leq \lambda_{n-1} \quad (n \in \mathbf{N}^j), \qquad \lambda_\infty := \sup_{n \in \mathbf{N}^j} \lambda_n \in L_p.$$

Denote by $\mathcal{P}_p$ the space of this kind of martingales and endow it with the following norm (or quasinorm):

$$\|f\|_{\mathcal{P}_p} := \inf \|\lambda_\infty\|_p \qquad (0 < p \leq \infty)$$

where the infimum is taken over all predictable sequences $(\lambda_n, n \in \mathbf{N}^j)$ having the above property.

If, in the previous definition, we replace the inequality $|f_n| \leq \lambda_{n-1}$ by

$$S_n(f) \leq \lambda_{n-1}$$

then the martingale f is said to be a martingale with *predictable quadratic variation in* L_p. The space containing these martingales is denoted by $\mathcal{Q}_p$ with the norm

$$\|f\|_{\mathcal{Q}_p} := \inf \|\lambda_\infty\|_p \qquad (0 < p \leq \infty)$$

where the infimum is taken over all predictable sequences again. It is clear that the infimums taken in the $\mathcal{P}_p$ and $\mathcal{Q}_p$ norms can be achieved. Indeed, let $(\lambda_n^{(k)}, n \in \mathbf{N}^j)$ $(j = 1,2)$ be a predictable sequence of (f_n) for every $k \in \mathbf{N}$ such that $\|\lambda_\infty^{(k)}\|_p \to \|f\|_{\mathcal{P}_p}$ whenever $k \to \infty$. Set

$$\lambda_n := \inf_{k \in \mathbf{N}} \lambda_n^{(k)} \qquad (n \in \mathbf{N}^j);$$

it is obvious that (λ_n) is a predictable sequence of (f_n) and

$$\|f\|_{\mathcal{P}_p} = \|\lambda_\infty\|_p.$$

The $\mathcal{Q}_p$ spaces for which the proof is similar were first introduced by the author for discrete time in [197] and for continuous time in [198]. These spaces can be handled like the spaces $\mathcal{P}_p$ and H_p^s can be handled and, moreover, with the help of these spaces a simple proof of Burkholder-Davis-Gundy's inequality is given.

Two normed or quasinormed spaces X and Y are *equivalent* (briefly $X \sim Y$) if and only if X and Y are isomorphic and their normes are equivalent, i.e. there exist constants $c, C > 0$ such that for all $x \in X$

$$c\|x\|_X \leq \|x\|_Y \leq C\|x\|_X.$$

The ***dual*** of an arbitrary normed or quasinormed space X is denoted by X'. We say that Y is the dual space of X if $X' \sim Y$.

The BMO and Lipschitz spaces for one-parameter are to be introduced. It will be shown later that these spaces are equivalent to the duals of the H_p spaces. BMO_q^- $(1 \leq q < \infty)$ denotes the space of those functions $f \in L_q$ for which

$$\|f\|_{BMO_q^-} := \sup_{n \in \mathbf{N}} \|(E_n|f - E_{n-1}f|^q)^{1/q}\|_\infty < \infty.$$

We remark that we have $E_0 f = 0$ for every $f \in L_1$. As a generalization of this space the Lipschitz spaces are obtained: $\Lambda_q^-(\alpha)$ $(1 \leq q < \infty, \alpha \geq 0)$ consists of functions $f \in L_q$ for which

$$\|f\|_{\Lambda_q^-(\alpha)} := \sup_{n \in \mathbf{N}} \sup_{A \in \mathcal{F}_n} P(A)^{-1/q-\alpha} \Big(\int_A |f - E_{n-1}f|^q \, dP\Big)^{1/q} < \infty.$$

Obviously, $\Lambda_q^-(0) = BMO_q^-$. The spaces $\Lambda_q(\alpha)$ can similarly be defined: $\Lambda_q(\alpha)$ $(1 \leq q < \infty, \alpha \geq 0)$ denotes the space of functions $f \in L_q$ for which

$$\|f\|_{\Lambda_q(\alpha)} := \sup_{n \in \mathbf{N}} \sup_{A \in \mathcal{F}_n} P(A)^{-1/q-\alpha} \Big(\int_A |f - E_n f|^q \, dP\Big)^{1/q} < \infty.$$

It can easily be shown that the $\Lambda_q(\alpha)$ norm is equal to the following one

$$\|f\|_q^\circ := \sup_{\nu \in \mathcal{T}_1} P(\nu \neq \infty)^{-1/q-\alpha} \|f - f^\nu\|_q \qquad (1 \leq q < \infty, \alpha \geq 0). \tag{1.4}$$

Indeed, for all stopping times $\nu \in \mathcal{T}_1$,

$$\begin{aligned} P(\nu \neq \infty)^{-1-q\alpha} \|f - f^\nu\|_q^q &= P(\nu \neq \infty)^{-1-q\alpha} \sum_{n=0}^{\infty} \int_{\{\nu = n\}} |f - E_n f|^q \, dP \\ &\leq \|f\|_{\Lambda_q(\alpha)}^q P(\nu \neq \infty)^{-1-q\alpha} \sum_{n=0}^{\infty} P(\nu = n)^{1+q\alpha}. \end{aligned}$$

Since $1 + q\alpha \geq 1$, one has $\|f\|_q^\circ \leq \|f\|_{\Lambda_q(\alpha)}$. On the other hand, let $n \in \mathbf{N}$ and $A \in \mathcal{F}_n$ be arbitrary and let $\nu(\omega) = n$ if $\omega \in A$, or else $\nu(\omega) = \infty$. Then ν is evidently a stopping time and

$$\|f\|_q^\circ \geq P(A)^{-1/q-\alpha} \Big(\int_A |f - E_n f|^q \, dP\Big)^{1/q}.$$

From this we can conclude that

$$\|\cdot\|_q^\circ = \|\cdot\|_{\Lambda_q(\alpha)}.$$

If $\alpha = 0$ then the $\Lambda_q(\alpha)$ space is denoted by BMO_q. Thus it can immediately be seen that the BMO_q norm $(1 \leq q < \infty)$ is equal to the following two ones

$$\|f\|_{BMO_q} = \sup_{n\in\mathbf{N}} \|(E_n|f - E_n f|^q)^{1/q}\|_\infty$$

and

$$\|f\|_{BMO_q} = \sup_{\nu\in\mathcal{T}_1} P(\nu \neq \infty)^{-1/q}\|f - f^\nu\|_q.$$

Note that in martingale theory the spaces BMO^-, BMO, $\Lambda^-(\alpha)$ and $\Lambda(\alpha)$ are usually denoted by BMO, BMO^+, $\Lambda(\alpha)$ and $\Lambda^+(\alpha)$, respectively. However, in our treatment it is more suitable to use new notations.

The form (1.4) of the BMO_q and the $\Lambda_q(\alpha)$ norms will be generalized for two parameters: $\Lambda_q(\alpha)$ $(1 \leq q < \infty)$ denotes the space of those functions $f \in L_q$ for which

$$\|f\|_{\Lambda_q(\alpha)} = \sup_{\nu\in\mathcal{T}_2} P(\nu \neq \infty)^{-1/q-\alpha}\|f - f^\nu\|_q < \infty \qquad (1 \leq q < \infty, \alpha \geq 0)$$

(see Bernard [9], Weisz [199]). Again, the $\Lambda_q(0)$ space will be denoted by BMO_q. An element of BMO is called a function of *bounded mean oscillation.*

1.2. VILENKIN ORTHONORMED SYSTEM

In this section special stochastic bases generated by orthonormed function systems are considered. The concept of the martingales can be regarded as a partly generalization of the Vilenkin-Fourier series. Vilenkin systems are dealt with in Section 2.6 as well as in Chapters 4 and 6. Some theorems are proved for Vilenkin martingales, only (see Section 3.3).

A Vilenkin system, which is a generalization of the Walsh system, was introduced by Vilenkin [186] in 1947. Since that time several authors have been dealing with the Vilenkin systems (e.g. Gosselin [84], Schipp [163], Simon [170], Young [203]). The extension of the results from Walsh system to Vilenkin systems is usually non-trivial. To demonstrate the difference we note that while Carleson's theorem, which says that for a function $f \in L_p$ $(1 < p < \infty)$ the Vilenkin-Fourier series of f converges a.e. to f, is proved for the one-parameter Walsh system (see Chapter 4), it is still unknown for one-parameter unbounded Vilenkin systems.

The σ-algebras generated by a Vilenkin system are easy to handle since they are all atomic (see (1.5) and (1.6)). Moreover, in case the Vilenkin system is bounded, these σ-algebras are samples of regular, otherwise, of non-regular stochastic basis.

Let us now introduce the Vilenkin systems. In this section $\Omega = [0,1)$ or $\Omega = [0,1) \times [0,1)$, $\mathcal{A}$ is the σ-algebra of the one- or two-dimensional Borel sets and P is the one- or two-dimensional Lebesgue measure. Let $(p_n, n \in \mathbf{N})$ and $(q_n, n \in \mathbf{N})$ be two sequences of natural numbers with entries at least 2. Introduce the notations $P_0 = Q_0 = 1$ and

$$P_{n+1} := \prod_{k=0}^{n} p_k, \quad Q_{n+1} := \prod_{k=0}^{n} q_k \qquad (n \in \mathbf{N}).$$

Every point $x \in [0,1)$ can be written in the following way:

$$x = \sum_{k=0}^{\infty} \frac{x_k}{P_{k+1}}, \qquad 0 \le x_k < p_k,\ x_k \in \mathrm{N}.$$

In case there are two different forms, we choose the one for which $\lim_{k\to\infty} x_k = 0$. With this convention the construction of x is unique.

The functions

$$r_n(x) := \exp\frac{2\pi \imath x_n}{p_n} \qquad (n \in \mathrm{N})$$

are called *generalized Rademacher functions* where $\imath := \sqrt{-1}$. It can easily be proved that these functions are independent.

The product system generated by these functions is a *one-dimensional Vilenkin system*:

$$w_n(x) := \prod_{k=0}^{\infty} r_k(x)^{n_k}$$

where $n = \sum_{k=0}^{\infty} n_k P_k$, $0 \le n_k < p_k$ and $n_k \in \mathrm{N}$.

The Kronecker product $(w_{n,m}; n, m \in \mathrm{N})$ of two different Vilenkin systems is said to be a *two-dimensional Vilenkin system.* Thus

$$w_{n,m}(x,y) := w_n(x)w'_m(y) = \prod_{k=0}^{\infty} r_k(x)^{n_k} \prod_{l=0}^{\infty} r'_l(y)^{m_l}$$

where

$$r'_n(y) := \exp\frac{2\pi \imath y_n}{q_n}$$

and $n = \sum_{k=0}^{\infty} n_k P_k$, $m = \sum_{l=0}^{\infty} m_l Q_l$, $0 \le n_k < p_k$, $0 \le m_l < q_l$ and $n_k, m_l \in \mathrm{N}$.

It is well known that a Vilenkin system is orthonormed. If $p_n = 2$ (and $q_n = 2$) for every $n \in \mathrm{N}$ then it is called the *one- (two-) dimensional Walsh system.*

Let $\mathcal{F}_n$ and $\mathcal{F}'_m$ be the σ-algebras generated by $\{r_0, \dots, r_{n-1}\}$ and by $\{r'_0, \dots, r'_{m-1}\}$, respectively, and let $\mathcal{F}_{n,m}$ be the one generated by $\mathcal{F}_n \times \mathcal{F}'_m$. It is easy to see that

$$\mathcal{F}_n = \sigma\{[kP_n^{-1}, (k+1)P_n^{-1}) : 0 \le k < P_n\} \tag{1.5}$$

and

$$\mathcal{F}_{n,m} = \sigma\{[kP_n^{-1}, (k+1)P_n^{-1}) \times [lQ_m^{-1}, (l+1)Q_m^{-1}) : 0 \le k < P_n, 0 \le l < Q_m\}. \tag{1.6}$$

The intervals and rectangles inside the curly brackets of (1.5) and (1.6) are said to be the *atoms* of the σ-algebras $\mathcal{F}_n$ and $\mathcal{F}_{n,m}$, respectively. If $p_n = 2$ (and $q_n = 2$) for every $n \in \mathrm{N}$ then the sequence $(\mathcal{F}_n)$ $[(\mathcal{F}_{n,m})]$ is called *one- (two-) parameter sequence of dyadic σ-algebras.*

After a simple consideration we can conclude that the sequences $(\mathcal{F}_n)$ and $(\mathcal{F}_{n,m})$ of σ-algebras satisfy the conditions of the previous section, amongst others, the condition F_4.

Denote by $I_n(x)$ and by $I_{n,m}(x,y)$ the atoms of the σ-algebras $\mathcal{F}_n$ and $\mathcal{F}_{n,m}$, respectively, for which $x \in I_n(x)$ and $(x,y) \in I_{n,m}(x,y)$ $(n,m \in \mathbf{N}, x,y \in [0,1))$, respectively. The *Dirichlet kernel function* of a Vilenkin system is defined by

$$D_n := \sum_{k=0}^{n-1} w_k, \qquad D_{n,m} := \sum_{k=0}^{n-1}\sum_{l=0}^{m-1} w_{k,l} \qquad (n,m \in \mathbf{N}).$$

Obviously, $D_{n,m}(x,y) = D_n(x)D_m(y)$. It is known (see e.g. Vilenkin [186]) that for $n,m \in \mathbf{N}$,

$$D_{P_n}(x) := \begin{cases} P_n & \text{if } x \in I_n(0) \\ 0 & \text{if } x \in [0,1) \setminus I_n(0) \end{cases}$$

and

$$D_{P_n,Q_m}(x,y) := \begin{cases} P_nQ_m & \text{if } (x,y) \in I_{n,m}(0,0) \\ 0 & \text{if } (x,y) \in [0,1)^2 \setminus I_{n,m}(0,0). \end{cases}$$

If $f \in L_1$ then the numbers

$$\hat{f}(n) := E(f\overline{w}_n), \qquad \hat{f}(n,m) := E(f\overline{w}_{n,m})$$

are said to be the n-th and the (n,m)-th *Vilenkin-Fourier coefficients* of f $(n,m \in \mathbf{N})$, respectively. Let us extend this definition to martingales as well. If $f = (f_k, k \in \mathbf{N}^j)$ is a martingale then let

$$\hat{f}(n) := \lim_{k\to\infty} E(f_k\overline{w}_n) \qquad (n \in \mathbf{N}^j).$$

Since w_n is $\mathcal{F}_m$ measurable for a large enough $m \in \mathbf{N}^2$, it can immediately be seen that this limit does exist. Note that if $f \in L_1$ then $E_kf \to f$ in L_1 norm as $k \to \infty$, hence

$$\hat{f}(n) = \lim_{k\to\infty} E((E_kf)\overline{w}_n) \qquad (n,k \in \mathbf{N}^j).$$

Thus the Vilenkin-Fourier coefficients of $f \in L_1$ are the same as the ones of the martingale (E_nf) obtained from f.

The *one-* and *two-parameter Vilenkin-Fourier series* of an integrable function f are given by

$$f(x) \sim \sum_{k=0}^{\infty} \hat{f}(k)w_k(x), \qquad f(x,y) \sim \sum_{k=0}^{\infty}\sum_{l=0}^{\infty} \hat{f}(k,l)w_{k,l}(x,y),$$

respectively. In line with the conditions of Section 1.1 we always suppose that $\hat{f}(0) = \hat{f}(k,0) = \hat{f}(0,k) = 0$ $(k \in \mathbf{N})$. Denote by R_nf and $R_{n,m}f$ the n-th and the (n,m)-th partial sum of the Vilenkin-Fourier series of f, respectively, namely,

$$R_nf := \sum_{k=0}^{n-1} \hat{f}(k)w_k, \qquad R_{n,m}f := \sum_{k=0}^{n-1}\sum_{l=0}^{m-1} \hat{f}(n,m)w_{n,m}.$$

If $f_n := R_{P_n} f$ and $f_{n,m} := R_{P_n,Q_m} f$ then it is easy to prove (see Schipp, Wade, Simon, Pál [167]) that

$$(1.7) \qquad f_n(x) := \sum_{k=0}^{P_n-1} \hat{f}(k) w_k(x) = P_n \int_{I_n(x)} f \, dP = E_n f(x)$$

and

$$(1.8) \qquad \begin{aligned} f_{n,m}(x,y) : &= \sum_{k=0}^{P_n-1} \sum_{l=0}^{Q_m-1} \hat{f}(k,l) w_{k,l}(x,y) \\ &= P_n Q_m \int_{I_{n,m}(x,y)} f \, dP \\ &= E_{n,m} f(x,y), \end{aligned}$$

that is to say, $(f_n, n \in \mathbb{N})$ resp. $(f_{n,m}; n, m \in \mathbb{N})$ is the martingale relative to $(\mathcal{F}_n)$ resp. $(\mathcal{F}_{n,m})$ obtained from f. Moreover, a sequence $f = (f_n, n \in \mathbb{N})$ resp. $f = (f_{n,m}; n, m \in \mathbb{N})$ of integrable and adapted functions is a martingale if and only if there exist complex numbers c_k resp. $c_{k,l}$ such that

$$f_n = \sum_{k=0}^{P_n-1} c_k w_k \qquad \text{resp.} \qquad f_{n,m} = \sum_{k=0}^{P_n-1} \sum_{l=0}^{Q_m-1} c_{k,l} w_{k,l}.$$

Of course, $c_k = \hat{f}(k)$ and $c_{k,l} = \hat{f}(k,l)$.

From the theorems mentioned in Section 1.1 it follows immediately that if $f \in L_p$ then in the one-parameter case $R_{P_n} f \to f$ in L_p norm and a.e. $(1 \le p < \infty)$ and in the two-parameter case $R_{P_{n_1},Q_{n_2}} f \to f$ in L_p norm $(1 \le p < \infty)$ and a.e. $(1 < p < \infty)$ whenever $n \to \infty$.

A one-dimensional Vilenkin system generated by a bounded sequence (p_n) (i.e. $p_n = O(1)$) as well as a two-dimensional one constructed from $p_n = O(1)$ and $q_n = O(1)$ are said to be *bounded*. It is easy to show that if $p_n \le N$ for every $n \in \mathbb{N}$ then the sequence $(\mathcal{F}_n, n \in \mathbb{N})$ of σ-algebras is regular. Indeed, applying (1.7) we get that

$$\begin{aligned} f_{n+1}(x) &= P_{n+1} \int_{I_{n+1}(x)} f_{n+2} \, dP \\ &\le \frac{P_{n+1}}{P_n} P_n \int_{I_n(x)} f_{n+2} \, dP \\ &= p_n f_n(x) \\ &\le N f_n(x). \end{aligned}$$

With (1.8) the regurality of the stochastic basis $(\mathcal{F}_n, n \in \mathbb{N}^2)$ can similarly be proved whenever both the sequences (p_k) and (q_k) are bounded.

The martingale difference sequence can also be given by

$$d_{n+1} f = \sum_{k=P_n}^{P_{n+1}-1} c_k w_k, \qquad d_{n+1,m+1} f = \sum_{k=P_n}^{P_{n+1}-1} \sum_{l=Q_m}^{Q_{m+1}-1} c_{k,l} w_{k,l}$$

which can be rewritten as

$$(1.9)\qquad d_{n+1}f = \sum_{i=1}^{p_n-1} v_n^{(i)} r_n^i, \qquad d_{n+1,m+1}f = \sum_{i=1}^{p_n-1}\sum_{j=1}^{q_m-1} v_{n,m}^{(i,j)} r_n^i r_m'^j$$

where every $v_n^{(i)}$ is $\mathcal{F}_n$- and every $v_{n,m}^{(i,j)}$ is $\mathcal{F}_{n,m}$-measurable.

Since for $i, l = 1, \ldots, p_n - 1$ we have

$$(1.10)\qquad \begin{aligned} &E_n(r_n^i) = 0,\\ &E_n(r_n^i \overline{r_n^l}) = \delta(i-l),\\ &|r_n^i| = 1, \end{aligned}$$

from which we obtain that

$$(1.11)\qquad s(f) = \Big(\sum_{n=0}^{\infty}\sum_{i=1}^{p_n-1} |v_n^{(i)}|^2\Big)^{1/2}, \quad s(f) = \Big(\sum_{n=0}^{\infty}\sum_{m=0}^{\infty}\sum_{i=1}^{p_n-1}\sum_{j=1}^{q_m-1} |v_{n,m}^{(i,j)}|^2\Big)^{1/2}.$$

CHAPTER 2

ONE-PARAMETER MARTINGALE HARDY SPACES

Martingale theory came into the limelight for about 40 years when Doob [61] proved that the composition of an analytic or harmonic function on the unit circle of the complex plain and the Brownian motion is a continuous parameter continuous martingale. Martingale theory and the theory of the martingale Hardy spaces can excellently be applied in the theory of complex functions and in the theory of the classical Hardy spaces (see Durrett [64]). In this book we deal with discrete parameter martingales. Many theorems relative to the discrete martingales can easily be extended to continuous parameter martingales. For the theory of continuous parameter martingales we refer to Dellacherie, Meyer [58].

Burkholder and Gundy [22], [19], [27] proved that the martingale Hardy spaces H_p^S and H_p^* are equivalent whenever $1 < p < \infty$. A few years later Davis [55] extended this result to the case $p = 1$ (see Theorem 2.12). In 1973 Garsia [82] and Herz [94] characterized the dual of H_1^S as a BMO space. Since that time many mathematicians have been investigating martingale inequalities and martingale Hardy spaces.

In this chapter the five martingale Hardy spaces introduced earlier are considered. In Section 2.1 the basic idea of this book, the idea of the atomic decomposition is given. An atom is a simple, easy to handle martingale. The martingales of the spaces H_p^s, $\mathcal{P}_p$ and $\mathcal{Q}_p$ are decomposed into a sum of atoms. Many theorems can be proved with the help of this method, that is to say, the statements need to be shown for atoms, only. In Section 2.2 with the atomic decomposition a new and simple proof of Burkholder-Davis-Gundy's inequality mentioned above is given. The equivalence of the five martingale Hardy spaces is proved in case the stochastic basis $\mathcal{F}$ is regular (Corollary 2.23). In Section 2.3, amongst others, it is shown that the dual of H_1^s resp. H_p^s resp. H_1^S is BMO_2 resp. H_q^s $(1/p + 1/q = 1)$ resp. BMO_2^- (Theorems 2.24, 2.26, Corollary 2.35). Moreover, it is proved that the duals of the so-called VMO subspaces of the BMO spaces are H_1^s and H_1^S (Theorems 2.39, 2.41, 2.43). In Section 2.4 a classical result due to Garsia [82] and Herz [94] is proved (Theorem 2.53): the BMO_q^- spaces are all equivalent for $1 \leq q < \infty$. A family of the BMO spaces that are equivalent to BMO_2 is given (Theorem 2.49). Moreover, three sharp functions are investigated and it is proved that the L_p norms of these sharp functions are equivalent to the H_p^s, H_p^* and H_p^S norms of the martingale, respectively. In Section 2.5 the boundedness of the martingale transforms between two Hardy spaces is in question and amongst other new theorems the equivalence between $\mathcal{P}_p$ and $\mathcal{Q}_p$ $(0 < p < \infty)$ is demonstrated. In Section 2.6 Vilenkin martingales are considered. The spaces H_1^S, BMO and VMO are characterized via conjugate martingale transforms.

2.1. ATOMIC DECOMPOSITIONS

The atomic decomposition is a useful characterization of Hardy spaces by the help of which some duality theorems and martingale inequalities can be proved.

With the help of stopping times three kinds of atoms are defined and the atomic decomposition of the spaces H_p^s, $\mathcal{P}_p$ and $\mathcal{Q}_p$ are verified. In harmonic analysis the concept of simple atoms the supports of which are $\mathcal{F}_n$ measurable sets for any $n \in \mathbf{N}$ is usually used. We show that the atomic decomposition of the spaces H_p^s, $\mathcal{P}_p$ and $\mathcal{Q}_p$ is valid for simple atoms, too. Many results of this section can be found for discrete time in Weisz [197] and for continuous time in Weisz [198].

Let us introduce first the concept of an atom:

Definition 2.1. *A measurable function a is a (p,∞) atom of the first category (briefly $(1,p,\infty)$ atom) if there exists a stopping time $\nu \in \mathcal{T}_1$ such that*

$$(i) \qquad a_n := E_n a = 0 \qquad \text{if} \qquad \nu \geq n$$

$$(ii) \qquad \|s(a)\|_\infty \leq P(\nu \neq \infty)^{-1/p}.$$

Replacing (ii) by the inequality

$$\|S(a)\|_\infty \leq P(\nu \neq \infty)^{-1/p}$$

and by

$$\|a^*\|_\infty \leq P(\nu \neq \infty)^{-1/p},$$

we get the concept of a (p,∞) atom of the second category ($(2,p,\infty)$ atom) and the concept of a (p,∞) atom of the third category ($(3,p,\infty)$ atom), respectively. Denote by $A_i(p,\infty)$ the set of (i,p,∞)-atoms $(i = 1,2,3)$.

Note that $(3,1,\infty)$ atoms and in the continuous case both $(2,p,\infty)$ and $(3,p,\infty)$ atoms have already been investigated, moreover, the atomic decomposition of $\mathcal{P}_p$ is known (see Bernard, Masisonneuve [10], Chevalier [46]). Atomic decomposition was first used by Coifman and Weiss [52] in the classical case and by Herz [94] for the discrete parameter $\mathcal{P}_1$ space.

The *atomic decomposition of the martingale Hardy spaces* H_p^s, $\mathcal{Q}_p$ and $\mathcal{P}_p$ is formulated.

Theorem 2.2. *If the martingale $f = (f_n; n \in \mathbf{N})$ is in H_p^s $(0 < p < \infty)$ then there exist a sequence $a^k \in A_1(p,\infty)$ $(k \in \mathbf{Z})$ of $(1,p,\infty)$ atoms and a sequence $\mu = (\mu_k, k \in \mathbf{Z}) \in l_p$ of real numbers such that for all $n \in \mathbf{N}$*

$$\sum_{k=-\infty}^{\infty} \mu_k E_n a^k = f_n \tag{2.1}$$

and

$$\Big(\sum_{k=-\infty}^{\infty} |\mu_k|^p\Big)^{1/p} \leq C_p \|f\|_{H_p^s}. \tag{2.2}$$

Moreover, the sum $\sum_{k=l}^{m} \mu_k a^k$ *converges to* f *in* H_p^s *norm as* $m \to \infty$, $l \to -\infty$, *too. Conversely, if* $0 < p \leq 1$ *and the martingale* f *has a decomposition of type (2.1) then* $f \in H_p^s$ *and*

$$\|f\|_{H_p^s} \sim \inf(\sum_{k=-\infty}^{\infty} |\mu_k|^p)^{1/p} \tag{2.3}$$

where the infimum is taken over all decompositions of f *of the form (2.1).*

Theorem 2.3. *If, in Theorem 2.2, we replace* H_p^s *resp.* $A_1(p,\infty)$ *by* $\mathcal{Q}_p$ *resp.* $A_2(p,\infty)$ *or by* $\mathcal{P}_p$ *resp.* $A_3(p,\infty)$ $(0 < p < \infty)$ *then (2.1) and (2.2) hold, too. The sum* $\sum_{k=l}^{m} \mu_k a^k$ *converges to* f *in the first case in* $\mathcal{Q}_p$ *norm* $(0 < p \leq 2)$ *and in the second case in* $\mathcal{P}_p$ *norm* $(0 < p \leq 1)$ *as* $m \to \infty$, $l \to -\infty$. *The converse and (2.3) are also valid in both cases when* $0 < p \leq 1$.

Proof of Theorem 2.2. Assume that $f \in H_p^s$. Let us consider the following stopping times for all $k \in \mathbf{Z}$:

$$\nu_k := \inf\{n \in \mathbf{N} : s_{n+1}(f) > 2^k\}.$$

The sequence of these stopping times is obviously non-decreasing. It is easy to see that

$$f_n = \sum_{k \in \mathbf{Z}} (f_n^{\nu_{k+1}} - f_n^{\nu_k}). \tag{2.4}$$

Let

$$\mu_k := 2^k 3 P(\nu_k \neq \infty)^{1/p}$$

and

$$a_n^k := \frac{f_n^{\nu_{k+1}} - f_n^{\nu_k}}{\mu_k}$$

(if $\mu_k = 0$ then let $a_n^k = 0$, $k \in \mathbf{Z}$, $n \in \mathbf{N}$). It is clear that, for a fixed k, (a_n^k) is a martingale. Since $s(f_n^{\nu_k}) \leq 2^k$,

$$s(a_n^k) \leq P(\nu_k \neq \infty)^{-1/p} \qquad (n \in \mathbf{N}),$$

consequently, (a_n^k) is L_2-bounded and so there exists $a^k \in L_2$ such that

$$E_n a^k = a_n^k$$

and

$$s(a^k) \leq P(\nu_k \neq \infty)^{-1/p}.$$

If $n \leq \nu_k$ then $a_n^k = 0$, thus we obtain that a^k is really a $(1,p,\infty)$ atom. By Abel rearrangement we get

$$\begin{aligned} \sum_{k \in \mathbf{Z}} |\mu_k|^p &= 3^p \sum_{k \in \mathbf{Z}} 2^{kp} P(\nu_k \neq \infty) \\ &= 3^p \sum_{k \in \mathbf{Z}} 2^{kp} P[s(f) > 2^k] \end{aligned} \tag{2.5}$$

$$= 3^p \sum_{k\in\mathbf{Z}} (2^k)^p P[s^p(f) > (2^k)^p]$$

$$= \frac{3^p}{2^p-1} \sum_{k\in\mathbf{Z}} [(2^p)^{k+1} - (2^p)^k] P[s^p(f) > (2^p)^k]$$

$$= \frac{3^p}{2^p-1} \sum_{k\in\mathbf{Z}} (2^p)^k P[(2^p)^{k-1} < s^p(f) \leq (2^p)^k]$$

$$\leq \frac{3^p}{2^p-1} E[s^p(f)]$$

which proves (2.1) and (2.2).

Obviously,

$$f - \sum_{k=l}^{m} \mu_k a^k = (f - f^{\nu_{m+1}}) + f^{\nu_l}. \tag{2.6}$$

Notice that $(f - f^{\nu_{m+1}}) \to 0$ in H_p^s norm as $m \to \infty$ because the a.e. limit of

$$s^p(f - f^{\nu_{m+1}}) = [s^2(f) - s^2(f^{\nu_{m+1}})]^{p/2}$$

is equal to zero and it can be majorized by the integrable function $s^p(f)$. Since $s(f^{\nu_l}) \leq 2^l$, we obtain that the series $\sum_{k=l}^{m} \mu_k a^k$ converges to f in H_p^s norm as $m \to \infty$, $l \to -\infty$.

The last statement of Theorem 2.2 is to be verified. Assume that $0 < p \leq 1$ and f has a decomposition of the form (2.1). It is easy to check that in this case

$$s(f) \leq \sum_{k=-\infty}^{\infty} |\mu_k| s(a^k).$$

Since $0 < p \leq 1$,

$$E[s^p(f)] \leq \sum_{k=-\infty}^{\infty} |\mu_k|^p E[s^p(a^k)].$$

If a is a $(1, p, \infty)$ atom then it comes from its definition that

$$\chi(\nu \geq k) E_{k-1} |d_k a|^2 = E_{k-1}[\chi(\nu \geq k)|d_k a|^2] = 0,$$

thus $s(a) = 0$ on the set $\{\nu = \infty\}$. Consequently,

$$E(s^p(a)) \leq 1, \tag{2.7}$$

in other words, the inequality

$$E(s^p(f)) \leq \sum_{k=-\infty}^{\infty} |\mu_k|^p$$

holds, which proves Theorem 2.1. ■

Proof of Theorem 2.3. The proof shall be given for $\mathcal{Q}_p$, only, because it is slightly more complex than the one for $\mathcal{P}_p$. Let $f \in \mathcal{Q}_p$ and $(\lambda_n, n \in \mathbf{N})$ be an adapted, non-decreasing sequence such that $S_n(f) \le \lambda_{n-1}$ and $\lambda_\infty \in L_p$. The stopping times ν_k are defined in this case by

$$\nu_k := \inf\{n \in \mathrm{N} : \lambda_n > 2^k\}.$$

Let a^k and μ_k $(k \in \mathrm{N})$ be defined as in the proof of the previous theorem. The formulas (2.1) and (2.2) can be proved in the same way as in Theorem 2.2. The equation

$$f - \sum_{k=l}^{m} \mu_k a^k = (f - f^{\nu_{m+1}}) + f^{\nu_l}$$

holds similarly to (2.6). We can show that $\sum_{k=l}^{m} \mu_k a^k$ converges to f in H_p^S norm for every p as $m \to \infty$, $l \to -\infty$. Furthermore,

$$\begin{aligned} S^2(f - f^{\nu_{m+1}}) &= S^2(f) - S^2(f^{\nu_{m+1}}) \\ &= \sum_{k=m+1}^{\infty} [S^2(f^{\nu_{k+1}}) - S^2(f^{\nu_k})] \\ &= \sum_{k=m+1}^{\infty} S^2(f^{\nu_{k+1}} - f^{\nu_k}) \\ &= \sum_{k=m+1}^{\infty} |\mu_k|^2 S^2(a^k). \end{aligned} \tag{2.8}$$

Set

$$\rho_n^k := \chi(\nu_k \le n)\|S(a^k)\|_\infty$$

and

$$(\rho_n)^2 := \sum_{k=m+1}^{\infty} |\mu_k|^2 (\rho_n^k)^2. \tag{2.9}$$

From (2.8) it is clear that (ρ_n) is adapted, non-decreasing and that

$$S_n(f - f^{\nu_{m+1}}) \le \rho_{n-1}.$$

If $p/2 \le 1$ then we get from (2.9) that

$$E(\rho_\infty^p) \le \sum_{k=m+1}^{\infty} |\mu_k|^p.$$

Using the analogue of (2.2) we obtain that $(f - f^{\nu_{m+1}})$ converges to zero in Q_p norm as $m \to \infty$. It is easy to see that $\|f^{\nu_l}\|_{Q_p} \to 0$ as $l \to -\infty$ in this case, too. The proof for the $\mathcal{P}_p$ spaces is similar.

To prove the last statement in the theorem we only have to show that if $a \in A_2(p, \infty)$ and $a \in A_3(p, \infty)$ then we have

$$\|a\|_{Q_p} \le 1 \tag{2.10}$$

and

$$\|a\|_{\mathcal{P}_p} \leq 1, \tag{2.11}$$

respectively (cf. (2.7)). To verify these, let

$$\lambda_n := \chi(\nu \leq n)\|S(a)\|_\infty$$

and

$$\lambda_n := \chi(\nu \leq n)\|a\|_\infty,$$

respectively, where ν is the stopping time in the definition of the atom a. In both cases (λ_n) is a predictable sequence because it is non-decreasing and $a_{n+1} = 0$ on the set $\{\nu > n\}$. The inequalities (2.10) and (2.11) follow immediately from this. The rest of the proof is similar to the one in Theorem 2.2. ∎

Note that the part of the last theorem relative to $\mathcal{P}_1$ appeared first in a disguised form in a proof due to Herz in [94]. Moreover, Theorem 2.2 and 2.3 answer partly the question put by Chao in [37], namely, whether there exists an atomic decomposition of H_p^* for a sequence of non-regular σ-algebras. (If $\mathcal{F}$ is regular then $H_p^* \sim \mathcal{P}_p$, see Corollary 2.23.)

We narrow the definition of an atom down to the stopping times of which ranges consist of an integer and/or ∞.

Definition 2.4. *A measurable function a is a $(1,p,\infty)$ simple atom if there exist $n \in \mathbf{N}$, $H \in \mathcal{F}_n$ such that*

$$\begin{array}{ll} (i) & a_n := E_n a = 0 \\ (ii) & \|s(a)\|_\infty \leq P(H)^{-1/p} \\ (iii) & \{a \neq 0\} \subset H. \end{array}$$

Replacing $s(a)$ in (ii) by $S(a)$ resp. by a^* we get the concept of a $(2,p,\infty)$ resp. of a $(3,p,\infty)$ simple atom.

If $H \in \mathcal{F}_n$ for any $n \in \mathbf{N}$ then the map

$$\nu_H(\omega) = \begin{cases} n & \text{if } \omega \in H \\ \infty & \text{if } \omega \notin H \end{cases}$$

is a stopping time. Since $H = \{\nu_H \neq \infty\}$, we can see that every (i,p,∞) simple atom is an (i,p,∞) atom $(i = 1,2,3; 0 < p < \infty)$.

In harmonic analysis the $(3,p,\infty)$ simple atoms are mainly applied (see e.g. Coifman, Weiss [52], Schipp, Wade, Simon, Pál [167]). In the following theorem we show that the martingales in H_p^s, $\mathcal{P}_p$ and $\mathcal{Q}_p$ can be decomposed into simple atoms as well.

Theorem 2.5. *Theorem 2.2 and 2.3 hold also for simple atoms.*

Proof. The theorem will be verified for the space H_p^s, only. The proofs for the other two spaces are similar. For the proof it is sufficient to construct a suitable atomic decomposition. Rewrite (2.4) in the form

$$f_n = \sum_{k\in\mathbf{Z}} \sum_{l=0}^{n-1} \chi(\nu_k = l)(f_n^{\nu_{k+1}} - f_n^{\nu_k}).$$

Let

$$\mu_{k,l} := 2^k 3P(\nu_k = l)^{1/p}$$

and

$$b_n^{k,l} := \chi(\nu_k = l)\frac{f_n^{\nu_{k+1}} - f_n^{\nu_k}}{\mu_{k,l}}.$$

The sequence $(b_n^{k,l})$ is an L_2-bounded martingale for fixed k and l, hence it can be identified with a function $b^{k,l} \in L_2$. To verify that $b^{k,l}$ is a $(1, p, \infty)$ simple atom, we only have to show (i) of Definition 2.4. Since

$$f_n^{\nu_{k+1}} - f_n^{\nu_k} = \sum_{i=0}^{n} d_i f \chi(\nu_k < i \le \nu_{k+1}), \tag{2.12}$$

we can conclude that

$$\chi(\nu_k = l)(f_n^{\nu_{k+1}} - f_n^{\nu_k}) = \chi(\nu_k = l) \sum_{i=l+1}^{n} d_i f \chi(\nu_k < i \le \nu_{k+1}).$$

From this $E_l(b^{k,l}) = 0$ follows immediately. The rest of the proof is similar to the one of Theorem 2.2. ∎

If every σ-algebra $\mathcal{F}_n$ is generated by finitely many (set) atoms then it can be assumed in Definition 2.4 that $H \in \mathcal{F}_n$ is an atom because in this case the set $\{\nu_k = l\}$ can be decomposed into the union of finitely many disjoint atoms (cf. the dyadic case in Coifman, Weiss [52] and Schipp, Wade, Simon, Pál [167]).

Note that Theorem 2.5 does not hold for two parameters. Since Definition 2.1 can be generalized for two parameters, it shall be used more frequently. For more about (p, q) atoms see Chapter 3.

2.2. MARTINGALE INEQUALITIES

In this section the connection between the five martingale Hardy spaces introduced earlier is considered. First the classical Doob's inequality and the well known fact that $H_p^* \sim L_p$ if $p > 1$ are proved. The convexity and concavity theorems (Garsia [82]) are generalized for an arbitrary countable index set. In Theorem 2.11 the relations $H_p^s \subset H_p^*, H_p^S$ $(0 < p \le 2)$; $H_p^*, H_p^S \subset H_p^s$ $(2 \le p < \infty)$ and $\mathcal{P}_p, \mathcal{Q}_p \subset H_p^*, H_p^S, H_p^s$ $(0 < p < \infty)$ are verified. The proof of the equivalence of $\mathcal{P}_p$ and $\mathcal{Q}_p$ $(0 < p < \infty)$ can be found in Section 2.5. The $\mathcal{Q}_p$ spaces were introduced to give a simple proof of Davis's inequality. Using the atomic decomposition and the relations mentioned above we give a new proof of Burkholder-Davis-Gundy's inequality which says that the spaces H_p^* and H_p^S are equivalent whenever $1 \le p < \infty$. A counterexample due to Marczinkievicz and Zygmund [122] shows that Burkholder-Davis-Gundy's inequality does not hold in general for $0 < p < 1$ (see Proposition

2.16). We obtain from Theorem 2.22 that the five martingale Hardy norms are equivalent for previsible martingales and for all parameters p. It is verified that the stochastic basis is regular if and only if every martingale is previsible. From this it follows that, in case the stochastic basis $\mathcal{F}$ is regular, all the five Hardy spaces are equivalent for every parameter p (see Corollary 2.23). The papers Weisz [197] and [198] are strongly utilized.

From Theorems 2.2 and 2.3 it follows immediately that

$$\|f\|_1 \leq \|f\|_H \tag{2.13}$$

where $H \in \{H_1^s, \mathcal{Q}_1, \mathcal{P}_1\}$ because $\|a\|_1 \leq 1$ if $a \in A_i(1,\infty)$ $(i = 1,2,3)$. The formula (2.13) holds obviously for $H = H_1^*$. We prove the well known Doob's inequality (see e.g. Neveu [142]):

Proposition 2.6. *Let $p > 1$. For every non-negative L_p-bounded submartingale $(f_n, n \in \mathrm{N})$ we have that $\sup_{n\in\mathrm{N}} f_n$ belongs to L_p, more precisely,*

$$\|\sup_{n\in\mathrm{N}} f_n\|_p \leq \frac{p}{p-1} \sup_{n\in\mathrm{N}} \|f_n\|_p.$$

Proof. For the proof we need the following lemma which, although its proof is very easy, is interesting in itself.

Lemma 2.7. *Every non-negative submartingale $(f_n, n \in \mathrm{N})$ satisfies the inequality*

$$\lambda P(f_n^* > \lambda) \leq \int_{\{f_n^* > \lambda\}} f_n \, dP \qquad (n \in \mathrm{N}, \lambda > 0).$$

Proof of Lemma 2.7. Consider the stopping time

$$\nu_\lambda := \inf\{n : f_n > \lambda\}.$$

Using the submartingale property we have

$$\begin{aligned}
\lambda P(\nu_\lambda \leq n) &= \lambda \sum_{k=0}^{n} P(\nu_\lambda = k) \\
&\leq \sum_{k=0}^{n} \int_{\{\nu_\lambda = k\}} f_k \, dP \leq \sum_{k=0}^{n} \int_{\{\nu_\lambda = k\}} E_k f_n \, dP \\
&= \sum_{k=0}^{n} \int_{\{\nu_\lambda = k\}} f_n \, dP = \int_{\{\nu_\lambda \leq n\}} f_n \, dP.
\end{aligned}$$

Since

$$\{\nu_\lambda \leq n\} = \{f_n^* > \lambda\},$$

the lemma is proved. ∎

To return to the proof of Proposition 2.6 write the inequality of the preceding lemma in the form

$$\lambda E(\chi(f_n^* > \lambda)) \leq E(f_n \chi(f_n^* > \lambda)) \qquad (n \in \mathrm{N}, \lambda > 0).$$

Integrating both sides with respect to the measure $p\lambda^{p-2}\,d\lambda$ $(p>1)$ and applying Fubini's theorem we get that

$$\begin{aligned}E(f_n^{*p}) &= \int_0^\infty p\lambda^{p-1}E(\chi(f_n^*>\lambda))\,d\lambda \\ &\le \int_0^\infty p\lambda^{p-2}E(f_n\chi(f_n^*>\lambda))\,d\lambda \\ &= \frac{p}{p-1}E(f_nf_n^{*\,p-1}).\end{aligned}$$

On the other hand, Hölder's inequality implies that

$$E(f_nf_n^{*\,p-1}) \le \|f_n\|_p\|f_n^{*\,p-1}\|_{p/(p-1)} = \|f_n\|_p\|f_n^*\|_p^{p-1}.$$

After division by $\|f_n^*\|_p^{p-1}<\infty$ the combination of the two preceding inequalities gives

$$\|f_n^*\|_p \le \frac{p}{p-1}\|f_n\|_p.$$

As f_n^* is non-decreasing, with taking the supremum over all $n\in\mathrm{N}$ the proposition becomes complete. ■

If $(f_n, n\in\mathrm{N})$ is a martingale then clearly $(|f_n|, n\in\mathrm{N})$ is a non-negative submartingale, so we get immediately that

$$\lambda P(f^*>\lambda) \le \int_{\{f^*>\lambda\}}|f|\,dP \qquad (f\in L_1, \lambda>0) \tag{2.14}$$

and

$$\|f\|_p \le \|f^*\|_p \le \frac{p}{p-1}\|f\|_p \qquad (f\in L_p, p>1), \tag{2.15}$$

in other words, $H_p^*\sim L_p$ if $p>1$.

We shall use the following generalization of the convexity and concavity theorem several times. In the one-parameter case this theorem for Young functions is due to Burkholder, Davis, Gundy [26], however, the idea of our proof is belonging to Garsia [82] and to Schipp, Wade, Simon, Pál [167]. For the proof we need the next well known definition and lemma.

Definition 2.8. *Let* $\mathbf{T}$ *be a countable index set.* $L_p(l_r)$ $(1\le p,r\le\infty)$ *denotes the space of the sequences* $\xi=(\xi_n, n\in\mathbf{T})$ *of measurable functions for which*

$$\|\xi\|_{L_p(l_r)} := \|(\sum_{n\in\mathbf{T}}|\xi_n|^r)^{1/r}\|_p < \infty.$$

Lemma 2.9. *The dual of* $L_p(l_r)$ *is* $L_q(l_s)$ *whenever* $1\le p,r<\infty$, $1/p+1/q=1$ *and* $1/r+1/s=1$. *The bounded linear functionals of* $L_p(l_r)$ *can be written in the form*

$$\Lambda(\xi) = \sum_{k\in\mathbf{T}}E(\xi_k\eta_k) \qquad (\xi\in L_p(l_r)), \tag{2.16}$$

furthermore,

$$\|\Lambda\| = \|\eta\|_{L_q(l_s)}$$

for any $\eta \in L_q(l_s)$.

The proof is similar to the one of the duality between L_p and L_q (see e.g. Benedek, Panzone [5]).

The convexity and concavity theorem is formulated.

Theorem 2.10. *Let* $\mathbf{T}$ *be a countable index set and* $(\mathcal{A}_t, t \in \mathbf{T})$ *be an arbitrary (not necessarily monotone) sequence of* σ*-algebras with the assumption* $\sigma(\cup_{t\in\mathbf{T}}\mathcal{A}_t) = \mathcal{A}$. *Suppose that for all* $h \in L_p$ *Doob's inequality*

$$\| \sup_{t\in\mathbf{T}} |E_t h| \|_p \le C_p \|h\|_p \qquad (p > 1) \tag{2.17}$$

holds where E_t *denotes the conditional expectation operator relative to* $\mathcal{A}_t$. *If* $(f_t, t \in \mathbf{T})$ *is a sequence of non-negative measurable functions then for* $1 \le p < \infty$ *we have*

$$E[(\sum_{t\in\mathbf{T}} E_t f_t)^p] \le C_q^p E[(\sum_{t\in\mathbf{T}} f_t)^p] \tag{2.18}$$

and

$$E[(\sum_{t\in\mathbf{T}} f_t)^{1/p}] \le B_p E[(\sum_{t\in\mathbf{T}} E_t f_t)^{1/p}] \tag{2.19}$$

where $1/p + 1/q = 1$, $B_p > 0$ *and* $C_q > 0$ *denotes the constant in (2.17).*

Proof. These inequalities are obvious for $p = 1$. Assume that $1 < p < \infty$. By Riesz's representation theorem

$$\| \sum_{t\in\mathbf{T}} E_t f_t \|_p = \sup_{\|g\|_q \le 1} |E[(\sum_{t\in\mathbf{T}} E_t f_t) g]|.$$

Hölder's inequality and (2.17) imply

$$\begin{aligned} |E[(\sum_{t\in\mathbf{T}} E_t f_t) g]| &\le \sum_{t\in\mathbf{T}} E(f_t |E_t g|) \\ &\le E\Big[(\sum_{t\in\mathbf{T}} f_t)(\sup_{t\in\mathbf{T}} |E_t g|)\Big] \\ &\le \| \sum_{t\in\mathbf{T}} f_t \|_p \| \sup_{t\in\mathbf{T}} |E_t g| \|_q \\ &\le C_q \| \sum_{t\in\mathbf{T}} f_t \|_p. \end{aligned}$$

The first inequality is established.

To verify the second one let r be the conjugate index to $2p$ (i.e. $1/r + 1/(2p) = 1$) and set $g_t := (f_t)^{1/(2p)}$ for $t \in \mathbf{T}$ and $g = (g_t, t \in \mathbf{T})$. It follows from Lemma 2.9

that

$$(E[(\sum_{t\in\mathbf{T}} f_t)^{1/p}])^{1/2} = \|g\|_{L_2(l_{2p})}$$

$$= \sup_{\|\lambda\|_{L_2(l_r)}\leq 1} |E(\sum_{t\in\mathbf{T}} g_t\lambda_t)|.$$

Moreover, it follows from Hölder's inequality that

$$|E(\sum_{t\in\mathbf{T}} g_t\lambda_t)| \leq \sum_{t\in\mathbf{T}} E[E_t|g_t\lambda_t|]$$

$$\leq \sum_{t\in\mathbf{T}} E\Big[(E_t g_t^{2p})^{1/(2p)}(E_t|\lambda_t|^r)^{1/r}\Big]$$

$$= \sum_{t\in\mathbf{T}} E\Big[(E_t f_t)^{1/(2p)}(E_t|\lambda_t|^r)^{1/r}\Big]$$

$$\leq E\Big[(\sum_{t\in\mathbf{T}} E_t f_t)^{1/(2p)}(\sum_{t\in\mathbf{T}} E_t|\lambda_t|^r)^{1/r}\Big]$$

$$\leq \Big(E[(\sum_{t\in\mathbf{T}} E_t f_t)^{1/p}]E[(\sum_{t\in\mathbf{T}} E_t|\lambda_t|^r)^{2/r}]\Big)^{1/2}.$$

Since $1 < 2/r = 2 - 1/p < 2$, it follows from the first inequality that

$$E\Big[(\sum_{t\in\mathbf{T}} E_t|\lambda_t|^r)^{2/r}\Big] \leq B_p E\Big[(\sum_{t\in\mathbf{T}} |\lambda_t|^r)^{2/r}\Big]$$

$$\leq B_p\|\lambda\|^2_{L_2(l_r)} \leq B_p$$

This completes the proof of the second inequality. ∎

Since (2.17) holds in the linear case, namely, if $\mathbf{T} = \mathbf{N}$ and $(\mathcal{A}_n, n \in \mathbf{N})$ is non-decreasing, the inequalities (2.18) and (2.19) are also valid. Note that in this case $C_q = p$ and $B_p \leq 4$.

The following two theorems formulate inclusion relations between Hardy spaces.

Theorem 2.11.

(i)

$$\|f\|_{H_p^\bullet} \leq C_p\|f\|_{H_p^\bullet}, \quad \|f\|_{H_p^S} \leq C_p\|f\|_{H_p^\bullet} \qquad (0 < p \leq 2)$$

(ii)

$$\|f\|_{H_p^\bullet} \leq C_p\|f\|_{H_p^\bullet}, \quad \|f\|_{H_p^\bullet} \leq C_p\|f\|_{H_p^S} \qquad (2 \leq p < \infty)$$

(iii)

$$\|f\|_{H_p^\bullet} \leq \|f\|_{\mathcal{P}_p}, \quad \|f\|_{H_p^S} \leq \|f\|_{\mathcal{Q}_p} \qquad (0 < p < \infty)$$

(iv)

$$\|f\|_{H_p^\bullet} \leq C_p\|f\|_{\mathcal{Q}_p}, \quad \|f\|_{H_p^S} \leq C_p\|f\|_{\mathcal{P}_p} \qquad (0 < p < \infty)$$

(v)

$$\|f\|_{H_p^\bullet} \leq C_p\|f\|_{\mathcal{P}_p}, \quad \|f\|_{H_p^\bullet} \leq C_p\|f\|_{\mathcal{Q}_p} \qquad (0 < p < \infty),$$

where the positve constants C_p depend only on p. (The symbol C_p may denote different constants in different contexts.)

Applying these results we can give a simple proof of the well-known Burkholder-Davis-Gundy's inequality which is one of the most fundamental theorems of martingale theory. Bernard and Maisonneuve [10] gave a very nice proof for the inequality $\|f\|_{H_1^s} \leq C\|f\|_{H_1^*}$. With the help of the $\mathcal{Q}_p$ spaces we can prove the previous inequality as well as its converse for every $1 \leq p < \infty$ with the same method as in [10]. The next theorem can be found for example in Burkholder [19], [22], Davis [55], Garsia [82] and, for continuous time, in Dellacherie, Meyer [58].

Theorem 2.12. **(Burkholder-Davis-Gundy's inequality)** *The spaces H_p^S and H_p^* are equivalent for $1 \leq p < \infty$, namely,*

$$c_p\|f\|_{H_p^S} \leq \|f\|_{H_p^*} \leq C_p\|f\|_{H_p^S} \qquad (1 \leq p < \infty). \tag{2.20}$$

Since we apply Theorem 2.12 in the proof of some parts of Theorem 2.11 and Theorem 2.11 in the proof of Theorem 2.12, to bypass misunderstandings we prove these two theorems together.

Proof of Theorem 2.11. Part I. First of all we show that if a is a $(1, p, \infty)$ atom then $\|a\|_{H_p^*} \leq 2$ and $\|a\|_{H_p^S} \leq 1$ $(0 < p \leq 2)$. Indeed,

$$E(a^{*p}) = E(a^{*p}\chi(\nu \neq \infty)) \leq E^{p/2}(a^{*2})P(\nu \neq \infty)^{1-p/2}.$$

By Doob's inequality

$$E(a^{*2}) \leq 4E(a^2) = 4E(s^2(a)).$$

So

$$E(a^{*p}) \leq 2^p\left(P(\nu \neq \infty)^{-2/p}P(\nu \neq \infty)\right)^{p/2}P(\nu \neq \infty)^{1-p/2} = 2^p.$$

The inequality $\|a\|_{H_p^S} \leq 1$ can be proved similarly. The first inequality of (i) follows immediately from Theorem 2.2 for $0 < p \leq 1$.

Applying (2.4) and (2.12) to the function d_nf instead of f_n we can conclude that

$$d_nf = \sum_{k \in \mathbf{Z}}(d_nf)\chi(\nu_k < n \leq \nu_{k+1}).$$

Since, for fixed n, the sets $\{\nu_k < n \leq \nu_{k+1}\}$ are disjoint, we have

$$|d_nf|^2 = \sum_{k \in \mathbf{Z}}|d_nf|^2\chi(\nu_k < n \leq \nu_{k+1}). \tag{2.21}$$

Consequently,

$$S^2(f) = \sum_{k \in \mathbf{Z}} S^2(\mu_k a^k);$$

the definition of μ_k and a^k $(k \in \mathbf{N})$ can be found in the proof of Theorem 2.2. As $p/2 \leq 1$,

$$E(S^p(f)) \leq \sum_{k \in \mathbf{Z}} |\mu_k|^p E(S^p(a^k)) \leq \sum_{k \in \mathbf{Z}} |\mu_k|^p.$$

The second inequality of (i) follows from Theorem 2.2. Though this inequality could be proved with the help of (2.19), too, we proved it this way because this method will be used several times later as well.

Replacing $(\mathcal{A}_t, t \in \mathrm{T})$ by $(\mathcal{F}_{k-1}, k \in \mathrm{N})$, p by $p/2$ and f_t by $|d_k f|^2$ in (2.18) we obtain the second inequality of (ii).

(iii) comes easily from the definition.

It is simple to verify (see (i)) that $\|a\|_{H_p^*} \le 2$ if $a \in A_2(p,\infty)$ and $\|a\|_{H_p^s} \le 1$ if $a \in A_3(p,\infty)$ $(0 < p \le 2)$. Similarly to the proof of (i) we can also show that the first inequality of (iv) holds for $0 < p \le 1$ and the second one holds for $0 < p \le 2$. Denote by K_0 the set of bounded and, for some n, $\mathcal{F}_n$ measurable functions. K_0 is evidently contained by all the five martingale Hardy spaces introduced before. We apply an idea due to Chevalier (see [46]) to prove the first inequality of (iv) for all $f \in K_0$ and for $p > 1$. Assume that this inequality holds for all $f \in K_0$ and for a fixed $p/2$ $(p > 0)$, namely,

$$\|f\|_{H_{p/2}^*} \le C_{p/2} \|f\|_{\mathcal{Q}_{p/2}} \qquad (f \in K_0) \tag{2.22}$$

and show that it holds for p, too. Let

$$g_n := f_n^2 - S_n^2(f). \tag{2.23}$$

It is easy to see that $g = (g_n)$ is a martingale and $g \in K_0$ because $f \in K_0$. Moreover,

$$S_n^2(g) \le 4 f_{n-1}^{*\,2} S_n^2(f) \le 4 f_{n-1}^{*\,2} \lambda_{n-1}^2$$

where (λ_n) is a predictable sequence in L_p of $(S_n(f))$. By Hölder's inequality we obtain

$$\|g\|_{\mathcal{Q}_{p/2}} \le 2 \|f^* \lambda_\infty\|_{p/2} \le 2 \|f^*\|_p \|\lambda_\infty\|_p,$$

that is to say

$$\|g\|_{\mathcal{Q}_{p/2}} \le 2 \|f\|_{H_p^*} \|f\|_{\mathcal{Q}_p}. \tag{2.24}$$

Since $f_n^2 = g_n + S_n^2(f)$, we have

$$|f_n|^p \le 2^{p/2} \left(|g_n|^{p/2} + S_n^p(f) \right)$$

and

$$|f^*|^p \le 2^{p/2} \left(|g^*|^{p/2} + S^p(f) \right).$$

In other words

$$\|f\|_{H_p^*}^p \le 2^{p/2} \|g\|_{H_{p/2}^*}^{p/2} + 2^{p/2} \|f\|_{\mathcal{Q}_p}^p .$$

The next inequality follows from (2.22) and from (2.24):

$$\|f\|_{H_p^*}^p - 2^{p/2} \left(\|f\|_{\mathcal{Q}_p}^p + C_{p/2}^{p/2} 2^{p/2} \|f\|_{H_p^*}^{p/2} \|f\|_{\mathcal{Q}_p}^{p/2} \right) \le 0.$$

Solving this second-degree inequality for $z = \|f\|_{H_p^*}^{p/2}$ we get that

$$\|f\|_{H_p^*} \le (4C_{p/2} + \sqrt{2}) \|f\|_{\mathcal{Q}_p}$$

for all $f \in K_0$. Hence this inequality holds for all $f \in K_0$ and for all $1 < p < \infty$ because it holds for $0 < p \le 1$ as we could see above. It would be complicated to go on with the proof for all $f \in \mathcal{Q}_p$, Theorem 2.12 will be applied instead.

As in (2.21) the set $\{\nu_k < n \le \nu_{k+1}\}$ is $\mathcal{F}_{n-1}$ measurable, the proofs of both inequalities in (v) for $0 < p \le 2$ are again similar to the one of (i). ∎

This part of Theorem 2.11 shall be used in the proof of Theorem 2.12. Furthermore, for the proof we shall need Davis's decomposition of the martingales of H_p^S and H_p^* and some additional definitions. Let us denote by $\mathcal{G}_p$ $(0 < p < \infty)$ the space of martingales f for which

$$\|f\|_{\mathcal{G}_p} := \|\sum_{n=0}^{\infty} |d_n f|\|_p < \infty.$$

Note that $\mathcal{G}_p$ was introduced by Garsia [82].

Lemma 2.13. *Let $f \in H_p^S$ $(1 \le p < \infty)$. Then there exist $h \in \mathcal{G}_p$ and $g \in \mathcal{Q}_p$ such that $f_n = h_n + g_n$ for all $n \in \mathbb{N}$ and*

$$\|h\|_{\mathcal{G}_p} \le (2+2p)\|f\|_{H_p^S}, \qquad \|g\|_{\mathcal{Q}_p} \le (7+2p)\|f\|_{H_p^S}.$$

Lemma 2.14. *Let $f \in H_p^*$ $(1 \le p < \infty)$. Then there exist $h \in \mathcal{G}_p$ and $g \in \mathcal{P}_p$ such that $f_n = h_n + g_n$ for all $n \in \mathbb{N}$ and*

$$\|h\|_{\mathcal{G}_p} \le (4+4p)\|f\|_{H_p^*}, \qquad \|g\|_{\mathcal{P}_p} \le (13+4p)\|f\|_{H_p^*}.$$

The proofs of Lemmas 2.13 and 2.14 are similar, therefore we verify the first one, only. The second one can be found in Garsia [82] and Herz [94].

Proof of Lemma 2.13. Suppose that $\lambda_0 \le \lambda_1 \le \ldots$ is an adapted sequence of functions such that

$$S_n(f) \le \lambda_n, \qquad \lambda_\infty := \sup_{n \in \mathbb{N}} \lambda_n \in L_p.$$

Clearly,

$$d_n f = d_n f \chi(\lambda_n > 2\lambda_{n-1}) + d_n f \chi(\lambda_n \le 2\lambda_{n-1}).$$

Let

$$h := \sum_{k=1}^{\infty} \Big[d_k f \chi(\lambda_k > 2\lambda_{k-1}) - E_{k-1}(d_k f \chi(\lambda_k > 2\lambda_{k-1})) \Big]$$

and

$$g := \sum_{k=1}^{\infty} \Big[d_k f \chi(\lambda_k \le 2\lambda_{k-1}) - E_{k-1}(d_k f \chi(\lambda_k \le 2\lambda_{k-1})) \Big].$$

On the set $\{\lambda_k > 2\lambda_{k-1}\}$ we have $\lambda_k \le 2(\lambda_k - \lambda_{k-1})$, henceforth

$$|d_k f| \chi(\lambda_k > 2\lambda_{k-1}) \le \lambda_k \chi(\lambda_k > 2\lambda_{k-1}) \le 2(\lambda_k - \lambda_{k-1}).$$

Thus

$$\sum_{k=1}^{n} |d_k h| \le 2\lambda_n + 2\sum_{k=1}^{n} E_{k-1}(\lambda_k - \lambda_{k-1}).$$

Theorem 2.10 gives immediately

$$\|h\|_{\mathcal{G}_p} \le (2+2p)\|\lambda_\infty\|_p.$$

On the other hand, we obtain that

$$|d_k f|\chi(\lambda_k \le 2\lambda_{k-1}) \le \lambda_k \chi(\lambda_k \le 2\lambda_{k-1}) \le 2\lambda_{k-1},$$

consequently,

$$|d_k g| \le 4\lambda_{k-1}.$$

Finally we can conclude that

$$\begin{aligned} S_n(g) &\le S_{n-1}(g) + |d_n g| \\ &\le S_{n-1}(f) + S_{n-1}(h) + 4\lambda_{n-1} \\ &\le \lambda_{n-1} + 2\lambda_{n-1} + 2\sum_{k=1}^{n-1} E_{k-1}(\lambda_k - \lambda_{k-1}) + 4\lambda_{n-1}. \end{aligned}$$

As an application of Theorem 2.10 we get

$$\|g\|_{\mathcal{Q}_p} \le (7+2p)\|\lambda_\infty\|_p.$$

Setting $\lambda_n := S_n(f)$ we get Lemma 2.13. ∎

Now we are ready to prove Burkholder-Davis-Gundy's inequality.

Proof of Theorem 2.12. First we prove the right hand side of (2.20) for all $f \in H_1^S$ if $p = 1$ and for all $f \in K_0$ if $1 < p < \infty$. It is easy to check that

$$\|h\|_{H_p^*} \le \|h\|_{\mathcal{G}_p}, \qquad \|h\|_{H_p^S} \le \|h\|_{\mathcal{G}_p}. \tag{2.25}$$

Let $f \in H_1^S$ if $p = 1$ and $f \in K_0$ if $1 < p < \infty$. Then there exist $h \in \mathcal{G}_p$ and $g \in \mathcal{Q}_p$ such that Lemma 2.13 holds. Since g is a finite sum of bounded differences, $g \in K_0$. Applying these results and Theorem 2.11 (iv) to $g \in K_0$ we really get the right hand side of (2.20):

$$\|f\|_{H_p^*} \le \|h\|_{H_p^*} + \|g\|_{H_p^*} \le \|h\|_{\mathcal{G}_p} + C_p\|g\|_{\mathcal{Q}_p} \le C_p\|f\|_{H_p^S}.$$

The left hand side of the inequality (2.20) can be proved in the same way for all $f \in H_p^*$ if $1 \le p \le 2$. Notice that the more difficult part of Theorem 2.12, the well known Davis's inequality (case $p = 1$) has already been proved.

Applying again Chevalier's idea we show that the left hand side of (2.20) holds for all $f \in K_0$ if $p > 2$. First of all we note that inequality

$$\|g\|_{H_{p/2}^S} \le 2\|f\|_{H_p^*}\|f\|_{H_p^S} \tag{2.26}$$

can be proved the same way as (2.24). The fact that $S_n^2(f) = f_n^2 - g_n$ comes from (2.23), consequently,

$$\|f\|_{H_p^S}^p \le 2^{p/2}\|g\|_{H_{p/2}^*}^{p/2} + 2^{p/2}\|f\|_{H_p^*}^p.$$

It follows from the right hand side of (2.20) (for $p/2$) and from (2.26) that

$$\|f\|_{H_p^S}^p - 2^{p/2}\left(\|f\|_{H_p^*}^p + C_{p/2}^{p/2} 2^{p/2} \|f\|_{H_p^*}^{p/2} \|f\|_{H_p^S}^{p/2}\right) \le 0.$$

Solving this inequality for $z = \|f\|_{H_p^S}^{p/2}$ we obtain the left hand side of (2.20) for $p > 2$ and for $f \in K_0$.

Up to this time (2.20) has been verified for $p = 1$ and, moreover, for $1 < p < \infty$, if $f \in K_0$. As $H_p^* \sim L_p$ $(1 < p < \infty)$, the space H_p^* is a Banach one and K_0 is dense in H_p^* $(1 \le p < \infty)$. It can easily be proved that the last two statements hold for H_p^S, too. Theorem 2.12 follows immediately from these and from (2.13). ■

The proof of Theorem 2.11 can be finished.

Continuation of the proof of Theorem 2.11. Part II. The first inequality of (i) for $1 < p \le 2$ resp. the first inequality of (ii) for $2 \le p < \infty$ is an easy consequence of Theorem 2.12 and of the second inequality of (i) resp. (ii). The rest of (iv) follows from Theorem 2.12 and from (iii) for every $1 < p < \infty$. For $2 \le p < \infty$ the inequalities in (v) follow from (ii) and (iii). The proof of Theorem 2.11 is complete. ■

Note that (i), (ii) and the first part of (v) for $p = 1$ can be found in Burkholder [19], Burkholder, Gundy [27], and Garsia [82], with another proofs.

In addition to this theorem, with another method, it is proved in Section 2.5 that $\mathcal{P}_p$ is equivalent to $\mathcal{Q}_p$ $(0 < p < \infty)$.

We can see from Example 2.17 that in general case neither (i) for $2 \le p < \infty$ nor (ii) for $0 < p \le 2$ hold.

Note that Chevalier [49] has proved a slightly sharper inequality than (2.20), namely,

$$\|\sup[f^*, S(f)]\|_p \le C_p \|\inf[f^*, S(f)]\|_p \qquad (1 \le p < \infty).$$

From Lemma 2.13 and 2.14 and from inequalities

$$\|f\|_{H_p^s} \le C_p \|f\|_{\mathcal{P}_p}, \quad \|f\|_{H_p^s} \le C_p \|f\|_{\mathcal{Q}_p} \qquad (0 < p < \infty)$$

(see Theorem 2.11 (v)) we get the next lemma that was first proved by Herz [94] for H_1^*.

Lemma 2.15. *Let $f \in X$ where $X \in \{H_p^*, H_p^S\}$ $(1 \le p < \infty)$. Then there exist $h \in \mathcal{G}_p$ and $g \in H_p^s$ such that $f_n = h_n + g_n$ for all $n \in \mathbb{N}$ and*

$$\|h\|_{\mathcal{G}_p} \le C_p \|f\|_X, \qquad \|g\|_{H_p^s} \le C_p \|f\|_X.$$

This statement is trivial for $2 \le p < \infty$. Lemma 2.15 will be used later to prove an inequality between $\|\cdot\|_{H_p^s}$ and $\|\cdot\|_{H_p^*}$ (see Corollary 2.36).

It is well known that some martingales can be obtained in a simple way as the sum of some independent random variables. More exactly, if $x_1, x_2, \ldots$ are independent random variables with zero mean then $\left(f_n := \sum_{k=0}^n x_n\right)_{n\in\mathbb{N}}$ is a martingale with respect to the stochastic basis $\left(\mathcal{F}_n := \sigma(x_1, x_2, \ldots, x_n)\right)_{n\in\mathbb{N}}$. Indeed, $(x_1, x_2, \ldots)$ is a martingale difference sequence because $E_{n-1}x_n = Ex_n = 0$.

Marcinkiewicz and Zygmund [122] have proved that $\|f\|_p$ is equivalent to $\|S(f)\|_p$ $(1 \le p < \infty)$ in case the martingale f is the sum of independent random variables with zero mean. They gave a counterexample for which this equivalence does not hold if $0 < p < 1$. The following counterexample of Burkholder-Davis-Gundy's inequality for $0 < p < 1$ is a slightly modified version of the one due to Marcinkiewicz and Zygmund [122]. It can be found in Burkholder, Gundy [27] without proof.

Proposition 2.16. *In general case neither $c_p > 0$ nor $C_p > 0$ exist such that the inequality*

$$c_p\|S(f)\|_p \le \|f^*\|_p \le C_p\|S(f)\|_p \tag{2.27}$$

holds for all martingales if $0 < p < 1$.

Proof. Let j be a positive integer and $d^j := (d_1^j, d_2^j, \ldots)$ be a sequence of independent, identically distributed functions such that

$$\begin{aligned} P(d_k^j = 1) &= 1 - (j+1)^{-1}, \\ P(d_k^j = -j) &= (j+1)^{-1}. \end{aligned} \tag{2.28}$$

Let $f^j := (f_1^j, f_2^j, \ldots)$ be the martingale defined by $f_n^j := \sum_{k=1}^n d_k^j$. If $j \ge 2n$ then obviously $|f_n^j| \ge n$, thus

$$E[(f_n^j)^{*p}] \ge E(|f_n^j|^p) \ge n^p.$$

The sum $\sum_{k=1}^n |d_k^j|^2$ can be estimated by n on a set the measure of which is $(1-(j+1)^{-1})^n$ and by nj^2 on a set the measure of which is $1-(1-(j+1)^{-1})^n$. So

$$E[S(f_n^j)^p] \le n^{p/2}(1-(j+1)^{-1})^n + n^{p/2}j^p[1-(1-(j+1)^{-1})^n].$$

If the right hand side of (2.27) holds then we have for all $j \ge 2n$ that

$$n^p \le C_p^p\Big(n^{p/2}(1-(j+1)^{-1})^n + n^{p/2}j^p[1-(1-(j+1)^{-1})^n]\Big). \tag{2.29}$$

It is easy to check that for a fixed n and $0 < p < 1$

$$\lim_{j\to\infty} n^{p/2}j^p[1-(1-(j+1)^{-1})^n] = 0.$$

So, taking the limit $j \to \infty$ in (2.29) we obtain that

$$1 \le C_p^p n^{-p/2}$$

which does not hold for all $n \in \mathbb{N}$. Consequently, the right hand side of (2.27) can not hold for all martingales.

To give a counterexample for the left hand side of (2.27) we have to modify slightly the definition of d_k^j. Let the independent sequence $d^j := (d_1^j, d_2^j, \ldots)$ be given for odd j as in (2.28) and for even $j = 2l$ as follows:

$$\begin{aligned} P(d_k^{2l} = -1) &= 1 - (2l+1)^{-1}, \\ P(d_k^{2l} = 2l) &= (2l+1)^{-1}. \end{aligned}$$

Let the martingale $f^j := (f_1^j, f_2^j, \ldots)$ be the same as above. The inequality

$$E[S(f_n^j)^p] \geq n^{p/2}$$

is trivial. The maximal function $(f_n^j)^*$ can be estimated by 1 on a set the measure of which is $(1-(j+1)^{-1})^n$ and by nj on a set the measure of which is $1-(1-(j+1)^{-1})^n$. So

$$E[(f_n^j)^{*p}] \leq (1-(j+1)^{-1})^n + n^p j^p [1-(1-(j+1)^{-1})^n].$$

From the left hand side of (2.27) it would follow that for every j, n

$$c_p^p n^{p/2} \leq (1-(j+1)^{-1})^n + n^p j^p [1-(1-(j+1)^{-1})^n].$$

Taking again the limit $j \to \infty$, we can prove, as we did above, that the left hand side of (2.27) can not hold for all martingales, either. ∎

From this it follows that Lemmas 2.13 and 2.14 can not hold for $0 < p < 1$, otherwise, with the previous method, we would have shown Burkholder-Davis-Gundy's inequality for every p.

It comes from the next Example that the other Hardy spaces are also different in general case.

Example 2.17. *Let $\mathcal{F}_0 := \{\emptyset, \Omega\}$ and $\mathcal{F}_1 = \mathcal{F}_2 = \ldots = \mathcal{A}$. Then $H_p^* = H_p^S = L_p \cap L_1$, $\|f\|_{H_p^*} = \|f\|_{H_p^S} = \|f\|_p$, $H_p^s = L_2$ and $\mathcal{P}_p = \mathcal{Q}_p = L_\infty$ $(0 < p < \infty)$.*

Nevertheless, for a regular stochastic basis, Hardy spaces are all equivalent with each other. We prove a slightly more general result. First let us generalize the definition of regularity. A martingale f is said to be *previsible* if there exists a real number $R > 0$ such that

$$|d_n f|^2 \leq R E_{n-1} |d_n f|^2 \tag{2.30}$$

for all $n \in \mathbb{N}$. The class of previsible martingales having the same constant R in (2.30) is denoted by $\mathcal{V}_R$. Note that Burkholder and Gundy [19], [27] considered a little bit more general condition.

The assumption (2.30) can also be defined with the exponent p instead of 2.

Lemma 2.18. *If (2.30) holds then there exists a positive number R_p such that for all $n \in \mathbb{N}$*

$$|d_n f|^p \leq R_p E_{n-1} |d_n f|^p \qquad (0 < p < \infty). \tag{2.31}$$

Proof. Let $0 < p \leq 2$. From (2.30) we obtain

$$\begin{aligned} E_{n-1}|d_n f|^2 &= E_{n-1}(|d_n f|^{2-p} |d_n f|^p) \\ &\leq E_{n-1}\left[R^{(2-p)/2}(E_{n-1}|d_n f|^2)^{(2-p)/2} |d_n f|^p\right] \\ &= R^{(2-p)/2}(E_{n-1}|d_n f|^2)^{(2-p)/2} E_{n-1}|d_n f|^p. \end{aligned}$$

Thus

$$(E_{n-1}|d_n f|^2)^{p/2} \leq R^{(2-p)/2} E_{n-1} |d_n f|^p. \tag{2.32}$$

Again, by (2.30)

$$|d_n f|^p \le R^{p/2}(E_{n-1}|d_n f|^2)^{p/2} \le RE_{n-1}|d_n f|^p.$$

Note that $R_p = R$ for $0 < p \le 2$. For $2 \le p < \infty$ the inequality (2.31) can be obtained from Hölder's inequality with $R_p = R^{p/2}$. ∎

The condition (2.31) for $p = 1$ is belonging to Garsia ([82] III.3.15). Now we show that the condition (2.30) is 'almost' equivalent to the definition of the regularity of the stochastic basis.

Proposition 2.19. *If (2.30) holds for all martingales with the same constant R then the stochastic basis $\mathcal{F}$ is regular. The converse is also valid.*

Proof. Let $f = (f_n)$ be a non-negative martingale. Then

$$E_{n-1}|f_n - f_{n-1}| = 2E_{n-1}[(f_n - f_{n-1})^-] \le 2f_{n-1}.$$

From (2.30) and (2.32) with $p = 1$ we obtain

$$\begin{aligned}|d_n f|^2 &\le RE_{n-1}|d_n f|^2\\ &\le R^2(E_{n-1}|f_n - f_{n-1}|)^2\\ &\le 4R^2 f_{n-1}^2.\end{aligned}$$

Therefore

$$f_n \le f_{n-1} + |d_n f| \le (1+2R)f_{n-1} \qquad (n \in \mathbb{N})$$

which yields that $\mathcal{F}$ is regular. The converse comes immediately from the definition of regularity. ∎

The following lemma will be used in the proof of the equivalence of Hardy spaces.

Lemma 2.20. *For an arbitrary martingale f and $0 < p < \infty$ we have*

$$E(\sup_{n\in\mathbb{N}} E_{n-1}|f_n|^p) \le 2E(f^{*p})$$

and

$$E(\sup_{n\in\mathbb{N}} E_{n-1}|S_n^p(f)|) \le 2E[S^p(f)].$$

Proof. We prove the first inequality, only, the second one is similar. Obviously,

$$\begin{aligned}\sup_{n\in\mathbb{N}} E_{n-1}|f_n|^p &\le \sup_{n\in\mathbb{N}} E_{n-1}(f_n^{*p})\\ &= \sup_{n\in\mathbb{N}} E_{n-1}[f_{n-1}^{*p} + (f_n^{*p} - f_{n-1}^{*p})]\\ &\le f^{*p} + \sum_{n=1}^{\infty} E_{n-1}(f_n^{*p} - f_{n-1}^{*p}).\end{aligned}$$

The lemma follows immediately from this. ∎

Similarly to this proof for $p = 1$, we can verify the next theorem with the convexity theorem (see Theorem 2.10).

Corollary 2.21. *For an arbitrary martingale f and for $1 \le p < \infty$*

$$\| \sup_{n \in \mathbf{N}} E_{n-1}|f_n| \|_p \le (1+p)\|f^*\|_p,$$

moreover, if $\mathcal{F}$ is regular then the converse inequality holds also with the constant R.

Now we are in the position of being able to prove the equivalence of the five Hardy spaces.

Theorem 2.22. *For a previsible martingale $f \in \mathcal{V}_R$ one has for every $0 < p < \infty$ that*

$$\|f\|_{H_p^s} \le C_p\|f\|_{H_p^S} \le C_p\|f\|_{H_p^*} \le$$
$$\le C_p\|f\|_{\mathcal{P}_p} \le C_p\|f\|_{\mathcal{Q}_p} \le C_p\|f\|_{H_p^s}$$

where the constants C_p are depending only on the previsibility constant R and on p.

Proof. The inequalities

$$\|f\|_{H_p^*} \le \|f\|_{\mathcal{P}_p}, \quad \|f\|_{H_p^S} \le \|f\|_{\mathcal{Q}_p} \qquad (0 < p < \infty)$$

come from Theorem 2.11 (iii). To prove the converse of the first inequality let $f \in H_p^* \cap \mathcal{V}_R$. Then, by (2.31),

$$\begin{aligned} |f_n|^p &\le C_p(|f_{n-1}|^p + |d_n f|^p) \\ &\le C_p(f_{n-1}^{*\,p} + E_{n-1}|d_n f|^p) \\ &\le C_p(f_{n-1}^{*\,p} + E_{n-1}|f_n|^p). \end{aligned}$$

Together with Lemma 2.20 this implies that

$$\|f\|_{\mathcal{P}_p} \le C_p\|f\|_{H_p^*} \qquad (0 < p < \infty). \tag{2.33}$$

Notice that

$$\begin{aligned} S_n^p(f) &\le C_p(S_{n-1}^p(f) + E_{n-1}|d_n f|^p) \\ &\le C_p(S_{n-1}^p(f) + E_{n-1}|S_n^p(f)|). \end{aligned}$$

So the inequality

$$\|f\|_{\mathcal{Q}_p} \le C_p\|f\|_{H_p^S} \qquad (0 < p < \infty)$$

can be proved similarly to (2.33). From these and from Theorem 2.11 (iv) it follows that, for a previsible martingale f,

$$\|f\|_{H_p^*} \le C_p\|f\|_{H_p^S} \le C_p\|f\|_{H_p^*} \qquad (0 < p < \infty).$$

By Theorem 2.11 (i) and (v) we have

$$\|f\|_{H_p^*} \le C_p\|f\|_{H_p^s} \le C_p\|f\|_{H_p^*} \qquad (0 < p \le 2).$$

We can establish $S(f) \leq R^{1/2}s(f)$ by the previsibility property, so the inequality

$$\|f\|_{H_p^S} \leq C_p\|f\|_{H_p^s} \leq C_p\|f\|_{H_p^S} \qquad (2 \leq p < \infty)$$

follows from Theorem 2.11 (ii). The proof of the theorem is complete. ■

The inequalities between the H_1^* and $\mathcal{P}_1$ norms can be found in Garsia [82]. The inequalities between H_p^*, H_p^S and H_p^s are proved with another argument in Burkholder [19], Burkholder, Gundy [27] and Gundy [86].

The following corollary follows immediately from Proposition 2.19 and from Theorem 2.22.

Corollary 2.23. *If $\mathcal{F}$ is regular then H_p^s, H_p^S, H_p^*, $\mathcal{P}_p$ and $\mathcal{Q}_p$ are all equivalent* $(0 < p < \infty)$.

2.3. DUALITY THEOREMS

It was proved by Herz in [94] that the dual space of H_1^s is BMO_2. Furthermore, in [95] he gave a description of the dual of H_p^s in case $0 < p < 1$, too, and considering a sequence of atomic σ-algebras he proved that its dual space is equivalent to $\Lambda_2(\alpha)$ $(\alpha = 1/p - 1)$. We shall show with the help of the atomic decomposition that it is also true for a sequence of arbitrary σ-algebras (see Weisz [197]). We follow the proof due to Pratelli [147] and verify that the dual of H_p^s is H_q^s $(1 < p < \infty, 1/p + 1/q = 1)$. Also with the atomic decomposition, it is proved that $\Lambda_1(\alpha)$ is equivalent to a subspace of the dual of $\mathcal{P}_p$ and, in the regular case, the dual of $\mathcal{P}_p$ is $\Lambda_1(\alpha)$ $(0 < p \leq 1, \alpha = 1/p - 1)$ (see Weisz [197] and for $p = 1$ Bernard, Maisonneuve [10], Herz [94]). As a consequence, we shall obtain that $\Lambda_1(\alpha)$ is equivalent to $\Lambda_2(\alpha)$ $(\alpha \geq 0)$ in the regular case. As generalizations of BMO_p and BMO_p^-, the spaces $\mathcal{K}_q^p$ and $\mathcal{K}_q^{p-}$ are introduced and Herz type inequalities are given between the H_p^s and $\mathcal{K}_q^2$, $\mathcal{Q}_p$ and $\mathcal{K}_q^2$, and between the $\mathcal{P}_p$ and $\mathcal{K}_q^1$ norms $(1 \leq p < \infty)$. The dual of the space $\mathcal{G}_p$ $(1 \leq p < \infty)$, that is defined by the L_p norm of the l_1 norm of the martingale differences, is characterized. With this result a new proof of the duality between H_1^* and BMO_2^- is given. This theorem was first proved by Garsia [82] and Herz [94] in 1973. As a corollary, we get an inequality due to Rosenthal [150] and Burkholder [19] in which the H_q^* norm is estimated by the sum of the H_q^s norm and the L_q norm of the supremum of the martingale differences. The spaces H_1^s, H_1^* and $\mathcal{P}_1$ are non-reflexive. It is interesting to ask whether it can be found a subspace of BMO, as in the classical case (see Coifman, Weiss [52]), the dual of which is one of the Hardy spaces. We define the VMO_p resp. the VMO_p^- space as the closure of the vectorspace of the step functions in the BMO_p resp. BMO_p^- norm. A characterization of the VMO spaces with limit is given in case every σ-algebra is generated by finitely many atoms. If every σ-algebra is generated by countably many atoms then the duals of VMO_2, VMO_2^- and VMO_1 are H_1^s, H_1^* and $\mathcal{P}_1$, respectively (Weisz [197], [189]). This result was first proved by Schipp [165] for dyadic martingales.

First the duality between H_p^s and $\Lambda_2(\alpha)$ is proved.

Theorem 2.24. *The dual space of H_p^s is $\Lambda_2(\alpha)$ $(0 < p \leq 1, \alpha = 1/p - 1)$.*

Proof. By Theorem 2.2 one can establish that L_2 is dense in H_p^s. We are going to prove that

$$l_\phi(f) := E(f\phi) \qquad (f \in L_2)$$

is a continuous linear functional on H_p^s where $\phi \in \Lambda_2(\alpha)$ is arbitrary. Take the same stopping times ν_k, atoms a^k and real numbers μ_k $(k \in \mathbf{Z})$ as we did in Theorem 2.2. Modifying slightly the proof of that theorem we get that besides (2.1) the equation

$$\sum_{k\in\mathbf{Z}} \mu_k a^k = \sum_{k\in\mathbf{Z}} (f^{\nu_{k+1}} - f^{\nu_k}) = f$$

holds a.e. and also in L_2 norm if $f \in L_2$.

Since $\phi \in L_2$, we have

$$l_\phi(f) = \sum_{k\in\mathbf{Z}} \mu_k E(a^k \phi).$$

By (i) of the definition of the atom a^k

$$E(a^k\phi) = E[a^k(\phi - \phi^{\nu_k})].$$

Using Hölder's inequality and (ii) of Definition 2.1 we can conclude that

$$\begin{aligned} |l_\phi(f)| &\le \sum_{k\in\mathbf{Z}} |\mu_k| \int_\Omega |a^k||\phi - \phi^{\nu_k}|\, dP \\ &\le \sum_{k\in\mathbf{Z}} |\mu_k| \|a^k\|_2 \|\phi - \phi^{\nu_k}\|_2 \\ &\le \sum_{k\in\mathbf{Z}} |\mu_k| P(\nu_k \ne \infty)^{-1/p+1/2} \|\phi - \phi^{\nu_k}\|_2 \\ &\le \sum_{k\in\mathbf{Z}} |\mu_k| \|\phi\|_{\Lambda_2(\alpha)}. \end{aligned}$$

Since $0 < p \le 1$, the inequality

$$|l_\phi(f)|^p \le \sum_{k\in\mathbf{Z}} |\mu_k|^p \|\phi\|^p_{\Lambda_2(\alpha)}$$

holds. Consequently, we obtain from Theorem 2.2 that

$$|l_\phi(f)| \le C_p \|f\|_{H_p^s} \|\phi\|_{\Lambda_2(\alpha)} \qquad (0 < p \le 1, f \in L_2). \tag{2.34}$$

Conversely, let $l \in (H_p^s)'$ be an arbitrary element of the dual space. We show that there exists $\phi \in \Lambda_2(\alpha)$ such that $l = l_\phi$ and

$$\|\phi\|_{\Lambda_2(\alpha)} \le \|l\|. \tag{2.35}$$

First we note that by

$$\|f\|_{H_p^s} \le \|s(f)\|_2 = \|f\|_2 \qquad (f \in L_2)$$

the space L_2 can be embedded continuously in H_p^s. Consequently, there exists $\phi \in L_2$ such that

$$l(f) = E(f\phi) \qquad (f \in L_2).$$

Let ν be an arbitrary stopping time and

$$g := \frac{\phi - \phi^\nu}{\|\phi - \phi^\nu\|_2 P(\nu \neq \infty)^{1/p-1/2}}. \tag{2.36}$$

The function g is not necessarily a $(1, p, \infty)$ atom, however, it satisfies the condition (i) of Definition 2.1, namely,

$$s(g) = s(g)\chi(\nu \neq \infty).$$

By Hölder's inequality we obtain

$$\begin{aligned}\|g\|_{H_p^s} &= \frac{\left(\int_\Omega s^p(\phi - \phi^\nu)\chi(\nu \neq \infty)\,dP\right)^{1/p}}{\|\phi - \phi^\nu\|_2 P(\nu \neq \infty)^{1/p-1/2}} \\ &\le \frac{\left(\int_\Omega |\phi - \phi^\nu|^2\,dP\right)^{1/2} P(\nu \neq \infty)^{1/p-1/2}}{\|\phi - \phi^\nu\|_2 P(\nu \neq \infty)^{1/p-1/2}} = 1.\end{aligned}$$

Thus

$$\|l\| \ge |l(g)| = E(g(\phi - \phi^\nu)) = P(\nu \neq \infty)^{-1/p+1/2}\|\phi - \phi^\nu\|_2$$

which proves (2.35) and the theorem as well. ∎

The dual of H_p^s $(1 < p < \infty)$ was characterized for discrete parameter by Herz [95] and for continuous parameter by Pratelli [147]. We follow the proof belonging to Pratelli. Another proof can be found in Chapter 5 (see Theorem 5.17). First we prove that the H_p^s spaces are uniformly convex.

Theorem 2.25. *If $2 \le q < \infty$ then the H_q^s spaces are uniformly convex.*

Proof. We remark that a Banach space X is uniformly convex if for all $\epsilon > 0$ there exists $\delta > 0$ such that the properties $x, y \in X$, $\|x\| \le 1$, $\|y\| \le 1$ and $\|x - y\| \ge \epsilon$ imply $\|x + y\| \le 2(1 - \delta)$.

We shall use the following well-known inequalities. For two positive real numbers x, y and for $1 \le p < \infty$ we have

$$(x + y)^p \le 2^{p-1}(x^p + y^p), \qquad (x + y)^p \ge (x^p + y^p).$$

If $f, g \in H_q^s$ then it is easy to see that

$$s^2(f + g) + s^2(f - g) = 2[s^2(f) + s^2(g)].$$

Since $q \ge 2$, we obtain

$$\begin{aligned}[s^2(f + g)]^{q/2} + [s^2(f - g)]^{q/2} &\le [s^2(f + g) + s^2(f - g)]^{q/2} \\ &\le 2^{q/2} 2^{q/2-1}[s^q(f) + s^q(g)],\end{aligned}$$

that is to say

$$s^q(f+g) \le 2^{q-1}[s^q(f)+s^q(g)] - s^q(f-g).$$

Using the properties $\|f\|_{H_q^s} \le 1$, $\|g\|_{H_q^s} \le 1$ and $\|f-g\|_{H_q^s} \ge \epsilon$ we get

$$\|f+g\|_{H_q^s} \le (2^q - \epsilon^q)^{1/q} = 2\Big(1-\frac{\epsilon^q}{2^q}\Big)^{1/q}.$$

With the choice

$$\delta = 1 - \Big(1-\frac{\epsilon^q}{2^q}\Big)^{1/q}$$

the theorem is proved. ■

By Milman's theorem (see Yosida [204] p. 127) uniform convexity implies reflexivity. Thus H_q^s $(2 \le q < \infty)$ is reflexive. Consequently, the following theorem is enough to be shown for $1 < p \le 2$.

Theorem 2.26. *If $1 < p < \infty$ then the dual of H_p^s is H_q^s where $1/p + 1/q = 1$.*

Proof. We can suppose that $1 < p \le 2$. Let $g \in H_q^s$ and

$$l_g(f) := E\Big(\sum_{n=0}^{\infty} d_n f d_n g\Big) \qquad (f \in H_p^s).$$

By Hölder's inequality

$$\begin{aligned} |l_g(f)| &\le \sum_{n=0}^{\infty} E\big[E_{n-1}(|d_n f||d_n g|)\big] \\ &\le E\Big[\sum_{n=0}^{\infty} (E_{n-1}|d_n f|^2)^{1/2} (E_{n-1}|d_n g|^2)^{1/2}\Big] \\ &\le \|f\|_{H_p^s} \|g\|_{H_q^s}. \end{aligned}$$

Thus

$$\|l_g\| \le \|g\|_{H_q^s}.$$

Reversely, if $l \in (H_p^s)'$ then $l \in L_2'$, consequently, there exists $g \in L_2$ such that

$$l(f) = l_g(f) = E(fg) \qquad (f \in L_2). \tag{2.37}$$

First assume that $g \in H_\infty^s$. Set

$$v_n := \frac{s_n^{q-2}(g)}{\|g\|_{H_q^s}^{q-1}}.$$

We define a martingale h as the martingale transform of g by (v_n), namely,

$$d_n h := v_n d_n g.$$

Since v_n is $\mathcal{F}_{n-1}$ measurable, h is really a martingale. Thus

$$s^2(h) = \sum_{n=0}^{\infty} \frac{s_n^{2q-4}(g)}{\|g\|_{H_q^s}^{2q-2}} E_{n-1}|d_n g|^2 \le \frac{s^{2q-2}(g)}{\|g\|_{H_q^s}^{2q-2}}.$$

Since $g \in H_\infty^s$, this implies that $h \in L_2$. On the other hand,

$$E[s^p(h)] \le \frac{E\left[s^{(2q-2)p/2}(g)\right]}{\|g\|_{H_q^s}^{(2q-2)p/2}} = 1.$$

Henceforth

$$\begin{aligned}\|l\| &\ge \sup_{f\in L_2, \|f\|_{H_p^s}\le 1} |l(f)| \ge |l(h)| \\ &= \frac{1}{\|g\|_{H_q^s}^{q-1}} E\left[\sum_{n=0}^{\infty} s_n^{q-2}(g) E_{n-1}|d_n g|^2\right] \\ &= \frac{1}{\|g\|_{H_q^s}^{q-1}} E\left[\sum_{n=0}^{\infty} s_n^{q-2}(g)\left(s_n^2(g) - s_{n-1}^2(g)\right)\right].\end{aligned}$$

Applying the classical inequality

$$x^\alpha - 1 \le \alpha(x-1)x^{\alpha-1} \qquad (1 \le \alpha, x)$$

to $x = s_n^2(g)/s_{n-1}^2(g)$ and to $\alpha = q/2 \ge 1$ we get that

$$\frac{2}{q}[s_n^q(g) - s_{n-1}^q(g)] \le [s_n^2(g) - s_{n-1}^2(g)]s_n^{q-2}(g).$$

From this we can conclude that

$$\begin{aligned}(2.38) \qquad \|l\| &\ge \sup_{f\in L_2, \|f\|_{H_p^s}\le 1} |E(fg)| \\ &\ge \frac{2}{q}\frac{1}{\|g\|_{H_q^s}^{q-1}} E(s^q(g)) \\ &= \frac{2}{q}\|g\|_{H_q^s}.\end{aligned}$$

Now we prove (2.38) for an arbitrary g satisfying (2.37). Let

$$\nu_k := \inf\{n \in \mathbb{N} : s_{n+1}(g) > k\}$$

and g^{ν_k} the corresponding stopped martingale. As $g \in L_2$, it is easy to see that $\nu_k \nearrow \infty$. Moreover, g^{ν_k} is an element of H_∞^s. It follows from (2.38) that

$$\|g^{\nu_k}\|_{H_q^s} \le \frac{q}{2} \sup_{f\in L_2, \|f\|_{H_p^s}\le 1} |E(fg^{\nu_k})|.$$

Since $E(fg^{\nu_k}) = E(f^{\nu_k}g)$, $f^{\nu_k} \in L_2$ and $\|f^{\nu_k}\|_{H_p^s} \leq 1$, we obtain

$$\|g^{\nu_k}\|_{H_q^s} \leq \frac{q}{2} \sup_{f \in L_2, \|f\|_{H_p^s} \leq 1} |E(fg)|.$$

Obviously, $s(g^{\nu_k}) \nearrow s(g)$ a.e. as $k \to \infty$. Therefore

$$\|g\|_{H_q^s} \leq \frac{q}{2}\|l\|$$

which proves the theorem. ■

The dual space of H_∞^s is considered in the dyadic case in Weisz [192].

Note that, similarly to (2.34), the inequality

$$(2.39) \qquad |l_\phi(f)| \leq C_p\|f\|_{\mathcal{Q}_p}\|\phi\|_{\Lambda_2(\alpha)} \qquad (0 < p \leq 1, \alpha = 1/p - 1; f \in H_\infty^S)$$

can be verified. However, by Example 2.17, in which $\mathcal{Q}_p = L_\infty$ and $\Lambda_2(\alpha) = L_2$ were obtained, $\Lambda_2(\alpha)$ is not equivalent to a subspace of $\mathcal{Q}_p'$ because the dual of L_∞ is the space of bounded, finitely additive measures.

Now we consider the dual of $\mathcal{P}_p$. Let us denote by $(\mathcal{P}_p')_1$ those elements l from the dual space of $\mathcal{P}_p$ for which there exists $\phi \in L_1$ such that

$$l(f) = E(f\phi) \qquad (f \in L_\infty).$$

In spite of the fact that Garsia claims in his book ([82], p. 130) that the dual of $\mathcal{P}_p$ is $\Lambda_1(\alpha)$, this does not hold since in Example 2.17 we have $\mathcal{P}_p = L_\infty$ and $\Lambda_1(\alpha) = L_1$ and the dual of L_∞ is usually not L_1. However, the following theorem is true:

Theorem 2.27. *$\Lambda_1(\alpha)$ is equivalent to a subspace of the dual of $\mathcal{P}_p$, more precisely, $(\mathcal{P}_p')_1 \sim \Lambda_1(\alpha)$ $(0 < p \leq 1, \alpha = 1/p - 1)$.*

Proof. The proof is similar to the one of Theorem 2.24, so we give it in sketch, only. If $\phi \in \Lambda_1(\alpha)$ and

$$l_\phi(f) := E(f\phi) \qquad (f \in L_\infty)$$

then the inequality

$$(2.40) \qquad |l_\phi(f)| \leq C_p\|f\|_{\mathcal{P}_p}\|\phi\|_{\Lambda_1(\alpha)} \qquad (f \in L_\infty)$$

can be proved as we did previously since

$$\sum_{k \in \mathbf{Z}} \mu_k a^k = \sum_{k \in \mathbf{Z}} (f^{\nu_{k+1}} - f^{\nu_k}) = f \qquad a.e.$$

and the partial sums of this series can be majorized by $2f^* \in L_\infty$.

To prove the converse, instead of the test function g defined in (2.36), take the $(3, p, \infty)$ atoms

$$a := \frac{1}{2}P(A)^{-1/p}\chi(A)(h - E_n h)$$

where $n \in \mathbf{N}$ and $A \in \mathcal{F}_n$ are arbitrary and

$$h := \operatorname{sign}(\phi - E_n\phi).$$

The proof of the theorem is complete. ∎

This theorem, for $p = 1$, can be found in Bernard, Maisonneuve [10] and Herz [94].

If L_2 can be embedded continuously in the space $\mathcal{P}_p$ then clearly $(\mathcal{P}'_p)_1 = \mathcal{P}'_p$. Hence, in this case, the dual of $\mathcal{P}_p$ is $\Lambda_1(\alpha)$. In Corollary 2.23 it was proved that $\mathcal{P}_p$ and H^*_p are equivalent in case $\mathcal{F}$ is regular. Since L_2 can be embedded continuously in the space H^*_p, the dual of $\mathcal{P}_p$ is $\Lambda_1(\alpha)$. In the regular case $\mathcal{P}_p$ is also equivalent to H^s_p, so their dual spaces are equivalent, too. Thus we obtain the following

Corollary 2.28. *If $\mathcal{F}$ is regular then the dual of $\mathcal{P}_p$ is $\Lambda_1(\alpha)$ $(0 < p \leq 1, \alpha = 1/p - 1)$, moreover, $\Lambda_1(\alpha) \sim \Lambda_2(\alpha)$ $(\alpha \geq 0)$.*

Now we extend the inequalities (2.34), (2.39) and (2.40) to the case $p \geq 1$. For this let us generalize the BMO_p spaces:

Definition 2.29. *The martingale spaces $\mathcal{K}^p_q$ and $\mathcal{K}^{p-}_q$ $(1 \leq p \leq q \leq \infty)$ are defined with the norm*

$$\|f\| := \inf_\gamma \|\gamma\|_q$$

where γ satisfies

$$E_n|f - f_n|^p \leq E_n\gamma^p \qquad (n \in \mathrm{N})$$

and

$$E_n|f - f_{n-1}|^p \leq E_n\gamma^p \qquad (n \in \mathrm{N}),$$

respectively.

By a short calculation it can be proved that $\mathcal{K}^p_\infty = BMO_p$, $\mathcal{K}^{p-}_\infty = BMO^-_p$ and, moreover, $\mathcal{K}^p_p \sim \mathcal{K}^{p-}_p \sim L_p$ $(1 < p < \infty)$. The following Herz type inequalities can be formulated.

Theorem 2.30. *Let $1/p + 1/q = 1$. For $1 \leq p < 2$ and for $\phi \in \mathcal{K}^2_q$ we have*

$$|l_\phi(f)| \leq 4\|f\|_{H^s_p}\|\phi\|_{\mathcal{K}^2_q} \qquad (f \in H^s_\infty)$$

and

$$|l_\phi(f)| \leq 4\|f\|_{\mathcal{Q}_p}\|\phi\|_{\mathcal{K}^2_q} \qquad (f \in H^S_\infty).$$

For $1 \leq p < \infty$ and for $\phi \in \mathcal{K}^1_q$ we have

$$|l_\phi(f)| \leq 8\|f\|_{\mathcal{P}_p}\|\phi\|_{\mathcal{K}^1_q} \qquad (f \in \mathcal{P}_\infty = L_\infty).$$

Moreover, H^s_∞ and H^S_∞ are dense in H^s_p and in $\mathcal{Q}_p$ $(1 \leq p < 2)$, respectively, and so is L_∞ in $\mathcal{P}_p$ $(1 \leq p < \infty)$.

For the proof see Weisz [198]. The third inequality was proved by Garsia [82]. Some other details about the $\mathcal{K}_q$ spaces can be found in the next section.

Independently of one another it was proved by Garsia [82] and Herz [94] (in the classical case by Fefferman and Stein [69]) that the dual of H^S_1 is BMO^-_2. A new proof of this result the idea of which is due to Bernard and Maisonneuve [10] is demonstrated. First the dual of $\mathcal{G}_p$ ($1 \le p < \infty$) will be characterized. Recall that the definition of the $\mathcal{G}_p$ norm is

$$\|f\|_{\mathcal{G}_p} := \|\sum_{n=0}^{\infty} |d_n f|\|_p.$$

Obviously, $\mathcal{G}_p$ is a subspace of $L_p(l_1)$. Similarly, we define some subspaces of $L_q(l_\infty)$ containing martingales.

Definition 2.31. *Denote by BD_q ($1 \le q \le \infty$) the space of martingales $f = (f_n, n \in \mathbf{N})$ for which*

$$\|f\|_{BD_q} := \|\sup_n |d_n f|\|_q < \infty.$$

Theorem 2.32. *The dual space of $\mathcal{G}_p$ is BD_q where $1 \le p < \infty$ and $1/p + 1/q = 1$.*

Proof. Setting $\phi \in BD_q$ and

$$l_\phi(f) := \sum_{k=1}^{\infty} E(d_k f d_k \phi) \qquad (f \in \mathcal{G}_p)$$

we obtain that

$$\begin{aligned} |l_\phi(f)| &\le E\Big(\sum_{k=1}^{\infty} |d_k f||d_k \phi|\Big) \\ &\le E\Big(\sum_{k=0}^{\infty} |d_k f| \sup_{k \in \mathbf{N}} |d_k \phi|\Big) \\ &\le \|f\|_{\mathcal{G}_p} \|\phi\|_{BD_q}, \end{aligned}$$

namely, $l_\phi \in (\mathcal{G}_p)'$ and $\|l_\phi\| \le \|\phi\|_{BD_q}$.

Conversely, let $l \in (\mathcal{G}_p)'$ be an arbitrary element of the dual space. We show that there exists $\phi \in BD_q$ such that $l = l_\phi$ and

$$\|\phi\|_{BD_q} \le \frac{2q}{q-1}\|l\|. \tag{2.41}$$

Let us embed $\mathcal{G}_p$ in the space $L_p(l_1)$ with the map $f \mapsto (d_k f, k \in \mathbf{N})$. Recall that the dual of $L_p(l_1)$ is $L_q(l_\infty)$ (see Lemma 2.9). By Banach-Hahn's theorem l can be extended onto $L_p(l_1)$ preserving its norm. Denoting by Λ the extension of l we have by Lemma 2.9 that there exists $\eta \in L_q(l_\infty)$ such that $\|\Lambda\| = \|l\| = \|\eta\|_{L_q(l_\infty)}$ and (2.16) hold. Thus

$$l(f_n) = \sum_{k=1}^{n} E[(d_k f)\eta_k] = \sum_{k=1}^{n} E\big[(d_k f)(E_k \eta_k - E_{k-1}\eta_k)\big]. \tag{2.42}$$

Defining

$$\phi_n := \sum_{k=1}^{n} (E_k \eta_k - E_{k-1}\eta_k) \qquad (\phi_0 := 0)$$

one can see that $\phi = (\phi_n, n \in \mathrm{N})$ is a martingale. Since

$$\sup_{k\in\mathrm{N}} |d_k\phi| \leq \sup_{k\in\mathrm{N}}(E_k|\eta_k| + E_{k-1}|\eta_k|) \leq 2\sup_{n\in\mathrm{N}} E_n(\sup_{k\in\mathrm{N}}|\eta_k|),$$

we derive (2.41) from Doob's inequality. Using the fact that $f_n \to f$ in $\mathcal{G}_p$ norm we have from (2.42) that $l = l_\phi$. ■

Note that if $\phi \in L_2 \cap BD_q$ and $f \in L_2 \cap \mathcal{G}_p$ then

$$l_\phi(f) = \lim_{n\to\infty} l_\phi(f_n) = \lim_{n\to\infty} E(f_n\phi_n) = E(f\phi). \tag{2.43}$$

It is worthy to emphasize the next consequence hidden in the proof of Theorem 2.32. A similar result for BMO_2^- can be found in Garsia [82].

Corollary 2.33. *If $\phi \in BD_q$ $(1 < q \leq \infty)$ then there exists a sequence $\eta = (\eta_n, n \in \mathrm{N}) \in L_q(l_\infty)$ of functions such that*

$$\phi_n = \sum_{k=1}^{n} d_k\eta_k \qquad (n \in \mathrm{N})$$

and

$$\|\eta\|_{L_q(l_\infty)} \leq \|\phi\|_{BD_q} \leq \frac{2q}{q-1}\|\eta\|_{L_q(l_\infty)}.$$

Using (2.65) of the next section and the equation

$$E_n|f - f_l|^2 = E_n\Big(\sum_{k=l+1}^{\infty} |d_kf|^2\Big) \qquad (l \geq n-1) \tag{2.44}$$

we obtain that

$$\sup\{\|\phi\|_{BMO_2}, \|\phi\|_{BD_\infty}\} \leq \|\phi\|_{BMO_2^-}. \tag{2.45}$$

On the other hand, it is easy to see that

$$\|\phi\|_{BMO_2^-} \leq \|\phi\|_{BMO_2} + \|\phi\|_{BD_\infty}. \tag{2.46}$$

Hence $BMO_2^- = BMO_2 \cap BD_\infty$. Though the dual of H_p^* $(1 < p < \infty)$ is known, together with the description of the one of H_1^*, it is worthy to characterize it, too.

Theorem 2.34. *The dual space of H_p^* $(1 \leq p \leq 2)$ can be given with the norm*

$$\|\phi\| := \|\phi\|_{H_q^s} + \|\phi\|_{BD_q} \qquad (2 \leq q \leq \infty)$$

where $1/p + 1/q = 1$ and with the only usage of the notation $H_\infty^s := BMO_2$.

Proof. Let $\phi \in H_q^s \cap BD_q$ be fixed. Note that, in this case, clearly $\phi \in L_2$. We shall prove that

$$l_\phi(f) := E(f\phi) \qquad (f \in L_2) \tag{2.47}$$

is a bounded linear functional of H_p^* $(1 \le p \le 2)$. Since L_2 is dense in H_p^*, functional l_ϕ is well defined. As $f_n \to f$ in L_2 norm $(n \to \infty)$, we have

$$l_\phi(f) = \lim_{n\to\infty} E(f_n\phi).$$

It comes from Davis's decomposition (see Lemma 2.15) that there exist martingales h and g such that $f_n = h_n + g_n$ and

$$\|h\|_{\mathcal{G}_p} \le C_p\|f\|_{H_p^*}, \qquad \|g\|_{H_p^s} \le C_p\|f\|_{H_p^*}.$$

If $f \in L_2$ then functions h_n and g_n are finite sums of square integrable differences, so they are in L_2, too. Henceforth

$$|E(f_n\phi)| \le |E(g_n\phi)| + |E(h_n\phi)|.$$

Applying Theorem 2.24, 2.26, 2.32 and (2.43) we can conclude that

$$\begin{aligned}|E(f_n\phi)| &\le C_p\|g_n\|_{H_p^s}\|\phi\|_{H_q^s} + \|h_n\|_{\mathcal{G}_p}\|\phi\|_{BD_q}\\ &\le C_p\|g\|_{H_p^s}\|\phi\|_{H_q^s} + \|h\|_{\mathcal{G}_p}\|\phi\|_{BD_q}.\end{aligned}$$

Now, from Lemma 2.15, we get that

$$|E(f\phi)| \le C_p\|f\|_{H_p^*}(\|\phi\|_{H_q^s} + \|\phi\|_{BD_q}), \tag{2.48}$$

namely, l_ϕ is really a bounded linear functional.

Conversely, assume that l is an arbitrary bounded linear functional on H_p^*. From Doob's inequality we have $\|f\|_{H_p^*} \le 2\|f\|_2$ $(1 \le p \le 2)$, thus l is also a bounded linear functional of L_2. Consequently, there exists $\phi \in L_2$ such that

$$l(f) = l_\phi(f) = E(f\phi) \qquad (f \in L_2).$$

On the other hand,

$$\|f\|_{H_p^s} \le C_p\|f\|_{H_p^*} \qquad (1 \le p \le 2)$$

(see Theorem 2.11 (i)) and, obviously,

$$\|f\|_{H_p^*} \le \|f\|_{\mathcal{G}_p} \qquad (1 \le p < \infty).$$

Henceforth, l is also bounded on H_p^s and on $\mathcal{G}_p$. It comes from Theorem 2.2 that L_2 is dense in H_p^s $(1 \le p \le 2)$. Moreover, it can easily be proved that $L_2 \cap \mathcal{G}_p$ is dense in $\mathcal{G}_p$. Consequently, we obtain from Theorem 2.24 and 2.26 that

$$\|\phi\|_{H_q^s} \le C_q\|l\| \qquad (2 \le q \le \infty)$$

and from Theorem 2.32 and (2.43) that

$$\|\phi\|_{BD_q} \le C_q\|l\| \qquad (2 \le q \le \infty).$$

Hence

$$\|\phi\|_{H_q^s} + \|\phi\|_{BD_q} \le C_q\|l\| \qquad (2 \le q \le \infty).$$

The proof of the theorem is complete. ∎

Single out this result for $p=1$. The inequalities (2.45) and (2.46) imply that

$$\|\cdot\|_{BMO_2^-} \sim \|\cdot\|_{BMO_2} + \|\cdot\|_{BD_\infty}.$$

So the following corollary follows.

Corollary 2.35. *The dual of H_1^* is BMO_2^-.*

We remark that if $1<p<\infty$ then the dual of H_p^* is H_q^* $(1/p+1/q=1)$. Thus the H_q^* norm and the norm given in Theorem 2.34 are equivalent.

Corollary 2.36. *For a martingale f we have*

$$\|f^*\|_q \le C_q\|s(f)\|_q + C_q\|\sup_{n\in\mathbf{N}}|d_nf|\|_q \qquad (2\le q<\infty).$$

Note that the converse of this inequality follows from Theorem 2.11 (ii), too. The statement of the above corollary is verified in Theorem 2.11 (i) for $0<q\le 2$. Corollary 2.36 was proved by Rosenthal [150] in case the differences (d_nf) are independent. Three years later Burkholder [19] proved it for arbitrary martingales. In Section 4.2 applying this inequality we shall prove the L_p $(1<p<\infty)$ norm convergence of Vilenkin-Fourier series.

In case $0<p<1$ both the duals of H_p^* and H_p^S are unknown, in general. However, without proof, we give a special result due to Herz [95].

Theorem 2.37. *Consider a sequence of atomic σ-algebras. Then the dual of H_p^S is $\Lambda_2^-(\alpha)$ $(0<p\le 1, \alpha=1/p-1)$.*

Of course, the duals of BMO_2 and BMO_2^- are not H_1^s and H_1^*, respectively. Nevertheless, H_1^s and H_1^* are equivalent to certain subspaces of the duals of BMO_2 and BMO_2^-, respectively. If $l_f(\phi)=l_\phi(f)$ then l_f is a bounded linear functional on BMO_2 resp. on BMO_2^- where $\phi\in BMO_2$ and $f\in H_1^s$ resp. $\phi\in BMO_2^-$ and $f\in H_1^*$. Moreover, the following inequalities hold also with the constant C_1 in (2.34) resp. (2.48):

$$\|f\|_{H_1^s} \le \|l_f\| \le C_1\|f\|_{H_1^s} \tag{2.49}$$

resp.

$$\|f\|_{H_1^*} \le \|l_f\| \le C_1\|f\|_{H_1^*}. \tag{2.50}$$

However, a kind of special subspaces of BMO_2 and BMO_2^- can be defined the duals of which are H_1^s and H_1^*, respectively. These subspaces will be denoted by VMO_2 and VMO_2^-, respectively. The relations between H_1^s, BMO_2 and VMO_2, and, moreover, between H_1^*, BMO_2^- and VMO_2^- are quite similar to the relation between l_1, l_∞ and the space c_0 of 0 sequences. It is known that the dual of the non-reflexive space l_1 is l_∞ and the dual of c_0 is l_1. The spaces H_1^s and H_1^* are two of the few examples of a separable, non-reflexive Banach space which is a dual space itself. Another example is the classical Hardy space (see Coifman, Weiss [52]).

From this time on to the end of this section let us suppose that every σ-algebra $\mathcal{F}_n$ is generated by *countably many (set) atoms*. Denote by $A(\mathcal{F}_n)$ the set of the atoms of the σ-algebra $\mathcal{F}_n$ and let

$$A(\mathcal{F}) := \bigcup_{n \in \mathbf{N}} A(\mathcal{F}_n).$$

Let us write L' and L to denote the linear envelope of the set

$$\{\chi(A) : A \in A(\mathcal{F})\} \tag{2.51}$$

and the vector space

$$\{\phi \in L' : E_0\phi = 0\}, \tag{2.52}$$

respectively. Let VMO_q and VMO_q^- be the closures of L in BMO_q and in BMO_q^- norm $(1 \le q < \infty)$, respectively. The elements of VMO are called the functions of *vanishing mean oscillation.* We shall see that the BMO_q norm is identical with the following one:

$$\|\phi\| := \sup_{n \in \mathbf{N}} \sup_{A \in A(\mathcal{F}_n)} P(A)^{-1/q} \Big(\int_A |\phi - \phi^A|^q \, dP\Big)^{1/q}$$

where $\phi^A := P(A)^{-1} \int_A \phi \, dP$. Indeed, suppose that $A \in A(\mathcal{F}_n)$ for some $n \in \mathbf{N}$, then $E_n\phi = \phi^A$ on the set A and

$$\|\phi\|_{BMO_q} \ge P(A)^{-1/q} \Big(\int_A |\phi - \phi^A|^q \, dP\Big)^{1/q}.$$

On the other hand, let $n \in \mathbf{N}$ and $A \in \mathcal{F}_n$ be arbitrary, so one has $A = \cup_{k=1}^{\infty} A_k$ where $A_k \in A(\mathcal{F}_n)$. Henceforth,

$$\begin{aligned} P(A)^{-1} \int_A |\phi - E_n\phi|^q \, dP &= P(A)^{-1} \sum_{k=1}^{\infty} \int_{A_k} |\phi - \phi^{A_k}|^q \, dP \\ &\le \|\phi\|^q P(A)^{-1} P\Big(\bigcup_{k=1}^{\infty} A_k\Big) = \|\phi\|^q. \end{aligned}$$

Similarly,

$$\|\phi\|_{BMO_q^-} = \sup_{n \in \mathbf{N}} \sup_{A \in A(\mathcal{F}_n)} P(A)^{-1/q} \Big(\int_A |\phi - \phi^{A^-}|^q \, dP\Big)^{1/q}$$

where, for an atom $A \in A(\mathcal{F}_n)$, A^- denotes the atom $A^- \in \mathcal{F}_{n-1}$ for which $A \subset A^-$.

If $\phi \in VMO_q$ and $\psi \in VMO_q^-$, it is obvious that

$$\lim_{n \to \infty} \sup_{A \in A(\mathcal{F}_n)} P(A)^{-1/q} \Big(\int_A |\phi - \phi^A|^q \, dP\Big)^{1/q} = 0$$

and

$$\lim_{n \to \infty} \sup_{A \in A(\mathcal{F}_n)} P(A)^{-1/q} \Big(\int_A |\psi - \psi^{A^-}|^q \, dP\Big)^{1/q} = 0.$$

If every σ-algebra $\mathcal{F}_n$ is generated by finitely many atoms then even the converse of the preceding statement holds.

Proposition 2.38. *If every σ-algebra $\mathcal{F}_n$ is generated by finitely many atoms then for the functions $\phi \in BMO_q$ and $\psi \in BMO_q^-$ we have $\phi \in VMO_q$ and $\psi \in VMO_q^-$ if and only if*

$$\lim_{n\to\infty} \|(E_n|\phi - E_n\phi|^q)^{1/q}\|_\infty = 0 \tag{2.53}$$

and

$$\lim_{n\to\infty} \|(E_n|\psi - E_{n-1}\psi|^q)^{1/q}\|_\infty = 0. \tag{2.54}$$

Proof. Assume that $\phi \in BMO_q$ satisfying (2.53). Let N be an index such that for all $n \geq N$

$$\|(E_n|\phi - E_n\phi|^q)^{1/q}\|_\infty < \epsilon$$

($\epsilon > 0$). Clearly, $E_N\phi \in L$ and the inequality

$$\|\phi - E_N\phi\|_{BMO_q} < \epsilon$$

follows from the equations

$$(\phi - E_N\phi) - E_n(\phi - E_N\phi) = (\phi - E_N\phi) \qquad (n < N)$$

and

$$(\phi - E_N\phi) - E_n(\phi - E_N\phi) = (\phi - E_n\phi) \qquad (n \geq N)$$

and from the inequality

$$\|(E_n|\phi - E_N\phi|^q)^{1/q}\|_\infty \leq \|(E_N|\phi - E_N\phi|^q)^{1/q}\|_\infty \qquad (n < N).$$

The formula (2.54) can be proved similarly. ∎

Now we can identify the dual of VMO_2.

Theorem 2.39. *If every σ-algebra $\mathcal{F}_n$ is generated by countably many atoms then the dual of VMO_2 is H_1^s.*

Proof. By Theorem 2.24 we have that

$$l_f(\phi) := \lim_{n\to\infty} E(f^n\phi) = E(f\phi) \qquad (\phi \in L)$$

is a bounded linear functional on VMO_2 ($f^n \in L_2$ and $f^n \to f$ in H_1^s norm).

Conversely, we can conclude that if $l \in VMO_2'$ then there exists $f \in H_1^s$ such that

$$l(\phi) = E(f\phi) \qquad (\phi \in L) \tag{2.55}$$

and

$$\|f\|_{H_1^s} \leq \|l\|. \tag{2.56}$$

To verify this, we embed the normed vector space $(L, \|\cdot\|_{VMO_2})$ isometrically in a space the dual of which can easily be found. Let

$$X_A := L_2(A, \mathcal{A} \cap A, P) =: L_2(A)$$

and

$$\|\xi\|_{X_A} := P(A)^{-1/2}\|\xi\|_{L_2(A)} \qquad (A \in A(\mathcal{F})). \tag{2.57}$$

It is obvious that

$$\left|\int_A \xi\zeta \, dP\right| \le P(A)^{-1/2}\|\xi\|_{L_2(A)} P(A)^{1/2}\|\zeta\|_{L_2(A)} \qquad (\xi, \zeta \in X_A).$$

On the other hand, it follows from Riesz's representation theorem that if $\Psi \in X_A'$ then there exists a unique $f \in L_2(A)$ such that

$$\Psi(\xi) = \int_A f\xi \, dP \qquad (\xi \in X_A)$$

and

$$\|\Psi\| = P(A)^{1/2}\|f\|_{L_2(A)}.$$

Let

$$X := \times_{A \in A(\mathcal{F})} X_A$$

with the norm

$$\|\xi\|_X := \sup_{A \in A(\mathcal{F})} \|\xi_A\|_{X_A} \qquad (\xi = (\xi_A, A \in A(\mathcal{F})) \in X).$$

Extend the functions of X_A from A to the whole Ω such that they take the value 0 outside A. Denote by X_0 those elements $\xi \in X$ for which $\xi_A = 0$ except for finitely many $A \in A(\mathcal{F})$. It is easy to see that if $\Lambda \in X_0'$ then there exists $f_A \in X_A$ $(A \in A(\mathcal{F}))$ such that

$$\Lambda(\xi) = \sum_{A \in A(\mathcal{F})} \int_A f_A \xi_A \, dP \qquad (\xi \in X_0)$$

and

$$\|\Lambda\| = \sum_{A \in A(\mathcal{F})} P(A)^{1/2}\|f_A\|_2 < \infty.$$

Now we embed $(L, \|\cdot\|_{VMO_2})$ in X_0 the following way:

$$R : L \longrightarrow X_0, \qquad R\phi := \Big((\phi - \phi^A)\chi(A), A \in A(\mathcal{F})\Big).$$

If $l \in (VMO_2)'$ then $l \circ R^{-1}$ is a bounded linear functional on the range of R, thus, by Banach-Hahn's theorem, $l \circ R^{-1}$ can be extended onto X_0 preserving its norm. Consequently, there exists $f_A \in X_A$ $(A \in A(\mathcal{F}))$ such that

$$\|l\| = \|l \circ R^{-1}\| = \sum_{A \in A(\mathcal{F})} P(A)^{1/2}\|f_A\|_2 \tag{2.58}$$

and

$$l(\phi) = \sum_{A \in A(\mathcal{F})} \int_A f_A(\phi - \phi^A)\, dP \qquad (\phi \in L).$$

It is easy to show that the last equation can be rewritten in the following form:

$$\begin{aligned} l(\phi) &= \sum_{A \in A(\mathcal{F})} \int_A (f_A - f_A^A)\phi\, dP \\ &= \sum_{k=0}^{\infty} \int_\Omega (f_k - E_k f_k)\phi\, dP \qquad (\phi \in L) \end{aligned} \tag{2.59}$$

where

$$\sum_{A \in A(\mathcal{F}_k)} f_A \chi(A) = f_k \qquad (k \in \mathbb{N}).$$

We show that the series

$$\sum_{A \in A(\mathcal{F})} (f_A - f_A^A \chi(A)) \tag{2.60}$$

converges a.e. and also in L_1 norm to a function $f \in H_1^s$ for which (2.56) holds. If $A \in A(\mathcal{F}_k)$ for some $k \in \mathbb{N}$, we have $\chi(A) f_A^A = E_k f_A$. Applying the inequality

$$\|E_k f_A\|_1 \le \|f_A\|_1 \le P(A)^{1/2} \|f_A\|_2$$

and considering (2.58) we get that the series (2.60) converges a.e. and also in L_1 norm to a function $f \in L_1$.

Furhermore, it is easy to see that

$$\|f\|_{H_1^s} \le \sum_{A \in A(\mathcal{F})} \|f_A - f_A^A \chi(A)\|_{H_1^s}.$$

We shall prove the inequality

$$\|f_A - f_A^A \chi(A)\|_{H_1^s} \le P(A)^{1/2} \|f_A\|_2. \tag{2.61}$$

Let $A \in A(\mathcal{F}_n)$ for a natural number $n \in \mathbb{N}$. Then

$$\begin{aligned} \|f_A - E_n f_A\|_{H_1^s} &= \int_A s(f_A - E_n f_A)\, dP \\ &\le \Big(\int_A s^2(f_A - E_n f_A)\, dP \Big)^{1/2} P(A)^{1/2}, \end{aligned}$$

that is to say,

$$\begin{aligned} \|f_A - E_n f_A\|_{H_1^s} &\le P(A)^{1/2} \Big(\int_A s^2(f_A)\, dP \Big)^{1/2} \\ &= P(A)^{1/2} \Big(\int_A |f_A|^2\, dP \Big)^{1/2}. \end{aligned}$$

Consequently, because of (2.58), the inequality (2.56) holds, indeed. The equation (2.55) follows easily from the L_1 convergence of the series (2.60). The proof of Theorem 2.39 has been completed. ■

Since

$$\sum_{A\in A(\mathcal{F})} P(A)^{1/2}\|f_A\|_2 = \sum_{n=0}^{\infty} \|(E_n|f_n|^2)^{1/2}\|_1, \tag{2.62}$$

using the same argument as before, we get the following

Corollary 2.40. *Let every σ-algebra $\mathcal{F}_n$ be generated by countably many atoms and $f \in H_1^s$. Then there exist functions $f_n \in L_2$ $(n \in \mathrm{N})$ such that*

$$f = \sum_{n=0}^{\infty}(f_n - E_n f_n)$$

a.e. and also in L_1 norm and, moreover,

$$C_1^{-1}\sum_{n=0}^{\infty} \|(E_n|f_n|^2)^{1/2}\|_1 \le \|f\|_{H_1^s} \le \sum_{n=0}^{\infty} \|(E_n|f_n|^2)^{1/2}\|_1$$

where C_1 is the constant appeared in (2.49).

The dual of VMO_1 can be identified in the same way.

Theorem 2.41. *If every σ-algebra $\mathcal{F}_n$ is generated by countably many atoms then the dual of VMO_1 is $\mathcal{P}_1$.*

Corollary 2.42. *Let every σ-algebra $\mathcal{F}_n$ be generated by countably many atoms and $f \in \mathcal{P}_1$. Then there exist functions $f_A \in L_\infty$ $(A \in A(\mathcal{F}))$ such that*

$$f = \sum_{A\in A(\mathcal{F})} (f_A - f_A^A \chi(A))$$

a.e. and also in L_1 norm and, moreover,

$$C_1^{-1} \sum_{A\in A(\mathcal{F})} P(A)\|f_A\|_\infty \le \|f\|_{\mathcal{P}_1} \le 2 \sum_{A\in A(\mathcal{F})} P(A)\|f_A\|_\infty$$

where C_1 is the constant appeared in (2.40).

Proof. As the proof is similar to the one of Theorem 2.39, we mention the essential differences, only. Instead of (2.57) we have to introduce the norm

$$\|\xi\|_{X_A} := P(A)^{-1}\|\xi\|_{L_1(A)}.$$

Instead of (2.58) we get that

$$\|l\| = \|l \circ R^{-1}\| = \sum_{A\in A(\mathcal{F})} P(A)\|f_A\|_\infty.$$

The statement corresponding to (2.61) can be proved as follows. It is easy to see that

$$\frac{f_A - f_A^A \chi(A)}{\|f_A - f_A^A \chi(A)\|_\infty} P(A)^{-1}$$

is a $(3,1,\infty)$ atom. Hence

$$\|f_A - f_A^A \chi(A)\|_{\mathcal{P}_1} \le \|f_A - f_A^A \chi(A)\|_\infty P(A) \le 2\|f_A\|_\infty P(A)$$

which proves both the theorem and the corollary. ∎

For other parameter q and under more general conditions the dual of the VMO_q space is given in Chapter 3. The dual of the closure of L in $\Lambda_2(\alpha)$ norm is investigated in Weisz [197].

Similar results hold for VMO_2^-. The following theorem was proved in the classical case by Coifman and Weiss [52] and for the one-parameter dyadic martingales by Schipp [165].

Theorem 2.43. *If every σ-algebra $\mathcal{F}_n$ is generated by countably many atoms then the dual of VMO_2^- is H_1^*.*

Proof. By Theorem 2.34, for a function $f \in H_1^*$, we have that

$$l_f(\phi) := E(f\phi) \qquad (\phi \in L)$$

is a bounded linear functional on VMO_2^-.

Since the proof of the converse is similar to the one in Theorem 2.39, we shall outline the main steps, only. The spaces X, X_0 and the functional Λ are defined the same way as there. The embedding of $(L, \|\cdot\|_{VMO_2^-})$ in X_0 is here the following:

$$R : L \longrightarrow X_0, \qquad R\phi := \left((\phi - \phi^{A^-})\chi(A), A \in A(\mathcal{F})\right).$$

In this case (2.59) can be written as

$$\begin{aligned} l(\phi) &= \sum_{A \in A(\mathcal{F})} \int_A (f_A - f_A^{A^-})\phi \, dP \\ &= \sum_{k=0}^{\infty} \int_\Omega (f_k - E_{k-1} f_k)\phi \, dP \qquad (\phi \in L). \end{aligned}$$

Applying the conditional Hölder's inequality we get that

$$\|E_{k-1} f_k\|_1 \le \|f_k\|_1 = \|E_k |f_k|\|_1 \le \|(E_k |f_k|^2)^{1/2}\|_1. \tag{2.63}$$

Because of (2.62) we obtain that the series

$$\sum_{k=0}^{\infty} (f_k - E_{k-1} f_k)$$

converges a.e. and also in L_1 norm to a function $f \in L_1$.

We shall show that $f \in H_1^*$ holds. It is easy to see that

$$\|f\|_{H_1^*} \leq \sum_{k=0}^{\infty} \|f_k - E_{k-1} f_k\|_{H_1^*}.$$

Obviously,

$$\begin{aligned} \|f_k - E_{k-1} f_k\|_{H_1^*} &= \| \sup_{n\geq k} |E_n(f_k - E_{k-1} f_k)| \|_1 \\ &\leq \| \sup_{n\geq k} |E_n f_k| \|_1 + \|E_{k-1} f_k\|_1. \end{aligned} \tag{2.64}$$

Moreover, by the conditional Hölder's inequality we have

$$\| \sup_{n\geq k} |E_n f_k| \|_1 = \|E_k(\sup_{n\geq k} |E_n f_k|)\|_1 \leq \|(E_k(\sup_{n\geq k} |E_n f_k|)^2)^{1/2}\|_1$$

for each $n \in \mathbb{N}$. Since

$$\chi(E) \sup_{n\geq k} |E_n f| = \sup_{n\geq k} |E_n(\chi(E) f)|$$

for all $f \in L_1$ and $E \in \mathcal{F}_k$, we can see by (2.15) that

$$E_k(\sup_{n\geq k} |E_n f|)^p \leq \left(\frac{p}{p-1}\right)^p E_k |f|^p \qquad (p > 1).$$

Applying this inequality to $p = 2$ we obtain

$$\| \sup_{n\geq k} |E_n f_k| \|_1 \leq 2 \|(E_k |f_k|^2)^{1/2}\|_1.$$

Therefore, it follows from (2.64), (2.63) and (2.62) that

$$\|f\|_{H_1^*} \leq 3 \sum_{k=0}^{\infty} \|(E_k |f_k|^2)^{1/2}\|_1 = 3\|l\|.$$

The proof can be finished as in Theorem 2.39. ∎

The proof of Theorem 2.43 contains the following information concerning the structure of H_1^*.

Corollary 2.44. *Let every σ-algebra $\mathcal{F}_n$ be generated by countably many atoms and $f \in H_1^*$. Then there exist functions $f_n \in L_2$ $(n \in \mathbb{N})$ such that*

$$f = \sum_{n=0}^{\infty} (f_n - E_{n-1} f_n)$$

a.e. and also in L_1 norm and, moreover,

$$C_1^{-1} \sum_{n=0}^{\infty} \|(E_n |f_n|^2)^{1/2}\|_1 \leq \|f\|_{H_1^*} \leq 3 \sum_{n=0}^{\infty} \|(E_n |f_n|^2)^{1/2}\|_1$$

where C_1 is the constant appeared in (2.50).

In the next section it is proved that the BMO_q^- spaces are all equivalent ($1 \leq q < \infty$), so the VMO_q^- spaces are equivalent, too.

2.4. *BMO* SPACES AND SHARP FUNCTIONS

In this section the connections between BMO, L_p and Hardy spaces are investigated. The equivalence of the BMO_p^- spaces ($1 \leq p < \infty$) was proved by John and Nirenberg [109] in the classical case and by Garsia [82] and Herz [94] for martingales. This theorem is extended to all $0 < p < \infty$. Two new BMO spaces, namely, BMO_r^s resp. BMO_r^S ($0 < r < \infty$) generated by the conditional quadratic resp. quadratic variation are introduced. Besides the equivalence of the BMO_p^- spaces ($1 \leq p < \infty$), we prove that $BMO_r^s \sim BMO_2$ and that $BMO_r^S \sim BMO_2^-$ for all $0 < r < \infty$. Our proof is similar to the one due to Garsia [82] and Bassily, Mogyoródi [3]. Note that the BMO_p spaces are usually not equivalent. The BMO_r^s spaces will be used in Chapter 5. As a consequence, we get that $s^2(f)$ and $S^2(f)$ are exponentially integrable if $\|f\|_{BMO_2} < 1$ resp. $\|f\|_{BMO_2^-} < 1$. If the stochastic basis is regular then all BMO_p^- and BMO_q spaces are equivalent. It was proved by Fefferman and Stein [69] in the classical case and by Garsia [82] in the martingale case that the L_p norm of the sharp function $f^\sharp$ is equivalent to the H_p^* norm of f ($1 < p < \infty$). We define two other sharp functions, namely, f^S resp. f^s with the help of the quadratic resp. conditional quadratic variation and prove that $\|f^S\|_p \sim \|f\|_{H_p^S}$ and $\|f^s\|_p \sim \|f\|_{H_p^s}$. Finally, some equivalences between the L_q norms of the sharp functions, the $\mathcal{K}_q$ and H_q norms are obtained.

It is easy to see that

$$\|f\|_{BMO_p} \leq 2\|f\|_\infty, \qquad \|f\|_{BMO_p^-} \leq 2\|f\|_\infty.$$

Moreover,

$$\begin{aligned}\|f\|_{BMO_2} &= \sup_{n\in\mathbf{N}} \|(E_n|f - f_n|^2)^{1/2}\|_\infty \\ &= \sup_{n\in\mathbf{N}} \|(E_n[s^2(f) - s_n^2(f)])^{1/2}\|_\infty \\ &\leq \sup_{n\in\mathbf{N}} \|(E_n[s^2(f)])^{1/2}\|_\infty \leq \|s(f)\|_\infty.\end{aligned}$$

It was verified in the previous sections that the dual of H_1^* is BMO_2^- and $L_p \subset H_1^*$ in case $1 < p \leq \infty$. The equivalence between the BMO_p^- spaces ($1 \leq p < \infty$) will be proved later, so we shall have

$$L_\infty \subset BMO_p^- \subset L_q \qquad (1 \leq q < \infty).$$

Furthermore, the dual of H_1^s is BMO_2, the dual of H_p^s is H_q^s and $H_p^s \subset H_1^s$ ($1 < p < \infty, 1/p + 1/q = 1$). Hence

$$L_\infty, H_\infty^s \subset BMO_2 \subset H_q^s \qquad (1 \leq q < \infty).$$

If $f \in BMO_p^-$ then $f_n \in L_\infty$ since

$$|f_n - f_{n-1}| \leq \|f\|_{BMO_p^-} \qquad (1 \leq p < \infty). \tag{2.65}$$

To prove (2.65) let us remark that the conditional expectations $(E_n|f_m - f_{n-1}|)_{m\geq n}$ increase as m increases and $n \in \mathrm{N}$ is fixed. This follows from the fact that the sequence $(|f_m - f_{n-1}|)_{m\geq n}$ is a submartingale. Since $f \in L_p$, this submartingale converges a.e. and also in L_1 norm to the function $|f - f_{n-1}|$. Thus

$$|f_m - f_{n-1}| \leq E_m|f - f_{n-1}|, \tag{2.66}$$

consequently,

$$E_n|f_m - f_{n-1}| \leq E_n|f - f_{n-1}| \qquad (m \geq n). \tag{2.67}$$

Setting $m = n$ in the last inequality and using Hölder's inequality we get (2.65). It is easy to see that

$$\|f\|_{BMO_p} \leq 2\|f\|_{BMO_p^-}. \tag{2.68}$$

Indeed, applying (2.65) and the inequality

$$(E_n|f - f_n|^p)^{1/p} \leq (E_n|f - f_{n-1}|^p)^{1/p} + |f_n - f_{n-1}|$$

we obtain (2.68). So the following relation holds:

$$L_\infty \subset BMO_p^- \subset BMO_p \subset L_p \qquad (1 \leq p < \infty).$$

Notice that the BMO_p spaces are usually not equivalent because in Example 2.17 we could see that $BMO_p = L_p$.

Now we introduce two new types of BMO spaces which are all equivalent to BMO_2 resp. to BMO_2^-. One of them, the so-called BMO_1^s space was first introduced by Garsia ([82] p. 159). In his book (p. 160) it is conjectured that BMO_1^s is not equivalent to BMO_2.

Definition 2.45. *BMO_r^s resp. BMO_r^S $(0 < r < \infty)$ denotes the set of the martingales $f = (f_n, n \in \mathrm{N})$ for which*

$$\|f\|_{BMO_r^s} := \sup_{n\in\mathrm{N}} \|(E_n[s^2(f) - s_n^2(f)]^{r/2})^{1/r}\|_\infty < \infty$$

resp.

$$\|f\|_{BMO_r^S} := \sup_{n\in\mathrm{N}} \|(E_n[S^2(f) - S_{n-1}^2(f)]^{r/2})^{1/r}\|_\infty < \infty.$$

Note that $BMO_2^s = BMO_2$, $H_\infty^s \subset BMO_r^s$ and $BMO_2^S = BMO_2^-$, $H_\infty^S \subset BMO_r^S$ are trivial. The BMO_r^s spaces will be used during the proofs of the interpolation theorems in Chapter 5.

To handle these two spaces together we introduce the following notation that will be used in this section, only. BMO_r^i $(0 < r < \infty; i \in \mathrm{N})$ consists of all sequences $(X_n, n \in \mathrm{N})$ of non-decreasing, non-negative functions for which X_n is $\mathcal{F}_{n-i}$ measurable, $X_n = 0$ for $0 \leq n < i \vee 1$ and

$$\|(X_n)\|_{BMO_r^i} := \sup_{n\in\mathrm{N}} \|(E_n[X^2 - X_{n-1+i}^2]^{r/2})^{1/r}\|_\infty < \infty$$

where $X := \sup_{n\in N} X_n$. Of course,

$$\|f\|_{BMO_r^s} = \|(s_n(f))\|_{BMO_r^1}$$

and

$$\|f\|_{BMO_r^S} = \|(S_n(f))\|_{BMO_r^0}.$$

The proofs of the equivalence of the BMO_q^- resp. BMO_r^s resp. BMO_r^S spaces are similar. For the last two equivalence theorems we give a common proof with the help of the BMO_r^i spaces.

Lemma 2.46. ([3]) *If γ is a function such that*

$$E_l|f - f_{l-1}| \le E_l\gamma$$

for all $l \in N$ then for arbitrary $\beta > \alpha > 0$

$$(\beta - \alpha)E_1[\chi(f_n^* > \beta)] \le E_1[\chi(f_n^* > \alpha)\gamma] \qquad (n \in N).$$

Proof. Let us introduce the following stopping times:

$$\nu_\lambda := \inf\{k \in N : |f_k| > \lambda\}$$

where $\lambda > 0$ is arbitrary. Obviously, $\nu_\beta \ge \nu_\alpha$. In this case

$$\begin{aligned} E_1[\chi(f_n^* > \beta)] &= E_1[\chi(\nu_\beta \le n)] \\ &= E_1\Big[\sum_{k=1}^{n}\sum_{l=1}^{k}\chi(\nu_\beta = k, \nu_\alpha = l)\Big]. \end{aligned}$$

On the set $\{\nu_\beta = k, \nu_\alpha = l\}$ we have $|f_k| > \beta$ and $|f_{l-1}| \le \alpha$, so $|f_k| - |f_{l-1}| > \beta - \alpha$. From this and from (2.66) we get that

$$\begin{aligned} E_1[\chi(f_n^* > \beta)] &\le E_1\Big[\sum_{k=1}^{n}\sum_{l=1}^{k}\chi(\nu_\beta = k, \nu_\alpha = l)\frac{|f_k| - |f_{l-1}|}{\beta - \alpha}\Big] \\ &\le E_1\Big[\sum_{l=1}^{n}\sum_{k=l}^{n}\chi(\nu_\beta = k, \nu_\alpha = l)\frac{E_k|f - f_{l-1}|}{\beta - \alpha}\Big]. \end{aligned}$$

Using the fact that the set $\{\nu_\beta = k, \nu_\alpha = l\}$ is $\mathcal{F}_k$ measurable we obtain

$$\begin{aligned} E_1[\chi(f_n^* > \beta)] &\le (\beta - \alpha)^{-1}E_1\Big[\sum_{l=1}^{n}\sum_{k=l}^{n}\chi(\nu_\beta = k, \nu_\alpha = l)|f - f_{l-1}|\Big] \\ &\le (\beta - \alpha)^{-1}E_1\Big[\sum_{l=1}^{n}\chi(\nu_\alpha = l)E_l|f - f_{l-1}|\Big]. \end{aligned}$$

Therefore, the inequality

$$\begin{aligned}(\beta-\alpha)E_1[\chi(f_n^*>\beta)] &\le E_1\Big[\sum_{l=1}^{n}\chi(\nu_\alpha=l)E_l\gamma\Big]\\ &\le E_1\Big[\sum_{l=1}^{n}\chi(\nu_\alpha=l)\gamma\Big]\\ &= E_1[\chi(\nu_\alpha\le n)\gamma]\end{aligned}$$

together with Lemma 2.46 are also proved. ∎

As the following lemma can be shown in the same way as Lemma 2.46, we sketch the proof, only (cf. Weisz [194]).

Lemma 2.47. *Let $0<r<\infty$ and $i\in\mathbf{N}$ be fixed. Assume that $(X_n, n\in\mathbf{N})$ is a sequence of non-decreasing, non-negative functions such that X_n is $\mathcal{F}_{n-i}$ measurable and $X_n=0$ for $0\le n<i\vee 1$. If γ is a function such that*

$$E_l(X_{m+i}^2-X_{l-1+i}^2)^{r/2}\le E_l\gamma^r$$

for all $m\ge l\ge(1-i)\vee 0$ then for arbitrary $\beta>\alpha>0$

$$(\beta-\alpha)^r E_{(1-i)\vee 0}[\chi(X_{n+i}>\beta)]\le E_{(1-i)\vee 0}[\chi(X_{n+i}>\alpha)\gamma^r] \qquad (n\in\mathbf{N}).$$

Proof. We have to work with the stopping times

$$\nu_\lambda:=\inf\{k\in\mathbf{N}: X_{k+i}>\lambda\} \qquad (\lambda>0).$$

In this case

$$E_{(1-i)\vee 0}[\chi(X_{n+i}>\beta)]=E_{(1-i)\vee 0}\Big[\sum_{k=(1-i)\vee 0}^{n}\ \sum_{l=(1-i)\vee 0}^{k}\chi(\nu_\beta=k,\nu_\alpha=l)\Big].$$

On the set $\{\nu_\beta=k,\nu_\alpha=l\}$ one has $X_{k+i}-X_{l+i-1}>\beta-\alpha$. Recalling that (X_n) is non-decreasing we have

$$\begin{aligned}&E_{(1-i)\vee 0}[\chi(X_{n+i}>\beta)]\\ &\quad=E_{(1-i)\vee 0}\Big[\sum_{l=(1-i)\vee 0}^{n}\ \sum_{k=l}^{n}\chi(\nu_\beta=k,\nu_\alpha=l)\frac{(X_{n+i}-X_{l+i-1})^r}{(\beta-\alpha)^r}\Big].\end{aligned}$$

Since

$$X_{n+i}-X_{l+i-1}\le(X_{n+i}^2-X_{l+i-1}^2)^{1/2}, \tag{2.69}$$

we obtain

$$(\beta-\alpha)^r E_{(1-i)\vee 0}[\chi(X_{n+i}>\beta)]$$
$$\leq E_{(1-i)\vee 0}\Big[\sum_{l=(1-i)\vee 0}^{n} \chi(\nu_\alpha = l) E_l (X_{n+i}^2 - X_{l+i-1}^2)^{r/2}\Big]$$
$$\leq E_{(1-i)\vee 0}[\chi(\nu_\alpha \leq n)\gamma^r].$$

The lemma is proved. ■

The following lemma is a simple generalization of the one due to Garsia [82] (p. 66).

Lemma 2.48. *Let $0 < r \leq 1$ and $i \in \mathbf{N}$ be fixed. Assume that $(A_n, n \in \mathbf{N})$ is a sequence of non-decreasing, non-negative functions such that A_n is $\mathcal{F}_{n-i}$ measurable, $A_n = 0$ for $0 \leq n < i \vee 1$ and*

$$[E_k(A_{n+i} - A_{k-1+i})^r]^{1/r} \leq B \qquad (n \geq k \geq (1-i)\vee 0). \tag{2.70}$$

Then for all $n \in \mathbf{N}$

$$E_{(1-i)\vee 0}(e^{tA_{n+i}^r}) \leq \frac{1}{1-tB^r}$$

provided that $tB^r < 1$ $(t > 0)$.

Proof. A slightly more general inequality will be proved. Let $m(u)$ be a non-decreasing function on $[0,\infty)$. Then

$$E_{(1-i)\vee 0}\Big[\int_0^{A_{n+i}^r} m(u)\,du\Big] = \sum_{k=(1-i)\vee 0}^{n} E_{(1-i)\vee 0}\Big[\int_{A_{k+i-1}^r}^{A_{k+i}^r} m(u)\,du\Big]$$
$$\leq \sum_{k=(1-i)\vee 0}^{n} E_{(1-i)\vee 0}[m(A_{k+i}^r)(A_{k+i}^r - A_{k+i-1}^r)].$$

Thus, setting $p_{(1-i)\vee 0} := m(A_{i+(1-i)\vee 0}^r)$ and

$$p_l := m(A_{l+i}^r) - m(A_{l+i-1}^r) \qquad (l > (1-i)\vee 0),$$

we obtain

$$E_{(1-i)\vee 0}\Big[\int_0^{A_{n+i}^r} m(u)\,du\Big] \leq \sum_{k=(1-i)\vee 0}^{n} \sum_{l=(1-i)\vee 0}^{k} E_{(1-i)\vee 0}[p_l(A_{k+i}^r - A_{k+i-1}^r)]$$
$$= \sum_{l=(1-i)\vee 0}^{n} E_{(1-i)\vee 0}[p_l(A_{n+i}^r - A_{l+i-1}^r)]$$
$$= \sum_{l=(1-i)\vee 0}^{n} E_{(1-i)\vee 0}[p_l E_l(A_{n+i}^r - A_{l+i-1}^r)].$$

The inequality $A^r_{n+i} - A^r_{l+i-1} \le (A_{n+i} - A_{l+i-1})^r$ and (2.70) imply

$$E_{(1-i)\vee 0}\Big[\int_0^{A^r_{n+i}} m(u)\,du\Big] \le B^r E_{(1-i)\vee 0}\Big(\sum_{l=(1-i)\vee 0}^{n} p_l\Big).$$

If A_n satisfies (2.70) then so does $A_n \wedge N$ ($N \in \mathbf{N}$) because

$$A_{n+i} \wedge N - A_{k+i-1} \wedge N \le A_{n+i} - A_{k+i-1}.$$

Consequently,

$$\begin{aligned} E_{(1-i)\vee 0}\Big[\int_0^{(A_{n+i}\wedge N)^r} m(u)\,du\Big] &\le B^r E_{(1-i)\vee 0}\Big(\sum_{l=(1-i)\vee 0}^{n} p_l\Big) \\ &= B^r E_{(1-i)\vee 0}\, m\big[(A_{n+i} \wedge N)^r\big]. \end{aligned}$$

If we choose $m(u) := e^{tu}$ then we get

$$\frac{1}{t} E_{(1-i)\vee 0}\Big(e^{t(A_{n+i}\wedge N)^r} - 1\Big) \le B^r E_{(1-i)\vee 0}\Big(e^{t(A_{n+i}\wedge N)^r}\Big).$$

Henceforth,

$$E_{(1-i)\vee 0}\Big(e^{t(A_{n+i}\wedge N)^r}\Big) \le \frac{1}{1 - tB^r}.$$

Tending with N to ∞ one can finish the proof. ■

The following remarkable consequence comes easily from Lemma 2.48.

Corollary 2.49. *Let $0 < r \le 1$ and $i \in \mathbf{N}$ be fixed. If $(X_n) \in BMO^i_r$ and $t\|(X_n)\|^r_{BMO^i_r} < 1$ for any $t > 0$ then*

$$E_{(1-i)\vee 0}(e^{tX^r}) \le \frac{1}{1 - t\|(X_n)\|^r_{BMO^i_r}}. \tag{2.71}$$

In special cases, for $X_n := s_n(f) =: s_n$ and $X_n := S_n(f) =: S_n$ one has

$$E_0(e^{ts^r}) \le \frac{1}{1 - t\|f\|^r_{BMO^s_r}}, \qquad E_1(e^{tS^r}) \le \frac{1}{1 - t\|f\|^r_{BMO^S_r}}, \tag{2.72}$$

respectively. Moreover,

$$E(e^{ts^2}) \le \frac{1}{1 - t\|f\|^2_{BMO_2}}, \qquad E(e^{tS^2}) \le \frac{1}{1 - t\|f\|^2_{BMO^-_2}}. \tag{2.73}$$

Proof. By (2.69) it is easy to see that X_n satisfies (2.70) with $B = \|(X_n)\|_{BMO^i_r}$. Lemma 2.48 proves (2.71) and (2.72). To show (2.73) check that

$$E_k(s^2_{n+1} - s^2_k) \le \|f\|^2_{BMO_2}$$

and

$$E_k(S^2_n - s^2_{k-1}) \le \|f\|^2_{BMO^-_2}.$$

So, the sequence (s_n^2) resp. (S_n^2) satisfies (2.70) with $r = 1$ and $B = \|f\|_{BMO_2}^2$ resp. $B = \|f\|_{BMO_2^-}^2$. Consequently, (2.73) follows from Lemma 2.48. ■

Note that the second inequality of (2.73) was proved by Burkholder [19] and by Garsia [82].

Now we are ready to prove the equivalence between the BMO_q^- spaces ($1 \leq q < \infty$). This result can be extended to all $0 < q < \infty$, too. More exactly, $BMO_r^S \sim BMO_2^-$ for every $0 < r < \infty$. Furthermore, opposed to the Garsia's conjecture we show that the BMO_r^s spaces ($0 < r < \infty$) are all equivalent to BMO_2 (see Weisz [194]).

Theorem 2.50. *The following equivalences hold: $BMO_p^- \sim BMO_1^-$ for $1 \leq p < \infty$, $BMO_r^S \sim BMO_2^S = BMO_2^-$ for $0 < r < \infty$ and $BMO_r^s \sim BMO_2^s = BMO_2$ for $0 < r < \infty$.*

Proof. From Hölder's inequality it follows that

$$\|f\|_{B_r} \leq \|f\|_{B_p} \qquad (r < p)$$

if $B_p \in \{BMO_p^-, BMO_p^s, BMO_p^S\}$ and in the first case $r = 1$.

To prove the converse let us apply Lemma 2.46 with $\gamma = \|f\|_{BMO_1^-} =: A$ and $\beta = \alpha + 2A$. Then

$$2E_1[\chi(f_n^* - 2A > \alpha)] \leq E_1[\chi(f_n^* > \alpha)].$$

Integrating it in α from 0 to ∞, we get

$$2E_1[(f_n^* - 2A)^+] \leq E_1(f_n^*)$$

where g^+ denotes the positive part of a measurable function g. Since

$$f_n^* \leq (f_n^* - 2A)^+ + 2A,$$

it follows from the previous inequality that

$$E_1(f_n^*) \leq 4A. \tag{2.74}$$

It is clear that $(f_m - f_{k-1})_{m\geq k-1}$ is a martingale relative to $(\mathcal{F}_m)_{m\geq k-1}$. Applying (2.74) to this martingale we obtain that

$$E_k\Big(\sup_{k\leq m\leq n} |f_m - f_{k-1}|\Big) \leq 4A$$

because

$$\|(f_m - f_{k-1})_{m\geq k-1}\|_{BMO_1^-} \leq \|f\|_{BMO_1^-}.$$

Since

$$f_n^* - f_{k-1}^* \leq \sup_{k\leq m\leq n} |f_m - f_{k-1}|,$$

the sequence (f_n^*) satisfies the condition (2.70) with $r = 1$, $i = 0$ and $B = 4A$. Hence, Lemma 2.48 implies

$$E_1(e^{tf_n^*}) \leq \frac{1}{1-4tA} \qquad (4tA < 1).$$

Applying this again to the martingale $(f_m - f_{k-1})_{m \geq k-1}$ we can see that

$$E_k\left[e^{t(|f-f_{k-1}|)}\right] \leq E_k\left[e^{t(\sup_{k \leq m \leq n} |f_m - f_{k-1}|)}\right] \leq \frac{1}{1-4tA}. \tag{2.75}$$

We are to prove similar inequalities for the other two types of BMO spaces. Assume that $0 < r \leq 1$. The inequality

$$E_0(e^{ts^r}) \leq \frac{1}{1-t\|f\|^r_{BMO_r^s}} \qquad (t\|f\|^r_{BMO_r^s} < 1)$$

was verified in Corollary 2.49 where $s = s(f)$. Applying this to the martingale $(f_m - f_k)_{m \geq k}$ and using the facts that

$$s^2[(f_m - f_k)_{m \geq k}] = s^2(f) - s_k^2(f)$$

and

$$\|(f_m - f_k)_{m \geq k}\|_{BMO_r^s} \leq \|f\|_{BMO_r^s}$$

we obtain

$$E_k[e^{t(s^2 - s_k^2)^{r/2}}] \leq \frac{1}{1-t\|f\|^r_{BMO_r^s}} \qquad (t\|f\|^r_{BMO_r^s} < 1). \tag{2.76}$$

The inequality

$$E_k[e^{t(S^2 - S_{k-1}^2)^{r/2}}] \leq \frac{1}{1-t\|f\|^r_{BMO_r^S}} \qquad (t\|f\|^r_{BMO_r^S} < 1) \tag{2.77}$$

can be proved similarly.

Set $D_1 := 4A$, $D_2 := \|f\|^r_{BMO_r^s}$, $D_3 := \|f\|^r_{BMO_r^S}$ and, moreover, $X_1 := |f - f_{k-1}|$, $X_2 := (s^2 - s_k^2)^{r/2}$ and $X_3 := (S^2 - S_{k-1}^2)^{r/2}$. It is easy to see that $\sup_{k \in \mathbb{N}} \|(E_k X_j)^{1/r}\|_\infty$ for $j = 1,2,3$ give the BMO_1^- $(r = 1)$ resp. BMO_r^s resp. BMO_r^S norms. By the conditional Markov inequality and by (2.75), (2.76), (2.77),

$$E_k[\chi(X_j > \lambda)] \leq E_k[\chi(e^{tX_j} > e^{\lambda t})] \leq \frac{1}{e^{\lambda t}} \frac{1}{1-tD_j} \qquad (j = 1,2,3).$$

Multiplying by $p/r\lambda^{p/r-1}$ $(p > r)$ and integrating it in λ from 0 to ∞ we get

$$\begin{aligned} E_k(X_j^{p/r}) &\leq \frac{p}{r}\frac{1}{1-tD_j} \int_0^\infty \lambda^{p/r-1} e^{-\lambda t}\, d\lambda \\ &= \frac{p}{r}\frac{1}{1-tD_j}\frac{1}{t^{p/r}} \int_0^\infty u^{p/r-1} e^{-u}\, du \end{aligned}$$

where $j = 1,2,3$ and in the first case $r = 1$. Choosing $t = 1/(2D_j)$ and using the fact that $\int_0^\infty u^{p/r-1}e^{-u}\,du$ is finite we obtain

$$E_k(X_j^{p/r}) \le C_p D_j^{p/r} \qquad (p > r; j = 1,2,3).$$

In other words,

$$\|f\|_{B_p} \le C_p \|f\|_{B_r} \qquad (p > r, 0 < r \le 1)$$

if $B_p \in \{BMO_p^-, BMO_p^s, BMO_p^S\}$ and in the first case $r = 1$. The proof of Theorem 2.50 is complete. ■

It was proved in Corollary 2.23 that, for a regular stochastic basis $\mathcal{F}$, the spaces H_1^* and $\mathcal{P}_1$ are equivalent. Henceforth, their dual spaces, BMO_2^- and BMO_1 are also equivalent. From this and from (2.68) we get immediately the next corollary.

Corollary 2.51. *The BMO_q spaces are usually not equivalent, however, if $\mathcal{F}$ is regular then all BMO_q and BMO_p^- spaces are equivalent* $(1 \le p, q < \infty)$.

Now the so-called sharp functions will be investigated. One of them, namely $f^\sharp$, was first investigated by Fefferman and Stein [69] in the classical case and by Garsia [82] in the martingale case. Garsia [82] has proved that the L_p norm of the sharp function $f^\sharp$ is equivalent to the H_p^* norm of f whenever $1 < p < \infty$. We introduce two other sharp functions and verify similar theorems for them.

Definition 2.52. *The sharp functions of a martingale f are defined as*

$$f^\sharp := \sup_{n\in\mathbb{N}} E_n|f - f_{n-1}|,$$

$$f_r^s := \sup_{n\in\mathbb{N}} [E_n(s^2 - s_n^2)^{r/2}]^{1/r} \qquad (0 < r < \infty, s = s(f)),$$

and

$$f_r^S := \sup_{n\in\mathbb{N}} [E_n(S^2 - S_{n-1}^2)^{r/2}]^{1/r} \qquad (0 < r < \infty, S = S(f)).$$

It is easy to see that $\|f^\sharp\|_\infty = \|f\|_{BMO_1^-}$, $\|f_r^s\|_\infty = \|f\|_{BMO_r^s}$ and $\|f_r^S\|_\infty = \|f\|_{BMO_r^S}$. Furthermore, the following theorem holds (Weisz [194]).

Theorem 2.53. *There exist constants $c_p > 0$ and $C_p > 0$ such that*

$$c_p\|f\|_{H_p^*} \le \|f^\sharp\|_p \le C_p\|f\|_{H_p^*} \qquad (1 < p < \infty),$$

$$c_p\|f\|_{H_p^s} \le \|f_r^s\|_p \le C_p\|f\|_{H_p^s} \qquad (0 < r < p < \infty)$$

and

$$c_p\|f\|_{H_p^S} \le \|f_r^S\|_p \le C_p\|f\|_{H_p^S} \qquad (0 < r < p < \infty).$$

Proof. Let us denote one of $|f_n|$, $s_n(f)$ and $S_n(f)$ by X_n. Write $\beta = 2\alpha$ in Lemmas 2.46 and 2.47. Multiplying by $p\alpha^{p-r-1}$ the inequality

$$\alpha^r E[\chi(\frac{X}{2} > \alpha)] \le E[\chi(X > \alpha)\gamma^r],$$

integrating it in α from 0 to ∞ and using Fubini's theorem we obtain

$$E\Big(|\frac{X}{2}|^p\Big) \le E\Big(p\gamma^r \int_0^\infty \alpha^{p-r-1}\chi(X > \alpha)\,d\alpha\Big) = c_p E(\gamma^r X^{p-r})$$

where $X := \sup_{n\in\mathbb{N}} X_n$ and in the first case, if $X_n = |f_n|$ then $r = 1$. By Hölder's inequality

$$\|X\|_p \le c_p \|\gamma\|_p \qquad (p > r).$$

If we choose $\gamma = f^\sharp$ $(r = 1)$ resp. $\gamma = f_r^s$ resp. $\gamma = f_r^S$ then the left hand sides of the inequalities in question are verified.

To prove the other side, let us estimate $f^\sharp$ $(r = 1)$ resp. f_r^s resp. f_r^S by $2\sup_{n\in\mathbb{N}}[E_n(X^r)]^{1/r}$. Using Doob's inequality we get

$$\|f^\sharp\|_p, \|f_r^s\|_p, \|f_r^S\|_p \le 2\big(E(\sup_{n\in\mathbb{N}} E_n[X^r])^{p/r}\big)^{1/p} \le C_p\|X\|_p \qquad (p > r)$$

and this completes the proof of the right hand sides of the inequalities. ∎

Let $(X_n, n \in \mathbb{N})$ be a sequence of non-decreasing, non-negative functions such that X_n is $\mathcal{F}_{n-i}$ measurable, $X_n = 0$ for $0 \le n < i \vee 1$ and let us define the sharp function of the sequence (X_n) by

$$X_{r,i}^\sharp := \sup_{n\in\mathbb{N}}[E_n(X^2 - X_{n+i-1}^2)^{r/2}]^{1/r} \qquad (0 < r < \infty, i \in \mathbb{N})$$

where $X := \sup_{n\in\mathbb{N}} X_n$. Then, obviously, $\|X_{r,i}^\sharp\|_\infty = \|(X_n)\|_{BMO_r^i}$ $(0 < r < \infty, i \in \mathbb{N})$. Furthermore, with the same proof as in Theorem 2.53, we can show that

$$c_p\|X\|_p \le \|X_{r,i}^\sharp\|_p \le C_p\|X\|_p \qquad (0 < r < p < \infty, i \in \mathbb{N}).$$

Let us deal with the connection between the L_q norm of the sharp functions and the $\mathcal{K}_q$ and H_q norms. By Definition 2.29 it can easily be shown that

$$\|f^\sharp\|_q \sim \|f\|_{\mathcal{K}_q^{1-}} \qquad (1 < q < \infty),$$

$$\|f_2^S\|_q \sim \|f\|_{\mathcal{K}_q^{2-}} \qquad (2 < q < \infty)$$

and

$$\|f_2^s\|_q \sim \|f\|_{\mathcal{K}_q^2} \qquad (2 < q < \infty).$$

Using the equation (2.44) we can verify that $\|f\|_{\mathcal{K}_2^2} \sim \|f\|_{\mathcal{K}_2^{2-}} \sim \|f\|_2$. This, together with Theorem 2.53, yields the next result.

Corollary 2.54. *We have the following equivalences:* $\mathcal{K}_q^{1-} \sim H_q^*$ $(1 < q < \infty)$, $\mathcal{K}_q^{2-} \sim H_q^S$ $(2 \le q < \infty)$ *and* $\mathcal{K}_q^2 \sim H_q^s$ $(2 \le q < \infty)$.

The first two equivalences are proved in Garsia [82], with another method. Though the inequality $\|f\|_{\mathcal{K}_1^{1-}} \le 2\|f\|_{H_1^*}$ is trivial neither its converse nor the equivalence between $\|f^\sharp\|_1$ and $\|f\|_{\mathcal{K}_1^{1-}}$ hold (see Garsia [82]).

2.5. MARTINGALE TRANSFORMS

A stopped martingale is a special martingale transform with the sequence $v_m = \chi(\nu \geq m)$ $(m \in \mathbb{N})$. It was shown by Burkholder [22] that if the multiplier sequence v is uniformly bounded then the martingale transform operator T_v is bounded from L_p to L_p for $1 < p < \infty$. In this section we study the boundedness of the operator T_v in case the maximal function of v is in L_p. Many of the theorems in this section are due to Chao and Long [42], [43]. We prove that T_v is bounded from H_q to H_r $(0 < q < \infty, v^* \in L_p, 1/r = 1/p + 1/q)$ if $H \in \{H^s, H^S, \mathcal{P}, \mathcal{Q}\}$ or $H = H^*$ $(1 \leq q < \infty)$. For the predictable spaces the converse holds as well: the elements of H_r are martingale transforms of the ones belonging to H_q where $r < q < \infty$ and $H \in \{H^s, \mathcal{P}, \mathcal{Q}\}$. This result for the $\mathcal{P}$ spaces can be found in Garsia [82]. In the endpoint case $q = \infty$, H_∞ must be replaced by a BMO space. We prove that a martingale is in $\mathcal{P}_p$ if and only if it is a transform of a martingale from BMO_2^-. Since the same holds also for $\mathcal{Q}_p$, as a consequence, we obtain that the spaces $\mathcal{P}_p$ and $\mathcal{Q}_p$ are equivalent for all $0 < p < \infty$.

We introduce the following classes of processes $v = (v_n, n \in \mathbb{N})$ adapted to $\mathcal{F}$:

$$V_p := \left\{v : \|v\|_{V_p} := \|v^*\|_p < \infty\right\}, \qquad 0 < p \leq \infty.$$

The *martingale transform* T_v for given martingale f and $v \in V_p$ is defined by $T_v f = (T_v f_n, n \in \mathbb{N})$ where

$$T_v f_n := \sum_{k=1}^{n} v_{k-1} d_k f.$$

Obviously, $T_v f$ is a martingale.

We say that an operator $T : X \longrightarrow Y$ is of type (X, Y) if it is bounded, namely,

$$\|Tx\|_Y \leq C\|x\|_X \qquad (x \in X).$$

The pointwise estimations

$$s(T_v f) \leq v^* s(f)$$

and

$$S(T_v f) \leq v^* S(f)$$

as well as Hölder's inequality imply the following result.

Theorem 2.55. *If $0 < p, q \leq \infty$, $v \in V_p$ and $1/r = 1/p + 1/q$ then T_v is of types (H_q^s, H_r^s) and (H_q^S, H_r^S) with $\|T_v\| \leq \|v\|_{V_p}$.*

Since by (2.15) and Theorem 2.12 we have $H_q^S \sim L_q$ $(1 < q < \infty)$, Burkholder's above result, namely, that T_v is bounded from L_q to L_q for $1 < q < \infty$ if $v \in V_\infty$, follows from the preceding theorem.

Using a duality argument and the fact that T_v is self-adjoint we obtain

Theorem 2.56. *Let $0 \leq \alpha < \infty$, $1/(1+\alpha) < p \leq \infty$ and $v \in V_p$. Then*

(i) T_v is of types $(\Lambda_2(\alpha), \Lambda_2(\beta))$ and $(\Lambda_2^-(\alpha), \Lambda_2^-(\beta))$ where $\beta = \alpha - 1/p \geq 0$ (i.e. $1/\alpha \leq p \leq \infty$)

(ii) T_v *is of types* $(\Lambda_2(\alpha), H_r^s)$, $(\Lambda_2^-(\alpha), H_r^S)$ *and* $(\Lambda_2^-(\alpha), H_r^*)$ *where* $0 < 1/r = 1/p - \alpha < 1$ *(i.e.* $1/(1+\alpha) < p < 1/\alpha$*).*

In both cases $\|T_v\| \le C\|v\|_{V_p}$.

Proof. Let $t = 1/(1+\alpha)$. If $1/p \le \alpha$ then choose $q \in (0,1]$ such that $1/p + 1/q = 1 + \alpha = 1/t$. Thus $\alpha = 1/t - 1$ and $\beta = 1/q - 1$. It follows from Theorem 2.24 that in case $\phi \in \Lambda_2(\alpha)$ and $f \in L_2$ we obtain

$$\begin{aligned}|E(fT_v\phi)| = |E(\phi T_v f)| &\le C\|\phi\|_{\Lambda_2(\alpha)}\|T_v f\|_{H_t^s} \\ &\le C\|v\|_{V_p}\|\phi\|_{\Lambda_2(\alpha)}\|f\|_{H_q^s}.\end{aligned}$$

Recall that L_2 is dense in H_q^s. Hence $T_v\phi \in (H_q^s)' \sim \Lambda_2(\beta)$ and

$$\|T_v\phi\|_{\Lambda_2(\beta)} \le C\|v\|_{V_p}\|\phi\|_{\Lambda_2(\alpha)}.$$

If $\alpha < 1/p < 1 + \alpha$ then choose again $q \in (1,\infty)$ such that $1/p + 1/q = 1/t$. In this case $1/q + 1/r = 1$. By the help of Theorem 2.26 we can prove with the same duality argument that $T_v\phi \in (H_q^s)' \sim H_r^s$ and $\|T_v\phi\|_{H_r^s} \le C\|v\|_{V_p}\|\phi\|_{\Lambda_2(\alpha)}$. The rest of the proof proceeds in the same way as above. ∎

Chao and Long have extended Theorem 2.56 (ii) to each $0 < p < 1/\alpha$ with an extrapolation method. We do not give its proof, it can be found in [43].

Theorem 2.57. *Let* $0 \le \alpha < \infty$, $0 < p < 1/\alpha$, $v \in V_p$ *and* $1/r = 1/p - \alpha$. *Then* T_v *is of types* $(\Lambda_2(\alpha), H_r^s)$, $(\Lambda_2^-(\alpha), H_r^S)$ *and* $(\Lambda_2^-(\alpha), H_r^*)$ *with* $\|T_v\| \le C\|v\|_{V_p}$.

In the special case $\alpha = 0$ we get the following

Corollary 2.58. *For* $0 < p < \infty$ *and* $v \in V_p$ *we have*

$$\|T_v f\|_{H_p^s} \le C_p\|v\|_{V_p}\|f\|_{BMO_2} \qquad (f \in BMO_2), \tag{2.78}$$

$$\|T_v f\|_{H_p^S} \le C_p\|v\|_{V_p}\|f\|_{BMO_2^-} \qquad (f \in BMO_2^-) \tag{2.79}$$

and

$$\|T_v f\|_{H_p^*} \le C_p\|v\|_{V_p}\|f\|_{BMO_2^-} \qquad (f \in BMO_2^-). \tag{2.80}$$

Now we consider the boundedness of T_v on the predictable spaces $\mathcal{P}_p$ and $\mathcal{Q}_p$. The following two theorems can be found in Chao, Long [42].

Theorem 2.59. *Let* $0 < p \le \infty$, $0 < q < \infty$, $v \in V_p$ *and* $1/r = 1/p + 1/q$. *Then* T_v *is of types* $(\mathcal{P}_q, \mathcal{P}_r)$ *and* $(\mathcal{Q}_q, \mathcal{Q}_r)$ *with* $\|T_v\| \le C\|v\|_{V_p}$.

Proof. Let (λ_n) be the predictable, non-decreasing least majorant of (f_n) and $f \in \mathcal{P}_q$. Then $|d_n f| \le 2\lambda_{n-1}$ and

$$|d_n(T_v f)| \le 2v_{n-1}^*\lambda_{n-1} =: \rho_{n-1}.$$

Applying twice Theorem 2.11 and Theorem 2.55 we get that

$$\|(T_v f)^*\|_r \le C\|s(T_v f)\|_r \le C\|v\|_{V_p}\|s(f)\|_q \le C\|v\|_{V_p}\|f\|_{\mathcal{P}_q}$$

for $0 < r \leq 2$. The same result holds for $2 < q < \infty$ if we simple write the operator S instead of s in the preceding inequality. Using this and the inequality

$$|T_v f_n| \leq (T_v f)^*_{n-1} + \rho_{n-1}$$

we can conclude that

$$\|T_v f\|_{\mathcal{P}_r} \leq C(\|(T_v f)^*\|_r + \|\rho_\infty\|_r) \leq C\|v\|_{V_p}\|f\|_{\mathcal{P}_q}.$$

The proof for the spaces $\mathcal{Q}_r$ is similar. ■

The boundedness of the operator T_v on the maximal Hardy spaces H^*_p can not be obtained in the same way as the one on the spaces H^S_p in Theorem 2.55. However, the following result holds.

Theorem 2.60. *Let $0 < p \leq \infty$, $1 \leq q < \infty$, $v \in V_p$ and $1/r = 1/p + 1/q$. Then T_v is of type (H^*_q, H^*_r) with $\|T_v\| \leq C\|v\|_{V_p}$.*

Proof. From Davis's decomposition with respect to H^*_q $(1 \leq q < \infty)$ (see Lemma 2.14) we get that $f = h + g$ and

$$\|\sum_{n=0}^{\infty} |d_n h|\|_q \leq C\|f\|_{H^*_q}, \qquad \|g\|_{\mathcal{P}_q} \leq C\|f\|_{H^*_q}.$$

Hence, by (2.25) and by Theorem 2.59,

$$\begin{aligned}\|(T_v f)^*\|_r &\leq C(\|(T_v h)^*\|_r + \|(T_v g)^*\|_r) \\ &\leq C(\|\sum_{n=0}^{\infty} |d_n(T_v h)|\|_r + \|T_v g\|_{\mathcal{P}_r}) \\ &\leq C\|v\|_{V_p}(\|\sum_{n=0}^{\infty} |d_n h|\|_q + \|g\|_{\mathcal{P}_q}).\end{aligned}$$

Consequently,

$$\|(T_v f)^*\|_r \leq C\|v\|_{V_p}\|f\|_{H^*_q}$$

which proves the theorem. ■

Martingale transforms are useful to study the relations between the spaces H^s_p, $\mathcal{P}_p$ and $\mathcal{Q}_p$. Garsia [82] has proved that for any given finite numbers r and q $(r < q)$ the elements in $\mathcal{P}_r$ are martingale transforms of those in $\mathcal{P}_q$ (see Theorem 2.61 (iii)). We shall show this result for the spaces H^s_r and $\mathcal{Q}_r$ as well as for the endpoint case $q = \infty$.

Theorem 2.61. *Let $0 < r < q < \infty$ and $1/r = 1/p + 1/q$.*

(i) For $f \in H^s_r$ there exist a function $g \in H^s_q$ and a sequence $v \in V_p$ such that $f = T_v g$ with

$$\|v\|^p_{V_p} \leq \|f\|^{r/2}_{H^s_{r/2}} \tag{2.81}$$

and

(2.82) $$\|g\|_{H_q^s} \leq \left(\frac{q}{r}\right)^{1/2} \|f\|_{H_r^s}^{r/q}.$$

Conversely, for any $v \in V_p$ and $g \in H_q^s$, the martingale $f = T_v g$ is in H_r^s and, moreover, $\|f\|_{H_r^s} \leq C\|v\|_{V_p}\|g\|_{H_q^s}$.

(ii) The same statements as in (i) hold when H^s is replaced by the $\mathcal{Q}$ space.

(iii) The statements in (i) hold also when H^s, (2.81) and (2.82) are replaced by $\mathcal{P}$,

(2.83) $$\|v\|_{V_p}^p \leq \|f\|_{\mathcal{P}_r}^r$$

and by

(2.84) $$\|g\|_{\mathcal{P}_q} \leq \frac{q}{r}\|f\|_{\mathcal{P}_r}^{r/q},$$

respectively.

Proof. The converse parts of (i), (ii) and (iii) come from Theorems 2.55 and 2.59.

Set $2t := 1 - r/q > 0$, $v_{n-1} := s_n^t(f)$ and

$$g_n := \sum_{k=1}^n s_k^{-t}(f) d_k f.$$

Then, obviously, $f = T_v g$ and

$$\|v\|_{V_p}^p = E(s^{tp}) = E(s^{r/2})$$

which proves (2.81). Moreover,

$$s_n^2(g) = \sum_{k=1}^n s_k^{-2t}(f) E_{k-1}|d_k f|^2.$$

It is easy to see that

$$\begin{aligned} s_n^2(g) &= \sum_{k=1}^n \frac{s_k^2(f) - s_{k-1}^2(f)}{s_k^{2t}(f)} \\ &\leq \int_0^{s_n^2(f)} \frac{1}{\alpha^{2t}}\, d\alpha = \frac{s_n^{2(-2t+1)}(f)}{-2t+1} = \frac{q}{r} s_n^{2r/q}(f). \end{aligned}$$

The inequality (2.82) follows from this.

To prove (ii) let (λ_n) be the predictable, non-decreasing least majorant of $(S_n(f))$, $v_n := \lambda_n^t$ and

$$g_n := \sum_{k=1}^n \lambda_{k-1}^{-t} d_k f.$$

Using Abel rearrangement we obtain

$$S_n^2(g) = \sum_{k=1}^{n} \frac{S_k^2(f) - S_{k-1}^2(f)}{\lambda_{k-1}^{2t}}$$
$$= \frac{S_n^2(f)}{\lambda_{n-1}^{2t}} + \sum_{k=1}^{n} S_k^2(f)\left(\frac{1}{\lambda_{k-1}^{2t}} - \frac{1}{\lambda_k^{2t}}\right).$$

The definition of the sequence (λ_n) implies

$$S_n^2(g) \le \lambda_{n-1}^{2-2t} + \sum_{k=1}^{n} \lambda_{k-1}^2\left(\frac{1}{\lambda_{k-1}^{2t}} - \frac{1}{\lambda_k^{2t}}\right).$$

Again by Abel rearrangement one can see that

$$S_n^2(g) \le \sum_{k=1}^{n} \frac{\lambda_{k-1}^2 - \lambda_{k-2}^2}{\lambda_{k-1}^{2t}(f)} \le \int_0^{\lambda_{n-1}^2} \frac{1}{\alpha^{2t}}\, d\alpha.$$

The proof can be finished as in (i).

The statements of (iii) can be shown similarly. ■

Theorem 2.59 and 2.60 do not hold if $q = \infty$. We can establish by Corollary 2.58 that, in this case, the spaces H_q^* and H_q^S could be replaced by BMO_2^- while the space H_q^s by BMO_2. We shall prove in the next theorem that the spaces $\mathcal{Q}_q$ and $\mathcal{P}_q$ could also be replaced by BMO_2^-, namely, that T_v is of types $(BMO_2^-, \mathcal{Q}_p)$ and $(BMO_2^-, \mathcal{P}_p)$. These results are stronger than (2.79) and (2.80) because, by Theorem 2.11, $\mathcal{Q}_p, \mathcal{P}_p \subset H_p^S \cap H_p^*$. In case $p = \infty$ and $q = \infty$ it was proved in Theorem 2.56 (i) that T_v is of type (BMO_2^-, BMO_2^-).

By the help of martingale transforms the next two theorems give a characterization of the spaces $\mathcal{Q}_p$, $\mathcal{P}_p$ and H_p^s.

Theorem 2.62. *If $0 < p < \infty$ and $f \in \mathcal{Q}_p$ then there exist $g \in BMO_2^-$ with $\|g\|_{BMO_2^-} \le \sqrt{2}$ and a (non-negative, non-decreasing) $v \in V_p$ with $\|v\|_{V_p} \le C_p \|f\|_{\mathcal{Q}_p}$ such that $f = T_v g$. Conversely, for any $v \in V_p$ and $g \in BMO_2^-$, the martingale $f = T_v g$ is in $\mathcal{Q}_p$ and, in addition to this, $\|f\|_{\mathcal{Q}_p} \le C_p \|v\|_{V_p} \|g\|_{BMO_2^-}$. Instead of $\mathcal{Q}$, the same statements hold for the $\mathcal{P}$ spaces with $\|g\|_{BMO_1^-} \le 4$.*

Proof. We give the proof for the $\mathcal{Q}_p$ spaces, only, because the one for the $\mathcal{P}_p$ spaces is similar and can be found in [42]. Let $f \in \mathcal{Q}_p$ and (λ_n) be the predictable, non-decreasing least majorant of $(S_n(f))$. Choose $0 < p_0 < p$ and define $v_0 := 1$ and for $n \ge 1$ let

$$v_n := \sup_{m \le n} (E_m[\lambda_\infty^{p_0}])^{1/p_0}.$$

If

$$g_n = \sum_{k=1}^{n} v_{k-1}^{-1} d_k f$$

then, clearly, $f = T_v g$. Furthermore, we have

$$\begin{aligned} S_N^2(g) - S_{n-1}^2(g) &= \sum_{k=n}^{N} v_{k-1}^{-2} |d_k f|^2 \\ &= \sum_{k=n}^{N} v_{k-1}^{-2}(S_k^2(f) - S_{k-1}^2(f)) \\ &= v_{N-1}^{-2}(S_N^2(f) - S_{n-1}^2(f)) \\ &\qquad + \sum_{k=n}^{N-1}(S_k^2(f) - S_{n-1}^2(f))(v_{k-1}^{-2} - v_k^{-2}). \end{aligned}$$

So we can conclude that

$$\begin{aligned} S_N^2(g) - S_{n-1}^2(g) &\le v_{N-1}^{-2}\lambda_{N-1}^2 + \sum_{k=n}^{N-1} \lambda_{k-1}^2(v_{k-1}^{-2} - v_k^{-2}) \\ &= v_{n-1}^{-2}\lambda_{n-1}^2 + \sum_{k=n}^{N-1} v_k^{-2}(\lambda_k^2 - \lambda_{k-1}^2). \end{aligned}$$

From Jensen's inequality

$$v_k^{-2} \le (E_k \lambda_\infty^{p_0})^{-2/p_0} \le E_k \lambda_\infty^{-2} \qquad (k \ge 1).$$

Therefore,

$$E_n(S_N^2(g) - S_{n-1}^2(g)) \le 1 + E_n\left(\sum_{k=n}^{N-1} E_k\left(\frac{\lambda_k^2 - \lambda_{k-1}^2}{\lambda_\infty^2}\right)\right) \le 2.$$

Hence, by Theorem 2.50, we obtain that $g \in BMO_2^-$ and $\|g\|_{BMO_2^-} \le \sqrt{2}$. Moreover, by Doob's inequality,

$$\|v\|_{V_p}^p = \|v^*\|_p^p = E((\lambda_\infty^{p_0})^{*\,p/p_0}) \le C_p E(\lambda_\infty^p) = C_p\|f\|_{\mathcal{Q}_p}^p.$$

For the converse assertion let $v \in V_p$ and $g \in BMO_2^-$. By (2.65) we have

$$\begin{aligned} S_n(T_v g) &\le S_{n-1}(T_v g) + |d_n(T_v g)| \\ &\le S_{n-1}(T_v g) + v_{n-1}^* \|g\|_{BMO_2^-}. \end{aligned}$$

Using (2.79) we can conclude that

$$\begin{aligned} \|T_v g\|_{\mathcal{Q}_p} &\le C_p \|S(T_v g)\|_p + C_p \|v^*\|_p \|g\|_{BMO_2^-} \\ &\le C_p \|v\|_{V_p} \|g\|_{BMO_2^-} \end{aligned}$$

which completes the proof. ∎

We remark that no similar characterization of the H_p^* spaces is obtainable. The reason is the following. For any $v \in V_p$ $(0 < p < \infty)$ and $g \in BMO_2^-$ we have

$T_v g \in \mathcal{P}_p$. Since, by Theorem 2.11, the space $\mathcal{P}_p$ is a proper subspace of H_p^*, in general, so $f \in H_p^*$ can not be represented as $f = T_v g$.

Theorem 2.63. *If $0 < p < \infty$ and $f \in H_p^s$ then there exist $g \in BMO_2$ with $\|g\|_{BMO_2} \le 1$ and a (non-negative, non-decreasing) $v \in V_p$ with $\|v\|_{V_p} \le C_p \|f\|_{H_p^s}$ such that $f = T_v g$. Conversely, for any $v \in V_p$ and $g \in BMO_2$, the martingale $f = T_v g$ is in H_p^s and, in addition to this, $\|f\|_{H_p^s} \le C_p \|v\|_{V_p} \|g\|_{BMO_2}$.*

The proof of this theorem can be found in Chao, Long [42] and, for two parameters, in Section 3.4.

The next result is a consequence of Theorem 2.62 and it is not obtainable with the method used in Section 2.2.

Corollary 2.64. *The spaces $\mathcal{Q}_p$ and $\mathcal{P}_p$ are equivalent for all $0 < p < \infty$, more exactly,*

$$c_p \|f\|_{\mathcal{P}_p} \le \|f\|_{\mathcal{Q}_p} \le C_p \|f\|_{\mathcal{P}_p}. \tag{2.85}$$

Proof. Assume that $f \in \mathcal{P}_p$. By Theorem 2.62 there exist $g \in BMO_2^-$ and $v \in V_p$ with $\|g\|_{BMO_1^-} \le 4$ and $\|v\|_{V_p} \le C_p \|f\|_{\mathcal{P}_p}$ such that $f = T_v g$. Applying again Theorem 2.62 to the $\mathcal{Q}_p$ spaces and using the equivalence of the BMO_q^- spaces we get that

$$\|f\|_{\mathcal{Q}_p} = \|T_v g\|_{\mathcal{Q}_p} \le C_p \|v\|_{V_p} \|g\|_{BMO_2^-} \le C_p \|f\|_{\mathcal{P}_p}.$$

The left hand side of (2.85) can be proved in the same way. ■

2.6. CONJUGATE TRANSFORMS OF VILENKIN MARTINGALES

The theory of the H_p spaces of conjugate harmonic functions on Euclidean spaces has been developed by Stein [177]. In particular case, $H_1(\mathbf{R}^n)$ can be characterized via Riesz's transforms:

$$H_1 = \{f \in L_1 : R_j f \in L_1, j = 1, \ldots, m\}. \tag{2.86}$$

Chao and Taibleson (see [39], [40], [41], [44], [45], [180]) have extended this theory to local fields. Moreover, for martingale spaces, Janson and Chao ([106], [41], [37]) studied transforms with matrix operators acting on the values of the difference sequences of q-martingales.

In this section conjugate martingale transforms with matrix operators acting on the generalized Rademacher series of the difference sequences of the Vilenkin martingales are investigated. These transforms are slightly more general than the ones considered in Section 2.5 and were first introduced by Gundy [86]. This section is based on the paper Weisz [190]. Similar theorems to the ones in Chao, Janson [106] and Janson [41] will be proved.

Contrary to the statement in Gundy [86], (2.86) was proved in case all matrices and martingales are real, only. This theorem is extended here to the complex case. More exactly, a necessary and sufficient condition is given for the transforms such that (2.86) holds whenever the martingale H_1 space is generated by a bounded

Vilenkin system. Note that, for a particular Vilenkin system, this space is identical with the H_1 space of q-martingales studied by Chao and Janson. A version of F. and M. Riesz's theorem will be proved. In the simplest case all matrices are diagonal the transforms used in this section are called multiplier transforms. It is in question whether H_1 can be characterized via a single multiplier transform if the multiplier has two values: -1 and 1. Moreover, a necessary and sufficient condition for (2.86) to hold for multiplier transforms is also given. A family of integrable functions for which $\|f\|_{L_1} \sim \|f\|_{H_1}$ is obtained. Similarly to Chao [36] we introduce also a transform in the dyadic case. A necessary and sufficient condition is given for $BMO = L_\infty + \sum_{i=1}^m T_i L_\infty$ and $VMO = C_W + \sum_{i=1}^m T_i C_W$ to hold where C_W denotes the continuous functions on a Vilenkin group. The first result was already known for q-martingales but for another type of transforms (see Janson [106]). In the classical case, for $BMO(\mathbf{T}^d)$ and for $VMO(\mathbf{T}^d)$ both results can be found in Fefferman, Stein [69] and in Janson [107].

Let $\mathcal{F}_n$ ($n \in \mathbf{N}$) be the σ-algebra generated by a Vilenkin system (see (1.5)) and suppose that $2 \le p_n \le N$ for all $n \in \mathbf{N}$. It was shown in Section 1.2 that in this case the stochastic basis $(\mathcal{F}_n)$ is regular. By Corollary 2.23 the Hardy spaces H_p^s, H_p^S, H_p^* and $\mathcal{P}_p$ are equivalent. Similarly, by Corollary 2.51, $BMO_p \sim BMO_q^-$. Denote by H_p one of the Hardy spaces and by BMO one of the BMO spaces. The following transform was introduced by Gundy [86].

Definition 2.65. *Let $A := (A_n, n \in \mathbf{N})$ be a sequence of matrices such that*

$$A_n : \mathbf{C}^{p_n - 1} \longrightarrow \mathbf{C}^{p_n - 1}.$$

With the differences of a martingale written in the form

$$d_{n+1} f = \sum_{j=1}^{p_n - 1} v_n^{(j)} r_n^j,$$

the differences of the conjugate martingale transform are defined by

$$d_{n+1}(Tf) := \sum_{j=1}^{p_n - 1} (A_n v_n)^{(j)} r_n^j$$

where $v_n := (v_n^{(j)})_{j=1}^{p_n - 1}$. Set $(Tf)_n := \sum_{k=1}^n d_k(Tf)$.

It is obvious that $((Tf)_n = Tf_n, n \in \mathbf{N})$ is a martingale.

The advantage of this transform is the fact that if the matrices are all diagonal then the so-called *multiplier transform* is obtained. Other martingale transforms with matrix operators are investigated by Janson and Chao [106], [41].

Assume that the (euclidean) norms of A_n ($n \in \mathbf{N}$) are uniformly bounded. The next proposition shows the boundedness of the operator T on Hardy and BMO spaces.

Proposition 2.66. *T is a bounded linear operator on BMO, on each H_p ($0 < p < \infty$) and, consequently, on each L_p ($1 < p < \infty$).*

Proof. Since

$$E_n|d_{n+1}(Tf)|^2 = \sum_{j=1}^{p_n-1} |(A_n v_n)^{(j)}|^2$$
$$\leq C \sum_{j=1}^{p_n-1} |v_n^{(j)}|^2 = CE_n|d_{n+1}f|^2,$$

one has $\|s(Tf)\|_p \leq \|s(f)\|_p$. By Corollary 2.23, we obtain that T is bounded on each H_p $(0 < p < \infty)$ and on each L_p $(1 < p < \infty)$.

The fact that $f \in BMO$ implies $f \in L_2$ and $Tf \in L_2$. Since

$$E_n|Tf - Tf_n|^2 = E_n(\sum_{k=n}^{\infty} E_k|d_{k+1}(Tf)|^2)$$
$$\leq CE_n(\sum_{k=n}^{\infty} E_k|d_{k+1}f|^2)$$
$$= CE_n|f - f_n|^2,$$

one has $\|Tf\|_{BMO} \leq \|f\|_{BMO}$. ∎

A characterization of H_1 can be given. Assume that $A^{(1)}, \ldots, A^{(m)}$ are sequences of matrices described in Definition 2.65 and let $T_1, \ldots, T_m$ be the corresponding conjugate martingale transforms. Proposition 2.66 shows that $f \in H_1$ implies that $T_1 f, \ldots, T_m f$ belong to H_1 and thus to L_1 as well. To prove the converse of this result we use the following very important lemma proved by Chao and Janson in [41] and in [106]. Set

$$U_q := \{x \in \mathbf{C}^q : \sum_{i=1}^{q} x_i = 0\}.$$

For a given martingale f we can regard $d_{n+1}f$ on an atom of $\mathcal{F}_n$ as an element of U_{p_n}.

Lemma 2.67. *Let W be a closed cone (i.e. $x \in W$ together with $t \geq 0$ imply $tx \in W$) consisting of elements of the form $x = (x^{(0)}, \ldots, x^{(m)})$ where $x^{(i)} = (x_1^{(i)}, \ldots, x_q^{(i)}) \in U_q$ such that if $x^{(i)} = \eta_i(\lambda_1, \ldots, \lambda_q)$ for some $\eta_i \in \mathbf{C}$, $i = 0, 1, \ldots, m$ and $\lambda_k \in \mathbf{R}$, $k = 1, 2, \ldots, q$ then $x = (0, 0, \ldots, 0)$. Then there exists a positive $p < 1$ such that*

$$\|a\|^p \leq \frac{1}{q} \sum_{k=1}^{q} \|(a^{(i)} + x_k^{(i)})_{i=0}^m\|^p \tag{2.87}$$

for $a = (a^{(i)})_{i=0}^m \in \mathbf{C}^{m+1}$ and $x = (x^{(i)})_{i=0}^m \in W$ where $\|\cdot\|$ denotes the euclidean norm of a vector.

Proof. Denote by

$$|||x||| := (\sum_{k=1}^{q} \sum_{i=0}^{m} |x_k^{(i)}|^2)^{1/2}$$

the euclidean norm of $x = (x^{(i)})_{i=0}^m \in W$. First we suppose that $|||x|||/\|a\|$ is small enough and use the binomial expansion:

$$\begin{aligned}
\sum_{k=1}^{q} \|(a^{(i)} + x_k^{(i)})_{i=0}^m\|^p &= \sum_{k=1}^{q}\Big(\sum_{i=0}^{m} |a^{(i)} + x_k^{(i)}|^2\Big)^{p/2} \\
&= \sum_{k=1}^{q}\Big(\sum_{i=0}^{m} |a^{(i)}|^2 + \sum_{i=0}^{m} 2\Re(\overline{a}^{(i)} x_k^{(i)}) + \sum_{i=0}^{m} |x_k^{(i)}|^2\Big)^{p/2} \\
&= \|a\|^p \sum_{k=1}^{q}\Big(1 + \frac{2\Re\sum_{i=0}^m \overline{a}^{(i)} x_k^{(i)} + \sum_{i=0}^m |x_k^{(i)}|^2}{\|a\|^2}\Big)^{p/2} \\
&= \|a\|^p \sum_{k=1}^{q}\Big(1 + \frac{p}{2}\frac{2\Re\sum_{i=0}^m \overline{a}^{(i)} x_k^{(i)}}{\|a\|^2} + \frac{p}{2}\frac{\sum_{i=0}^m |x_k^{(i)}|^2}{\|a\|^2} \\
&\quad + \frac{1}{2}\frac{p}{2}\Big(\frac{p}{2} - 1\Big)\Big(\frac{2\Re\sum_{i=0}^m \overline{a}^{(i)} x_k^{(i)}}{\|a\|^2}\Big)^2 + O\Big(\frac{|||x|||^3}{\|a\|^3}\Big)\Big).
\end{aligned}$$

The second term is equal to zero since $\sum_{k=1}^q x_k^{(i)} = 0$. To estimate the fourth term let α be the maximum of the continuous function $\sum_{k=1}^q \big(\Re\sum_{i=0}^m \overline{a}^{(i)} x_k^{(i)}\big)^2$ on the compact set

$$K_1 := \{(a, x) \in \mathbf{C}^{m+1} \times W : \|a\| = 1 \text{ and } |||x||| = 1\}.$$

Applying Schwarz's inequality we have

$$\begin{aligned}
\sum_{k=1}^{q}\Big(\Re\sum_{i=0}^{m} \overline{a}^{(i)} x_k^{(i)}\Big)^2 &\le \sum_{k=1}^{q} \Big|\sum_{i=0}^{m} \overline{a}^{(i)} x_k^{(i)}\Big|^2 \\
&= \sum_{k=1}^{q}\sum_{i=0}^{m}\sum_{j=0}^{m} \overline{a}^{(i)} x_k^{(i)} a^{(j)} \overline{x}_k^{(j)} \\
&\le \sum_{k=1}^{q}\sum_{i=0}^{m}\sum_{j=0}^{m} |a^{(j)} x_k^{(i)}|^2 = \|a\|^2 |||x|||^2 = 1
\end{aligned}$$

on K_1. In case an equality stood above, it would be implied that $\sum_{i=0}^m \overline{a}^{(i)} x_k^{(i)} \in \mathbf{R}$ and $a^{(i)} x_k^{(j)} = \lambda a^{(j)} x_k^{(i)}$ for some $(a, x) \in K_1$ and $\lambda \in \mathbf{R}$. Thus we would have

$$x_k^{(j)} = \sum_{i=0}^{m} \overline{a}^{(i)} a^{(i)} x_k^{(j)} = \lambda a^{(j)} \sum_{i=0}^{m} \overline{a}^{(i)} x_k^{(i)},$$

or, in other words

$$x^{(j)} = \lambda a^{(j)} \sum_{i=0}^{m} \overline{a}^{(i)} x^{(i)}$$

which contradicts the hypothesis. Hence $\alpha < 1$. Homogenity shows that, in general,

$$\sum_{k=1}^{q}\Big(\Re\sum_{i=0}^{m} \overline{a}^{(i)} x_k^{(i)}\Big)^2 \le \alpha \|a\|^2 |||x|||^2.$$

So we can conclude that

$$\|a\|^{-p}\sum_{k=1}^{q}\|(a^{(i)}+x_k^{(i)})_{i=0}^m\|^p$$
$$\geq q+\frac{p}{2}\frac{|||x|||^2}{\|a\|^2}+\frac{p}{2}(p-2)\alpha\frac{|||x|||^2}{\|a\|^2}+O\Big(\frac{|||x|||^3}{\|a\|^3}\Big)\geq q$$

if $\alpha < p < 1$ and $|||x||| < \epsilon\|a\|$ for some $\epsilon > 0$. Thus (2.87) is verified in this case.

To finish the proof we use another compactness argument. Set

$$K_2 := \{(a,x)\in \mathbf{C}^{m+1}\times W : \frac{1}{q}\sum_{k=1}^{q}\|(a^{(i)}+x_k^{(i)})_{i=0}^m\| = 1, |||x|||\geq \epsilon\|a\|\}.$$

Clearly, $a = q^{-1}\sum_{k=1}^{q}(a^{(i)}+x_k^{(i)})_{i=0}^m$. Consequently, $\|a\|\leq 1$ on K_2 and $\|a\| = 1$ only if $a^{(i)}+x_k^{(i)} = \lambda_k a^{(i)}$ with $\lambda_k \geq 0$. Thus we would have $x_k^{(i)} = (\lambda_k-1)a^{(i)}$, or,

$$x^{(i)} = a^{(i)}(\lambda_k-1)_{k=1}^q$$

which contradicts again our hypothesis.

Therefore, $\|a\|\leq\beta<1$ on K_2. From this we get that

$$\|a\|\leq\frac{\beta}{q}\sum_{k=1}^{q}\|(a^{(i)}+x_k^{(i)})_{i=0}^m\|$$

for $|||x|||\geq\epsilon\|a\|$. Thus

$$\|a\|^p\leq\Big(\frac{\beta}{q}\Big)^p\Big(\sum_{k=1}^{q}\|(a^{(i)}+x_k^{(i)})_{i=0}^m\|\Big)^p\leq\frac{1}{q}\sum_{k=1}^{q}\|(a^{(i)}+x_k^{(i)})_{i=0}^m\|^p$$

whenever $1 > p \geq \log q/\log(q/\beta)$ and $|||x|||\geq\epsilon\|a\|$, which proves the lemma. ■

Returning to the martingale H_1 space we obtain

Theorem 2.68. *Assume that, for each $n\in\mathbf{N}$, the matrices $A_n^{(1)},\dots,A_n^{(m)}$ have no common eigenvector $(z_1,\dots,z_{p_n-1})$ with $\overline{z_j} = z_{p_n-j}$ for each $j = 1,\dots,p_n-1$ and, moreover, that (f_n) is a martingale such that $\|f_n\|_1$ and $\|T_if_n\|_1$ $(i = 1,\dots,m)$ are uniformly bounded. Then (f_n) and (T_if_n) are martingales belonging to H_1. Furthermore,*

$$c\sum_{i=0}^{m}\|T_if\|_1\leq\|f\|_{H_1}\leq C\sum_{i=0}^{m}\|T_if\|_1 \tag{2.88}$$

where $T_0f_n := f_n$.

Proof. We are going to apply Lemma 2.67. Denote by $E_{i_0,\dots,i_{n-1}}$ the atoms of $\mathcal{F}_n$ such that

$$E_{i_0,\dots,i_{n-1}} = \bigcup_{i_n=1}^{p_n}E_{i_0,\dots,i_n}.$$

Set

$$a^{(i)} := T_i f_n(E_{i_0,\ldots,i_{n-1}}) \qquad \text{and} \qquad x_k^{(i)} := d_{n+1}(T_i f)(E_{i_0,\ldots,i_{n-1},k}).$$

It is easy to check that in this case W is a closed cone. Regard r_n^j as a p_n dimensional vector. Since, by (1.10), $(r_n^j)_{j=0}^{p_n-1}$ is an orthogonal basis in $\mathbf{C}^{p_n}$, we deduce that the real, nonzero vector $(\lambda_1,\ldots,\lambda_{p_n})$ can uniquely be written in the following form:

$$(\lambda_1,\ldots,\lambda_{p_n}) = \sum_{j=1}^{p_n-1} z_j r_n^j. \tag{2.89}$$

Recall that $\sum_{i=1}^{p_n}\lambda_i = 0$. Since $\overline{r_n^j} = r_n^{p_n-j}$, we obtain that $\overline{z_j} = z_{p_n-j}$ $(j = 1,\ldots,p_n-1)$. If, for every $0 \le i \le m$,

$$x^{(i)} = \eta_i(\lambda_1,\ldots,\lambda_{p_n}) \qquad (\eta_0 \ne 0)$$

then, by the definition and by (2.89), we get that

$$x^{(i)} = \eta_0 \sum_{j=1}^{p_n-1} (A_n^{(i)} z)^{(j)} r_n^j = \sum_{j=1}^{p_n-1} (\eta_i z_j) r_n^j$$

and, consequently, $z = (z_j)_{j=1}^{p_n-1}$ is a common eigenvector of $A_n^{(i)}$ $(0 \le i \le m)$ which is a contradiction. Hence $\eta_0 = 0$ and so $x = (0,0,\ldots,0)$ in Lemma 2.67. Thus the conditions in Lemma 2.67 are satisfied, so (2.87) holds. A usual martingale majorant argument is applied (see e.g. Chao, Janson [106], [41]). Set

$$g_n := \|(T_i f_n)_{i=0}^m\|.$$

Then the inequality (2.87) shows that $g_n^p \le E_n(g_{n+1}^p)$ for some $p < 1$, thus g_n^p is a positive submartingale. From the second condition of Theorem 2.68 we get that for every $n \in \mathbf{N}$

$$\|g_n^p\|_{1/p} = \|g_n\|_1^p \le (\sum_{i=0}^m \|T_i f_n\|_1)^p \le C. \tag{2.90}$$

Using Doob's inequality we can conclude that $\sup_{n\in\mathbf{N}} g_n^p \in L_{1/p}$, thus $\sup_{n\in\mathbf{N}} g_n \in L_1$. Since $|T_i f_n| \le g_n$, we get that $T_i f \in H_1$. The right hand side of (2.88) is obtained from

$$\|f\|_{H_1}^p \le \|\sup_{n\in\mathbf{N}} g_n\|_1^p = \|\sup_{n\in\mathbf{N}} g_n^p\|_{1/p} \le C_p \sup_{n\in\mathbf{N}} \|g_n^p\|_{1/p}$$

and from (2.90). The other side of (2.88) comes trivially from Proposition 2.66. The proof of Theorem 2.68 is complete. ∎

Note that we can verify Theorem 2.68 for all H_p spaces with $p > p_0$ in the same way whenever $0 < p_0 < 1$ is a fixed number. However, (2.88) does not hold for a small p. More precisely, according to Uchiyama [185], if $p_n = d$ $(n \in \mathbf{N})$ then there exists a $p_1 > 0$ such that (2.88) does not hold for $p < p_1$.

A finite measure ν on $(\Omega, \mathcal{A})$ defines a martingale (f_n) by

$$f_n(E_{i_0,\dots,i_{n-1}}) = P_n\nu(E_{i_0,\dots,i_{n-1}}).$$

Conversely, if (f_n) is a martingale then ν is a finite measure. Since $\|f_n\|_1 \leq \|\nu\|$, we get the following result that is analogous to F. and M. Riesz's theorem.

Corollary 2.69. *Assume that $A_n^{(1)}, \dots, A_n^{(m)}$ satisfy the assumptions of Theorem 2.68. If ν and $T_i\nu$ $(i = 1, \dots, m)$ are bounded measures then ν is absolutely continuous and belongs to H_1.*

In a special case the converse of Theorem 2.68 can also be proved. $H_1 \subset \{f \in L_1 : T_i f \in L_1, i = 1, \dots, m\}$ follows from Proposition 2.66. If $A_n^{(1)}, \dots, A_n^{(m)}$ do not have a common eigenvector with the property set in Theorem 2.68 then the reverse inclusion is proved there.

Theorem 2.70. *Assume that, for each $n \in \mathrm{N}$, the matrices $A_n^{(1)}, \dots, A_n^{(m)}$ have a common eigenvector $(z_{n;1}, \dots, z_{n;p_n-1})$ with $\overline{z_{n;j}} = z_{n;p_n-j}$ $(j = 1, \dots, p_n - 1)$ and $\sigma_n^{(i)} = \sigma^{(i)}$ $(n \in \mathrm{N},\ i = 1, \dots, m)$ for the corresponding eigenvalues. Then $H_1 \neq \{f \in L_1 : T_i f \in L_1, i = 1, \dots, m\}$.*

Proof. Since

$$(x_{n;1}, \dots, x_{n;p_n}) := \sum_{j=1}^{p_n-1} z_{n;j} r_n^j$$

is a real vector we can assume that $\min_k x_{n;k} = -1$ for all $n \in \mathrm{N}$.

Modifying the proof of Lemma 6 in Janson [106] we get

Lemma 2.71. *If $(x_{n;1}, \dots, x_{n;p_n})$ are real numbers such that $\sum_{k=1}^{p_n} x_{n;k} = 0$ and $\min_k x_{n;k} = -1$ for all $n \in \mathrm{N}$ then there exists a function $f \in L_1$ such that $f \notin H_1$ and*

$$d_{n+1} f(E_{i_0,\dots,i_{n-1},k}) = \lambda_{i_0,\dots,i_{n-1}} x_{n;k}$$

where $\lambda_{i_0,\dots,i_{n-1}}$ are real numbers.

Proof of Lemma 2.71. Define g_{n+1} by

$$g_{n+1}(E_{i_0,\dots,i_n}) := \prod_{k=0}^{n} (1 + x_{k;i_k}), \quad g_0 := 1.$$

The sequence (g_n) is a non-negative martingale since

$$g_{n+1}(E_{i_0,\dots,i_n}) = (1 + x_{n;i_n}) g_n(E_{i_0,\dots,i_{n-1}}).$$

So $\|g_n\|_1 = E(g_n) = E(g_0) = 1$. Suppose that $x_{n;1} = -1$ for all $n \in \mathrm{N}$. Then g_{n+1} is 0 if any $i_k = 1$ $(0 \leq k \leq n)$. Set

$$F_{n+1} := \bigcup_{i_0,\dots,i_{n-1} \neq 1} E_{i_0,\dots,i_{n-1},1}.$$

It is easy to see that the sets F_n are disjoint and

$$\int_{F_{n+1}} g_n \, dP = \frac{1}{p_n} \int_{\cup_{i_0,\ldots,i_{n-1}\neq 1} E_{i_0,\ldots,i_{n-1},}} g_n \, dP = \frac{1}{p_n}.$$

Let

$$f := \sum_{k=1}^{\infty} \frac{g_k}{k^2}.$$

In this case $f \in L_1$ and

$$f_n = \sum_{k=1}^{n-1} \frac{g_k}{k^2} + \sum_{k=n}^{\infty} \frac{g_n}{k^2} \geq \frac{g_n}{n+1}.$$

Hence

$$\int_{F_{n+1}} f^* \geq \int_{F_{n+1}} f_n \geq \int_{F_{n+1}} \frac{g_n}{n+1} = \frac{1}{(n+1)p_n}.$$

Since $p_n \leq N$, we obtain

$$Ef^* \geq \sum_{k=0}^{\infty} \int_{F_{n+1}} f^* = \infty.$$

Consequently, $f - E_0 f \in L_1$, nevertheless, $f - E_0 f \notin H_1$. ∎

To continue the proof of Theorem 2.70 let us take the function f constructed in Lemma 2.71. For this function we have obviously that $T_i f = \sigma^{(i)} \cdot f \in L_1$, however, $f \notin H_1$ which shows the theorem. ∎

From this it follows that if every p_n and $A_n^{(i)}$ ($n \in \mathbf{N}$) are equal then the conditions in Theorem 2.68 are necessary as well.

Corollary 2.72. *Suppose that $p_n = d$ and $B^{(i)} = A_n^{(i)}$ for every $n \in \mathbf{N}$ and $i = 1, \ldots, m$. Then $H_1 = \{f \in L_1 : T_i f \in L_1, i = 1, \ldots, m\}$ if and only if $B^{(1)}, \ldots, B^{(m)}$ have no common eigenvector $(z_1, \ldots, z_{d-1})$ for which $\overline{z_j} = z_{d-j}$ ($j = 1, \ldots, d-1$).*

Some examples of such transforms are to be presented. Denote by $a_{n;k,l}^{(i)}$ the elements of $A_n^{(i)}$. Simon asked in [170] whether in case $a_{n;k,l} := 0$ ($k \neq l$), $a_{n;k} := a_{n;k,k} = -1$ ($1 \leq k \leq [(p_n - 1)/2]$) and $a_{n;k} := a_{n;k,k} = 1$ ($[(p_n - 1)/2] < k \leq p_n - 1$) one has $H_1 = \{f \in L_1 : Tf \in L_1\}$. These transforms are used in [173] to prove that the Vilenkin-Fourier series of a function $f \in L_p$ converges to f in L_p norm ($1 < p < \infty$). The next results are more general and follow from Theorem 2.68 and Corollary 2.72.

Corollary 2.73. *Assume that T_i ($i = 1, \ldots, m$) are multiplier transforms and for each $n \in \mathbf{N}$ and each j with $1 \leq j \leq p_n - 1$ there is an i such that $a_{n;j}^{(i)} \neq a_{n;p_n-j}^{(i)}$. Then $H_1 = \{f \in L_1 : T_i f \in L_1, i = 1, \ldots, m\}$.*

From this it follows that if every p_n is odd then the answer to the question put by Simon is yes.

Corollary 2.74. *Suppose that* $p_n = d$ *and* $B^{(i)} = A_n^{(i)}$ *are diagonal* ($n \in \mathbf{N}$). *Then* $H_1 = \{f \in L_1 : T_i f \in L_1, i = 1, \ldots, m\}$ *if and only if for each* j *with* $1 \le j \le d-1$ *there is an* i *such that* $a_j^{(i)} \ne a_{d-j}^{(i)}$.

If d is even then for $j = d/2$ we get that $a_j^{(i)} = a_{d-j}^{(i)}$ for all $1 \le i \le m$. Hence, in this case, H_1 can not be characterized by any finite number of multiplier transforms. Thus the answer to the question under discussion is no.

Assume that every p_n ($n \in \mathbf{N}$) is even. For the eigenvalues σ_n and eigenvectors $z_n = (z_{n;1}, \ldots, z_{n;p_n-1})$ of Simon's matrices A_n mentioned in Theorem 2.70 we have $\sigma_n = a_{n;p_n/2} = 1$ and $z_{n;k} = 0$ ($k = 1, \ldots, p_n - 1,\ k \ne p_n/2$), $z_{n;p_n/2} \in \mathbf{R}$. It follows from Theorem 2.70 that, in this case, $H_1 \ne \{f \in L_1 : Tf \in L_1\}$, so the answer to the question above is no again.

If p_n is odd then let A_n be the diagonal matrix given by Simon. Let us modify this matrix for every even p_n. Set $a_{n;1,p_n/2} = -1$, $a_{n;k,k} = -1$ ($2 \le k < p_n/2$), $a_{n;p_n/2,1} = 1$, $a_{n;k,k} = 1$ ($p_n/2 < k \le p_n - 1$), and else $a_{n;k,l} = 0$. It is easy to check that this matrix does not have any eigenvector having the property as in Theorem 2.68 whenever $p_n > 2$ ($n \in \mathbf{N}$). If T denotes the corresponding transform then H_1 is characterized by an only transform.

Corollary 2.75. *If* $p_n > 2$ *for all* $n \in \mathbf{N}$ *and* T *denotes the lattest transform then* $H_1 = \{f \in L_1 : Tf \in L_1\}$.

The same corollary holds also for the following modification of Simon's matrix A_n for every even p_n: $a_{n;k,k} = -1$ ($1 \le k < p_n/2$), $a_{n;p_n/2,p_n-1} = 1$, $a_{n;k,k} = 1$ ($p_n/2 < k < p_n - 1$), $a_{n;p_n-1,p_n/2} = -1$, and else $a_{n;k,l} = 0$.

Note that this transform for $p_n = 3$ is identical with the so-called Hilbert transform H^3, however, for other $p_n = 2d+1$ it is different from H^{2d+1} (see Banuelos [2]).

The next corollary follows easily.

Corollary 2.76. *Suppose that for* $f \in L_1$ *one has* $\hat{f}(k) = 0$ *if for any* $n \in \mathbf{N}$

$$P_n \le k \le P_n + (p_n - 1 - [(p_n - 1)/2])P_n - 1.$$

Then $\|f\|_{H_1} \le C\|f\|_1$. *The same can be stated with the condition that* $\hat{f}(k) = 0$ *if for any* $n \in \mathbf{N}$

$$P_n + [(p_n - 1)/2]P_n \le k \le P_{n+1} - 1.$$

Set $f = f_1 + f_2$ such that $\hat{f}_1(k) = 0$ if $P_n \le k \le P_n + (p_n - 1 - [(p_n - 1)/2])P_n - 1$ and $\hat{f}_2(k) = 0$ if $P_n + [(p_n - 1)/2]P_n \le k \le P_{n+1} - 1$ ($n \in \mathbf{N}$). Thus, in case each p_n ($n \in \mathbf{N}$) is odd, $f \in H_1$ if and only if $f_1 \in L_1$ and $f_2 \in L_1$, namely,

$$c(\|f_1\|_1 + \|f_2\|_1) \le \|f\|_{H_1} \le C(\|f_1\|_1 + \|f_2\|_1). \tag{2.91}$$

With the help of Corollary 2.75 a similar result can also be obtained if not every p_n is odd. The inequality (2.91) is analogous to a result due to Gundy and Varopoulos (see Theorem 2 in Gundy, Varopoulos [88] and Corollary 4 in Chao, Janson [41]).

Note that all the results of this paper can also be proved in the same way for the trigonometric modell considered by Gundy and Varopoulos [88].

If $p_n = 2$ for some $n \in \mathbb{N}$ then the results above are meaningless. In this case the transforms are defined like in Chao [36]. If $p_n = 2$ but $p_{n-1} \neq 2$ and $p_{n+1} \neq 2$ then p_n as well as f_{n+1} and $\mathcal{F}_{n+1}$ are dropped and let $p'_{n-1} := 2p_{n-1}$. If $p_{n-1} \neq 2$ and $p_n = p_{n+1} = \ldots = 2$ then $p_n, p_{n+2}, \ldots$ as well as $f_{n+1}, f_{n+3}, \ldots$ and $\mathcal{F}_{n+1}, \mathcal{F}_{n+3}, \ldots$ are dropped and let $p'_{n+2k-1} = 2p_{n+2k-1}$ $(k = 0, 1, \ldots)$. In any other case let $p'_n = p_n$. Hence we get a martingale $(F_k := f_{n_k}, \mathcal{F}_{n_k})$ and a new sequence (p'_{n_k}) such that $p'_{n_k} > 2$. If f_{n+1} is dropped then $f_{n_{k_0}} := f_{n+2}$ is kept. Since

$$|f_{n+1}| \leq N \cdot E_n|f_{n+1}| \leq N \cdot E_n|f_{n+2}|,$$

we have

$$F^* \leq f^* \leq N \sup_{k \in \mathbb{N}} E_{n_{k-1}}|f_{n_k}|.$$

Using Corollary 2.21 we obtain

$$\| \sup_{k \in \mathbb{N}} E_{n_{k-1}}|f_{n_k}| \|_1 \leq 2\|F^*\|_1.$$

This yields that $f \in H_1$ if and only if $F \in H_1(\mathcal{F}_{n_k})$. Let us transform the martingale F. From $T_i F$ we define $T_i f$ by

$$T_i f_l := E_l(T_i F_{n_k})$$

for $l \leq n_k$. From Theorem 2.68 and 2.70 we obtain

Theorem 2.77. *Assume that $p_n = 2$ for some $n \in \mathbb{N}$, $(F_k, \mathcal{F}_{n_k})$ is the martingale corresponding to (f_n) and the transforms $T_1, \ldots, T_m$ of (F_k) have the same property as in Theorem 2.68. Then $f \in H_1$ (or, equivalently, $F \in H_1(\mathcal{F}_{n_k})$) if $\|T_i f_n\|_1$ (or, equivalently, if $\|T_i F_n\|_1$) are uniformly bounded $(i = 0, 1, \ldots, m)$. If every p_n is equal to 2 and $B^{(i)} = A_n^{(i)}$ then the preceding condition is necessary, too.*

Note that Gundy [86] has proved some results that are similar to Theorem 2.68 and 2.70. However, those results hold in the only case every $r_n^{(j)}$ in (1.10) is real. He has obtained that Theorem 2.68 and 2.70 hold if $B^{(i)} = A_n^{(i)}$ have no common real eigenvector $(p_n = d)$. He claims on p. 289 that, for complex $r_n^{(j)}$, the space H_1 is characterized by

$$\begin{pmatrix} 0 & 0 & 1 & 0 \\ 0 & 0 & 0 & 1 \\ -1 & 0 & 0 & 0 \\ 0 & -1 & 0 & 0 \end{pmatrix}$$

$(p_n = 5)$. This matrix does not have any real eigenvector, though it has an eigenvector having the property as in Corollary 2.72 (e.g. $[1, -\imath, \imath, 1]$). Consequently, by Corollary 2.72, the space H_1 can not be characterized by this matrix.

Now we come to the characterization of BMO and VMO. Denote by A_n^* the adjoint matrix of A_n and by T^* the corresponding martingale transform. It is easy to see that $E[(Tf)\bar{g}] = E(f\overline{T^* g})$. It is proved in Theorem 2.24 that the dual of H_1 is BMO and the bounded linear functionals can be given by

$$l_\phi(f) = E(f\bar{\phi}) \qquad (f \in L_2)$$

where $\phi \in BMO$ is arbitrary. From this we obtain

Theorem 2.78. *Assume that, for each $n \in \mathbf{N}$, the matrices $A_n^{(1)*}, \ldots, A_n^{(m)*}$ have no common eigenvector $(z_1, \ldots, z_{p_n-1})$ with $\overline{z_j} = z_{p_n-j}$ $(j = 1, \ldots, p_n - 1)$. Then $BMO = L_\infty + \sum_{i=1}^m T_i L_\infty$ and*

$$c\|g\|_{BMO} \leq \inf \sup_{0 \leq i \leq m} \|g_i\|_\infty \leq C\|g\|_{BMO} \tag{2.92}$$

where $g = \sum_{i=0}^m T_i g_i$. If $p_n = d$ and $B^{(i)} = A_n^{(i)}$ $(n \in \mathbf{N}, i = 0, 1, \ldots, m)$ then the preceding condition is necessary, too.

Proof. The proof is similar to that of Corollary 2 in Janson [106]. It can easily be proved by Hahn-Banach's theorem that the dual of the space

$$\{f \in L_1 : T_i^* f \in L_1, i = 1, \ldots, m\}$$

with the norm

$$\|(f, T_1^* f, \ldots, T_m^* f)\| := \sum_{i=0}^m \|T_i^* f\|_1$$

is the space $\sum_{i=0}^m T_i L_\infty$ with the norm

$$\|g\| := \inf \sup_{0 \leq i \leq m} \|g_i\|_\infty$$

where $g = \sum_{i=0}^m T_i g_i$. The continuous linear functionals are given by

$$l_g(f) := E(f\overline{g}) = \sum_{i=0}^m E(f\overline{T_i g_i}) = \sum_{i=0}^m E[(T_i^* f)\overline{g_i}]. \tag{2.93}$$

Theorem 2.68 proves the first part of the assertion.

To prove the necessarity suppose that the condition of Theorem 2.78 does not hold, though $BMO = \sum_{i=0}^m T_i L_\infty$ and (2.92) are true. Denote by L the vectorspace of the Vilenkin step functions with zero mean, more exactly, the vectorspace

$$\{f : f \text{ is } \mathcal{F}_n \text{ measurable for any } n \in \mathbf{N} \text{ and } Ef = 0\}.$$

L is dense in $\{f \in L_1 : T_i^* f \in L_1, i = 0, 1, \ldots, m\}$ because $T_i^* f_n = (T_i^* f)_n \to T_i^* f$ in L_1 norm as $n \to \infty$ and $T_i^* f_n \in L$ for every $i = 0, 1, \ldots, m$ and $n \in \mathbf{N}$. Since

$$\sum_{i=0}^m \|T_i^* f\|_1 \leq C\|f\|_{H_1}$$

(see (2.88)), we obtain from the next lemma that

$$H_1 = \{f \in L_1 : T_i^* f \in L_1, i = 0, 1, \ldots, m\}$$

which is a contradiction to Corollary 2.72. ∎

This theorem was first proved by Fefferman and Stein [69] for $BMO(\mathbf{R}^n)$. The following result is very interesting even in itself.

Lemma 2.79. *Suppose that X is a normed space, Y is a Banach space and their dual spaces are equivalent and, moreover, there exists a set U which is dense in X and also in Y. If $Y \subset X$ and*

$$c\|x\|_X \leq \|x\|_Y \tag{2.94}$$

then $X \sim Y$.

Proof. For an arbitrary $x \in U$, by Hahn-Banach's theorem, there exists $z \in Y'$ such that $\|z\|_{Y'} = 1$ and $z(x) = \|x\|_Y$. Thus

$$\|x\|_Y = |z(x)| \leq \|z\|_{X'}\|x\|_X \leq C\|x\|_X \tag{2.95}$$

for all $x \in U$. The space U is dense in X, so, for every $x \in X$, there exists a sequence $x_n \in U$ $(n \in \mathbf{N})$ which converges to x in X. By (2.95), (x_n) is a Cauchy sequence in Y, too. From (2.94) it follows that the limit of (x_n) in Y equals x as well. Taking the limit in (2.95) one completes the proof. ∎

It was proved in Theorem 2.39 that the dual of VMO is H_1 and the bounded linear functionals can be given by

$$l_f(\phi) = E(\phi\overline{f}) \qquad (\phi \in L)$$

where $f \in H_1$.

Let C_W represent the collection of functions $g : [0,1) \longrightarrow \mathbf{C}$ that are continuous at every Vilenkin irrational point (i.e. at every point that can not be written in the form k/P_n), continuous from the right on $[0,1)$, and have a finite limit from the left on $(0,1]$, all this in the usual topology. There is an isomorphism between $[0,1)$ and a Vilenkin group G (see e.g. Schipp, Wade, Simon, Pál [167]). G is compact and C_W is isomorphic to the space of the continuous functions on G. It is well known that the dual of C_W is the space M of all bounded measures on $\mathcal{A}$.

Now the characterization of the VMO space comes.

Theorem 2.80. *Assume that, for each $n \in \mathbf{N}$, the matrices $A_n^{(1)^*}, \ldots, A_n^{(m)^*}$ have no common eigenvector $(z_1, \ldots, z_{p_n-1})$ with $\overline{z_j} = z_{p_n-j}$ $(j = 1, \ldots, p_n - 1)$. Then $VMO = C_W + \sum_{i=1}^m T_i C_W$ and*

$$c\|g\|_{VMO} \leq \inf \sup_{0\leq i\leq m} \|g_i\|_\infty \leq C\|g\|_{VMO} \tag{2.96}$$

where $g_i \in C_W$ and $g = \sum_{i=0}^m T_i g_i$. If $p_n = d$ and $B^{(i)} = A_n^{(i)}$ $(n \in \mathbf{N}, i = 1, \ldots, m)$ then the preceding condition is necessary, too.

Proof. First we show that the dual of the space $\sum_{i=0}^m T_i C_W$ with the norm

$$\|g\| := \inf \sup_{0\leq i\leq m} \|g_i\|_\infty \qquad (g_i \in C_W, g = \sum_{i=0}^m T_i g_i)$$

is the space

$$\{(f_n) : \sup_{n\in\mathbf{N}} \sum_{i=0}^m \|T_i^* f_n\|_1 < \infty\}$$

where (f_n) is a martingale. From Stone-Weierstrass's theorem it comes that the Vilenkin polinomials are dense in C_W, thus L is dense in C_W and hence also in $\sum_{i=0}^m T_i C_W$. If a linear functional l has the form

$$l_f(g) = \lim_{n\to\infty} E(\overline{f_n} g) = \lim_{n\to\infty} \sum_{i=0}^{m} E(\overline{f_n} T_i g_i) = \lim_{n\to\infty} \sum_{i=0}^{m} E[\overline{(T_i^* f_n)} g_i] \tag{2.97}$$

where $g, g_i \in L$ $(i = 0, \dots, m)$ and $g = \sum_{i=0}^m T_i g_i$ then l is continuous and

$$\|l\| \le \sup_{n\in\mathbb{N}} \sum_{i=0}^{m} \|T_i^* f_n\|_1 .$$

Observe that the limit in (2.97) exists for all $g \in L$.

Conversely, if l is a continuous linear functional on $\sum_{i=0}^m T_i C_W$ then it is bounded on C_W as well. Thus there exists $\nu \in M$ such that

$$l(g) = \int_\Omega g \, d\overline{\nu} \qquad (g \in L)$$

and $\|\nu\| \le \|l\|$. So

$$l(T_i g_i) = \int_\Omega T_i g_i \, d\overline{\nu} = \int_\Omega g_i \, d\overline{T_i^* \nu} \qquad (g_i \in L).$$

Consequently, $\|T_i^* \nu\| \le \|l\|$. If (f_n) is the martingale defined by ν then the proof of the statement is complete.

If $B^{(i)*}$ have a common eigenvector as in Theorem 2.80 then, by Corollary 2.72,

$$H_1 \ne \{(f_n) : \sup_{n\in\mathbb{N}} \sum_{i=0}^{m} \|T_i^* f_n\|_1 < \infty\}$$

where (f_n) is a martingale, hence VMO is not equivalent to $\sum_{i=0}^m T_i C_W$.

Assume that the condition of Theorem 2.80 is satisfied. Then, by Theorem 2.68, the dual of VMO and the dual of $\sum_{i=0}^m T_i C_W$ are equivalent. C_W is a Banach space, thus the same holds for $\sum_{i=0}^m T_i C_W$ as well. Since $T_i : C_W \longrightarrow VMO$ are bounded, we obtain

$$\|g\|_{VMO} \le \sum_{i=0}^{m} \|T_i g_i\|_{VMO} \le \sum_{i=0}^{m} \|g_i\|_\infty .$$

Finally, from Lemma 2.79, we get (2.96) which proves Theorem 2.80. ∎

Note that if $B^{(i)*}$ have a common eigenvector as in Theorem 2.80 then $\sum_{i=0}^m T_i C_W$ can not be closed in BMO.

It is an open question whether these results can be extended to unbounded Vilenkin systems.

CHAPTER 3

TWO-PARAMETER MARTINGALE HARDY SPACES

Two-parameter martingales have been being investigated for about 20 years. One of the first results is belonging to Cairoli [29]; he proved in 1970 that Doob's inequality holds for two parameters, too. Then Burkholder-Gundy's inequality was proved by Metraux [125]. Davis's inequality is due to Brossard [14], [15] in case the stochastic basis is regular and to Frangos and Imkeller [76] for strong martingales. It is yet unknown whether it holds, in general. Bernard [9] has identified the dual of the dyadic H_1^S space.

In most cases the proofs of the one-parameter results can not simply be extended to two parameters, entirely new methods are needed. We point out only two fundamental reasons of this. The first one is that we can not apply the stopping time technique in the two-parameter case because of the difference of the one- and the two-parameter stopped martingales mentioned in Section 1.1. The other is that, for a one-parameter non-decreasing sequence of functions, the differences of the functions are all non-negative which does not hold for two parameters. Nevertheless, the method of the atomic decomposition can be used in both cases.

In this chapter some results of Chapter 2 are generalized. In Section 3.1 the definition of the atoms is modified and the atomic decomposition of the space H_p^s is formulated (see Weisz [199]). If the stochastic basis is regular then the martingales in H_p^S can also be decomposed into a sum of atoms. This result is due to Bernard [9] for dyadic martingales. In Section 3.2 several known martingale inequalities are given. Some of them are proved with the help of the atomic decomposition. Burkholder-Gundy's inequality and, for a regular stochastic basis, the equivalence of the five martingale Hardy spaces introduced in Section 1.1 are verified. The proof of the equivalence $H_p^* \sim H_p^S$ in case the stochastic basis is regular was found out by Brossard [14], [15]. In Section 3.3, amongst others, we prove that the dual of H_1^s is BMO_2 and, for Vilenkin martingales, the dual of H_p^s is H_q^s $(1 < p < \infty, 1/p + 1/q = 1)$. Moreover, if every σ-algebra is generated by finitely many (set) atoms then the dual of VMO_2 is H_1^s. For a regular stochastic basis the equivalence of the BMO_q spaces $(2 \leq q < \infty)$ is proved. A counterexample shows that BMO can not be defined simpler. Furthermore, opposed to the one-parameter case, the martingales in H_p^s can not be decomposed into a sum of simple atoms. In Section 3.4 the two-parameter martingale transforms are investigated and some results of Section 2.5 are extended. In Section 3.5 strong martingales are considered. This type of martingales is more similar to the one-parameter martingales. Almost every one-parameter result is extended to two-parameter strong martingales. With the help of Davis decomposition a new proof of Davis's inequality is given like in the one-parameter case in Section 2.2. In addition, the duality between H_1^* and BMO_2^- and the equivalence of the BMO_q^- spaces $(2 \leq q < \infty)$ are also verified.

3.1. ATOMIC DECOMPOSITIONS

The concept of an atom can be generalized for two parameters. However, in this case, instead of the (p,∞) atoms, the $(p,2)$ atoms are usable. In the two-parameter case the three different atoms in Definition 2.1 are all the same. This fact comes from the difference between the one- and the two-parameter stopped martingales.

In this section we formulate the atomic decomposition of the space H_p^s and, for regular stochastic basis, the one of H_p^S (see Weisz [199]). First the concept of the atoms is generalized.

Definition 3.1. *A function $a \in L_q$ is called a (p,q) atom if there exists a stopping time $\nu \in \mathcal{T}_2$ such that*

$$(i) \qquad a_n := E_n a = 0 \qquad \text{if} \qquad \nu \not\leq n$$

$$(ii) \qquad \|a^*\|_q \leq P(\nu \neq \infty)^{1/q-1/p} \qquad (0 < p \leq q, 1 < q \leq \infty).$$

Another equivalent definition of the atoms see in Bernard [9].

Now the analogue of Theorem 2.2 can be given for two parameters for $0 < p \leq 2$.

Theorem 3.2. *If the martingale $f = (f_n; n \in \mathbb{N}^2)$ is in H_p^s $(0 < p \leq 2)$ then there exist a sequence $(a^k, k \in \mathbb{Z})$ of $(p,2)$ atoms and a sequence $\mu = (\mu_k, k \in \mathbb{Z}) \in l_p$ of real numbers such that for all $n \in \mathbb{N}^2$*

$$\sum_{k=-\infty}^{\infty} \mu_k E_n a^k = f_n \tag{3.1}$$

and

$$\Big(\sum_{k=-\infty}^{\infty} |\mu_k|^p\Big)^{1/p} \leq C_p \|f\|_{H_p^s}. \tag{3.2}$$

Moreover, the sum $\sum_{k=l}^{m} \mu_k a^k$ converges to f in H_p^s norm as $m \to \infty$, $l \to -\infty$, too. Conversely, if $0 < p \leq 1$ and the martingale f has a decomposition of type (3.1) then $f \in H_p^s$ and

$$\|f\|_{H_p^s} \sim \inf\Big(\sum_{k=-\infty}^{\infty} |\mu_k|^p\Big)^{1/p} \tag{3.3}$$

where the infimum is taken over all decompositions of f of the form (3.1).

Proof. Assume that $f \in H_p^s$. Here finer stopping times than in the proof of Theorem 2.2 are needed to be introduced. Let

$$F_k := \{s(f) > 2^k\}$$

and consider the following stopping times for all $k \in \mathbb{Z}$:

$$\nu_k := \inf\{n \in \mathbb{N}^2 : E_n \chi(F_k) > 1/2\}.$$

It is easy to see that the equations (2.4) and (2.12) hold also in the two-parameter case:

(3.4) $$f_n = \sum_{k\in\mathbb{Z}} (f_n^{\nu_{k+1}} - f_n^{\nu_k})$$

and

(3.5) $$f_n^{\nu_{k+1}} - f_n^{\nu_k} = \sum_{m\leq n} (d_m f)\chi(\nu_k \ll m \not\gg \nu_{k+1}).$$

Set

$$\mu_k := 2^{k+3}\sqrt{2}P(\nu_k \neq \infty)^{1/p}$$

and

$$a_n^k := \frac{f_n^{\nu_{k+1}} - f_n^{\nu_k}}{\mu_k}.$$

For a fixed k, (a_n^k) is a martingale. We show that a^k is a $(p,2)$ atom corresponding to the stopping time ν_k. If $\nu_k \not\ll n$ then obviously $f_n^{\nu_{k+1}} = f_n^{\nu_k}$, thus (i) of Definition 3.1 holds. For (ii) we have to verify the inequality

$$E[(a^{k*})^2] \leq P(\nu_k \neq \infty)^{1-2/p}.$$

By Cairoli's maximal inequality (3.12) and by the definition of a^k the last inequality is equivalent to

(3.6) $$E(f_n^{\nu_{k+1}} - f_n^{\nu_k})^2 \leq 2\cdot 4^{k+1}P(\nu_k \neq \infty).$$

Using (3.5) and the fact that L_2 is isometric to H_2^s this inequality follows from the following one:

(3.7) $$E\Big(\sum_{n\in\mathbb{N}^2} E_{n-1}|d_n f|^2\chi(\nu_k \ll n \not\gg \nu_{k+1})\Big) \leq 2\cdot 4^{k+1}P(\nu_k \neq \infty).$$

Remember that the set $\{\nu_k \ll n \not\gg \nu_{k+1}\}$ is $\mathcal{F}_{n-1}$ measurable. Decompose the left side of (3.7) into two parts:

$$(A) = \sum_{n\in\mathbb{N}^2} E\Big(E_{n-1}|d_n f|^2\chi(\nu_k \ll n \not\gg \nu_{k+1})\chi(s(f) \leq 2^{k+1})\Big)$$

and

$$(B) = \sum_{n\in\mathbb{N}^2} E\Big(E_{n-1}|d_n f|^2\chi(\nu_k \ll n \not\gg \nu_{k+1})\chi(s(f) > 2^{k+1})\Big).$$

Then

(3.8) $$E\Big(\sum_{n\in\mathbb{N}^2} E_{n-1}|d_n f|^2\chi(\nu_k \ll n \not\gg \nu_{k+1}\Big) = (A) + (B).$$

Clearly,

(3.9) $$(A) \leq 4^{k+1}P(\nu_k \neq \infty).$$

Taking in (B) the conditional expectation with respect to $\mathcal{F}_{n-1}$ we obtain

$$(B) = \sum_{n\in\mathbf{N}^2} E\Big(E_{n-1}|d_n f|^2 \chi(\nu_k \ll n \not\gg \nu_{k+1}) E_{n-1}\chi(F_{k+1})\Big).$$

It follows from the definition of ν_{k+1} that, if $\nu_{k+1} \not\ll n$, then $E_{n-1}\chi(F_{k+1}) \leq 1/2$. So

$$(B) \leq \frac{1}{2} E\Big(\sum_{n\in\mathbf{N}^2} E_{n-1}|d_n f|^2 \chi(\nu_k \ll n \not\gg \nu_{k+1})\Big).$$

From this together with (3.8) and (3.9) we get (3.7), hence a^k is really a $(p,2)$ atom. Consequently, (3.1) holds. Of course, there exists a function $a^k \in L_2$ such that $E_n a^k = a_n^k$ $(n \in \mathbf{N}^2)$.

Applying Tsebisev's inequality and the equivalence between H_2^* and L_2 (see (3.12)) we obtain

$$\begin{aligned}
\sum_{k\in\mathbf{Z}} |\mu_k|^p &\leq C_p \sum_{k\in\mathbf{Z}} (2^k)^p P(\nu_k \neq \infty) \\
&= C_p \sum_{k\in\mathbf{Z}} (2^k)^p P\Big(\sup_{n\in\mathbf{N}^2} E_n\chi(F_k) > 1/2\Big) \\
&\leq C_p \sum_{k\in\mathbf{Z}} 4(2^k)^p E\Big[\sup_{n\in\mathbf{N}^2} (E_n\chi(F_k))^2\Big] \\
&\leq C_p \sum_{k\in\mathbf{Z}} (2^k)^p P(F_k) \\
&= C_p \sum_{k\in\mathbf{Z}} (2^k)^p P(s(f) > 2^k).
\end{aligned}$$

Similarly to the proof of Theorem 2.2 we can prove (3.2) by Abel rearrangement.

We obtain the inequality

$$\|f^{\nu_l}\|_{H_p^s} \leq \|f^{\nu_l}\|_2 \leq \sqrt{2}\cdot 2^l$$

with the method (3.6) was verified. One can show the same way as we did in Theorem 2.2 that the series $\sum_{k=l}^m \mu_k a^k$ converges to f in H_p^s norm as $m \to \infty$, $l \to -\infty$ $(0 < p \leq 2)$.

For the converse we prove that if a is a $(p,2)$ atom then

$$\|a\|_{H_p^s} \leq 1 \qquad (0 < p \leq 2).$$

Indeed, from the definition of the atom it follows that

$$\chi(\nu \not\ll n) E_{n-1}|d_n a|^2 = E_{n-1}[\chi(\nu \not\ll n)|d_n a|^2] = 0.$$

Hence $s(a) = 0$ on the set $\{\nu = \infty\}$. Applying the equation $\|s(a)\|_2 = \|a\|_2$ and Hölder's inequality we have

$$E(s^p(a)) \leq [E(s^2(a))]^{p/2} P(\nu \neq \infty)^{1-p/2} \leq 1.$$

If we assume that $0 < p \leq 1$ and that the martingale f has a decomposition of the form (3.1) then the inequality

$$E(s^p(f)) \leq \sum_{k=-\infty}^{\infty} |\mu_k|^p$$

can be proved again the way as in Theorem 2.2. The proof of the theorem is complete. ∎

If $\mathcal{F}$ is regular then the previous theorem can be shown for the H_p^S spaces, too.

Theorem 3.3. *If the stochastic basis $\mathcal{F}$ is regular then, instead of H_p^s, Theorem 3.2 holds for the H_p^S spaces.*

Proof. Let

$$F_k := \{S(f) > 2^k\}$$

and instead of $1/2$ write $1/(2R^2)$ in the definition of ν_k where R is the regularity constant. Since L_2 and H_2^S are isometric, the inequality (3.7) is also valid in the following form:

$$E\Big(\sum_{n\in\mathbf{N}^2} |d_n f|^2 \chi(\nu_k \ll n \not\gg \nu_{k+1}\Big) \leq 2\cdot 4^{k+1} P(\nu_k \neq \infty).$$

The formulas (A) and (B) can be defined as follows:

$$(A) = \sum_{n\in\mathbf{N}^2} E\Big(|d_n f|^2 \chi(\nu_k \ll n \not\gg \nu_{k+1})\chi(S(f) \leq 2^{k+1})\Big)$$

and

$$(B) = \sum_{n\in\mathbf{N}^2} E\Big(|d_n f|^2 \chi(\nu_k \ll n \not\gg \nu_{k+1})\chi(S(f) > 2^{k+1})\Big).$$

Applying the inequality $|d_n f|^2 \leq R^2 E_{n-1}|d_n f|^2$, which holds by regularity, we get

$$\begin{aligned}(B) &\leq \sum_{n\in\mathbf{N}^2} E\Big(R^2 E_{n-1}|d_n f|^2 \chi(\nu_k \ll n \not\gg \nu_{k+1})\chi(S(f) > 2^{k+1})\Big)\\ &= R^2 \sum_{n\in\mathbf{N}^2} E\Big(E_{n-1}|d_n f|^2 \chi(\nu_k \ll n \not\gg \nu_{k+1}) E_{n-1}\chi(F_{k+1})\Big)\\ &\leq \frac{1}{2} \sum_{n\in\mathbf{N}^2} E\Big(|d_n f|^2 \chi(\nu_k \ll n \not\gg \nu_{k+1})\Big).\end{aligned}$$

The rest can be proved similarly to Theorem 3.2. ∎

Note that, for dyadic martingales in which case $H_1^s = H_1^S$ and $|d_n f|^2$ is $\mathcal{F}_{n-1}$ measurable, this theorem was proved by Bernard [9].

3.2. MARTINGALE INEQUALITIES

In this section inequalities relative to the martingale Hardy spaces are given. Cairoli's maximal inequality is proved, namely, the equivalence $H_p^* \sim L_p$ $(p > 1)$.

Similarly to Theorem 2.11, in Theorem 3.5 the following relations are verified with the help of the atomic decomposition: $H_p^s \subset H_p^*, H_p^S$ $(0 < p \leq 2)$; $H_p^*, H_p^S \subset H_p^s$ $(2 \leq p < \infty)$; $\mathcal{P}_p, \mathcal{Q}_p \subset H_p^*, H_p^S$ $(1 < p < \infty)$ and $\mathcal{P}_p, \mathcal{Q}_p \subset H_p^s$ $(2 \leq p < \infty)$ (see Weisz [199]). Based on the one-parameter results Burkholder-Gundy's inequality is proved (see Metraux [125]). Neither a proof nor a counterexample for Davis's inequality is known. However, we show that all the five martingale Hardy spaces are equivalent if the stochastic basis is regular (see also Weisz [199]). The proof of the most complicated part of this result, more exactly, the proof of the inequality $\|f\|_{H_p^S} \leq C_p \|f\|_{H_p^*}$ $(0 < p \leq 1)$ is due to Brossard [14].

By Theorem 3.2 we obtain (2.13) for two parameters, too:

$$\|f\|_1 \leq \|f\|_{H_1^s}. \tag{3.10}$$

Of course, the inequality (3.10) holds also for the spaces H_1^* and $\mathcal{P}_1$ instead of H_1^s.

Doob's inequality for two parameters is to be proved. In the next theorem let

$$f_n^\bullet := \sup_{k \leq n_1} |f_{k,n_2}|.$$

Proposition 3.4. *For an arbitrary martingale f, $\lambda > 0$ and $n \in \mathbb{N}^2$ we have*

$$\lambda P(f_n^* > \lambda) \leq \int_{\{f_n^* > \lambda\}} f_n^\bullet \, dP \tag{3.11}$$

and if $f \in L_p$ $(1 < p \leq \infty)$ then

$$\|f_n\|_p \leq \|f_n^*\|_p \leq \left(\frac{p}{p-1}\right)^2 \|f_n\|_p. \tag{3.12}$$

Thus $H_p^* \sim L_p$ also in the two-parameter case if $1 < p \leq \infty$.

Proof. Observe that

$$f_n^* := \sup_{l \leq n_2} f_{n_1,l}^\bullet.$$

We show that $(f_{n_1,l}^\bullet, 0 \leq l \leq n_2)$ is a submartingale. By the condition F_4 one has

$$f_{k,l-1} = E_{n_1,l-1} f_{k,l}.$$

Taking the absolute value and the supremum one can see that

$$f_{n_1,l-1}^\bullet = \sup_{k \leq n_1} |f_{k,l-1}| \leq \sup_{k \leq n_1} E_{n_1,l-1} |f_{k,l}| \leq E_{n_1,l-1} f_{k,l}^\bullet.$$

Hence $(f_{n_1,l}^\bullet)$ is really a non-negative submartingale. The inequality (3.11) follows from Lemma 2.7. Applying twice Proposition 2.6 we get (3.12). ∎

The inequality (3.12) was first proved by Cairoli in [29].

The next extension of Theorem 2.11 follows from Theorems 3.2 and 3.6.

Theorem 3.5.

(i)

$$\|f\|_{H_p^*} \leq C_p \|f\|_{H_p^s}, \quad \|f\|_{H_p^S} \leq C_p \|f\|_{H_p^s} \qquad (0 < p \leq 2)$$

(ii)

$$\|f\|_{H_p^S} \le C_p \|f\|_{H_p^*}, \quad \|f\|_{H_p^S} \le C_p \|f\|_{H_p^s} \qquad (2 \le p < \infty)$$

(iii)

$$\|f\|_{H_p^*} \le \|f\|_{\mathcal{P}_p}, \quad \|f\|_{H_p^s} \le \|f\|_{\mathcal{Q}_p} \qquad (0 < p < \infty)$$

(iv)

$$\|f\|_{H_p^*} \le C_p \|f\|_{\mathcal{Q}_p}, \quad \|f\|_{H_p^s} \le C_p \|f\|_{\mathcal{P}_p} \qquad (1 < p < \infty)$$

(v)

$$\|f\|_{H_p^*} \le C_p \|f\|_{\mathcal{P}_p}, \quad \|f\|_{H_p^S} \le C_p \|f\|_{\mathcal{Q}_p} \qquad (2 \le p < \infty).$$

Proof. The verification proceeds like the one of Theorem 2.11. We could see in the proof of Theorem 3.2 that if a is a $(p, 2)$ atom then $\|a\|_{H_p^S} \le 1$ $(0 < p \le 2)$. Analogously to this one can show the inequalities $\|a\|_{H_p^*} \le 1$ and $\|a\|_{H_p^s} \le 1$ $(0 < p \le 2)$ for two parameters as well. The first inequality of (i) follows immediately from Theorem 3.2 for $0 < p \le 1$ while for $1 < p \le 2$ from the second inequality of (i) and from Theorem 3.6. If we apply (3.4) and (3.5) to the function $d_n f$ instead of f_n then the equation

$$d_n f = \sum_{k \in \mathbf{Z}} (d_n f) \chi(\nu_k \ll n \not\gg \nu_{k+1}) \tag{3.13}$$

is obtained. Since, for fixed n, the sets $\{\nu_k \ll n \not\gg \nu_{k+1}\}$ are disjoint, we have

$$|d_n f|^2 = \sum_{k \in \mathbf{Z}} |d_n f|^2 \chi(\nu_k \ll n \not\gg \nu_{k+1}).$$

Hence

$$S^2(f) = \sum_{k \in \mathbf{Z}} S^2(\mu_k a^k)$$

where the numbers μ_k and atoms a^k are defined in the proof of Theorem 3.2. As $p/2 \le 1$,

$$E(S^p(f)) \le \sum_{k \in \mathbf{Z}} |\mu_k|^p E(S^p(a^k)) \le \sum_{k \in \mathbf{Z}} |\mu_k|^p$$

holds. The second inequality of (i) follows from Theorem 3.2.

Using (3.12) and Theorem 2.10 we get that, if $(g_n, n \in \mathbf{N}^2)$ is a sequence of non-negative measurable functions and $1 \le r < \infty$, then

$$E[(\sum_{n \in \mathbf{N}^2} E_n g_n)^r] \le r^{2r} E[(\sum_{n \in \mathbf{N}^2} g_n)^r].$$

The second inequality of (ii) follows from this like in Theorem 2.11.

The rest of the theorem comes easily from the definition and from Theorem 3.6. The proof is complete. ■

Note that, with another proof, (i) can be found in Brossard [15].

The two-parameter Burkholder-Gundy's inequality was proved by Metraux [125] with applying the boundedness of the one-parameter martingale transforms investigated in Section 2.5 (see the next proof).

Theorem 3.6. (Burkholder-Gundy's inequality) *H_p^S and H_p^* are equivalent if $1 < p < \infty$, namely,*

$$c_p\|f\|_{H_p^S} \le \|f\|_{H_p^*} \le C_p\|f\|_{H_p^S} \qquad (1 < p < \infty). \tag{3.14}$$

Proof. Let

$$g_{n,m} := \sum_{k=1}^{n}\sum_{l=1}^{m} u_k v_l d_{k,l}f$$

and

$$h_{n,m} := \sum_{k=1}^{n}\sum_{l=1}^{m} v_l d_{k,l}f$$

where $u = (u_k, k \in \mathbb{N})$ and $v = (v_l, l \in \mathbb{N})$ are bounded sequences of real numbers. The one-parameter martingale $(g_{n,m}; n \in \mathbb{N})$ is the martingale transform of the one-parameter martingale $(h_{n,m}; n \in \mathbb{N})$ by the sequence u and, moreover, $(h_{n,m}; m \in \mathbb{N})$ is the one-parameter martingale transform of $(f_{n,m}; m \in \mathbb{N})$ by the sequence v. Applying the remark after Theorem 2.55 we have for $1 < p < \infty$ that

$$E|g|^p \le C_p E|h|^p \tag{3.15}$$

and

$$E|h|^p \le C_p E|f|^p. \tag{3.16}$$

We prove the inequality (3.14) like Burkholder did in the one-parameter case (see [22]). The two-parameter Khintchine's inequality (see Paley [145] p. 257) will be used: if $0 < p < \infty$ then there are positive constants A_p and B_p such that if $a = (a_{k,l}; k, l \in \mathbb{N})$ is a real number sequence then

$$\begin{aligned} A_p(\sum_{k=1}^{\infty}\sum_{l=1}^{\infty} |a_{k,l}|^2)^{p/2} &\le \int_0^1\int_0^1 |\sum_{k=1}^{\infty}\sum_{l=1}^{\infty} a_{k,l} r_k(x) r_l(y)|^p \,dx\,dy \\ &\le B_p(\sum_{k=1}^{\infty}\sum_{l=1}^{\infty} |a_{k,l}|^2)^{p/2} \end{aligned}$$

where r_k is the Rademacher function. Applying this inequality together with (3.15) and (3.16) to $1 < p < \infty$ with $a_{k,l} = d_{k,l}f$, $u_k = r_k(x)$ and $v_l = r_l(y)$ we obtain

$$\begin{aligned} A_p E(S(f)^p) &\le E(\int_0^1\int_0^1 |\sum_{k=1}^{\infty}\sum_{l=1}^{\infty} (d_{k,l}f) r_k(x) r_l(y)|^p \,dx\,dy) \\ &= \int_0^1\int_0^1 E|\sum_{k=1}^{\infty}\sum_{l=1}^{\infty} (d_{k,l}f) r_k(x) r_l(y)|^p \,dx\,dy \\ &\le C_p E|f|^p. \end{aligned}$$

Since f is also a martingale transform of $(\sum_{k=1}^{n}\sum_{l=1}^{m}(d_{k,l}f)r_k(x)r_l(y); n,m \in \mathbb{N})$, we get the other side of Burkholder-Gundy's inequality similarly. ∎

It is unknown whether (3.14) holds for $p = 1$.

However, if the stochastic basis is regular then (3.14) is true for every p, moreover, all the five Hardy spaces are equivalent. First we prove the equivalence of H_p^* and $\mathcal{P}_p$ and the one of H_p^S and $\mathcal{Q}_p$. The next theorem and corollary correspond to Lemma 2.20 and Corollary 2.21.

Theorem 3.7. *If $\mathcal{F}$ is regular then $H_p^* \sim \mathcal{P}_p$ and $H_p^S \sim \mathcal{Q}_p$ for all $0 < p < \infty$.*

Proof. We are going to prove the first equivalence, only. The inequality

$$\|f\|_{H_p^*} \leq \|f\|_{\mathcal{P}_p}$$

is obvious. To prove the converse let

$$\lambda_{n-1,m-1} := R^{2/p} \sup_{k\leq n} \sup_{l\leq m} (E_{k-1,l-1}|f_{k,l}|^p)^{1/p}.$$

In this case $\lambda_{n,m}$ is $\mathcal{F}_{n,m}$ measurable and, by the regularity of $\mathcal{F}$,

$$|f_{n,m}|^p \leq R^2 E_{n-1,m-1}|f_{n,m}|^p \leq \lambda_{n-1,m-1}^p.$$

Then

$$E\lambda_\infty^p = R^2 E(\sup_{n,m\in\mathbb{N}} E_{n-1,m-1}|f_{n,m}|^p).$$

By the F_4 hypotesis

$$\begin{aligned}\sup_{n,m\in\mathbb{N}} E_{n-1,m-1}|f_{n,m}|^p &\leq \sup_{n,m\in\mathbb{N}} E_{n-1,\infty}E_{\infty,m-1}f_{n,m}^{*\,p} \\ &\leq \sup_{n\in\mathbb{N}} E_{n-1,\infty} \sup_{m\in\mathbb{N}} E_{\infty,m-1}f_{n,m}^{*\,p}.\end{aligned}$$

If $\mu_n := \sup_{m\in\mathbb{N}} E_{\infty,m-1}f_{n,m}^{*\,p}$ then μ_n is $\mathcal{F}_{n,\infty}$ measurable. So

$$\begin{aligned}\sup_{n,m\in\mathbb{N}} E_{n-1,m-1}|f_{n,m}|^p &\leq \sup_{n\in\mathbb{N}} E_{n-1,\infty}\mu_n^* \\ &= \sup_{n\in\mathbb{N}} E_{n-1,\infty}[\mu_{n-1}^* + (\mu_n^* - \mu_{n-1}^*)] \\ &\leq \mu_\infty^* + \sum_{n=1}^{\infty} E_{n-1,\infty}(\mu_n^* - \mu_{n-1}^*).\end{aligned}$$

Hence

$$E(\sup_{n,m\in\mathbb{N}} E_{n-1,m-1}|f_{n,m}|^p) \leq 2E(\mu_\infty^*).$$

As

$$E(\mu_\infty^*) = E(\sup_{n\in\mathbb{N}} \sup_{m\in\mathbb{N}} E_{\infty,m-1}f_{n,m}^{*\,p}) \leq E(\sup_{m\in\mathbb{N}} E_{\infty,m-1}f_{\infty,m}^{*\,p}),$$

by the previous method we get that

$$E(\mu_\infty^*) \leq 2E(f^{*\,p}).$$

Consequently,

$$\|f\|_{\mathcal{P}_p} \le (4R^2)^{1/p} \|f\|_{H_p^s}$$

which proves the theorem. ∎

Similarly to the previous proof in case $p = 1$ we can verify the next result with the help of the convexity theorem (see Theorem 2.10).

Corollary 3.8. *For an arbitrary martingale f and for $1 \le p < \infty$*

$$\| \sup_{n,m \in \mathbb{N}} E_{n-1,m-1} |f_{n,m}| \|_p \le (1+p)^2 \|f^*\|_p,$$

moreover, if $\mathcal{F}$ is regular then the converse inequality holds also with the constant R^2.

The next result shows that, for a regular stochastic basis, the Hardy spaces generated by the quadratic and by the conditional quadratic variation are equivalent, too.

Theorem 3.9. *If $\mathcal{F}$ is regular then $H_p^s \sim H_p^S$ for all $0 < p < \infty$.*

Proof. Similarly to the proof of Theorem 3.5 (i) one can show the inequality

$$\|f\|_{H_p^s} \le C_p \|f\|_{H_p^S} \qquad (0 < p \le 2)$$

by Theorem 3.3. So the equivalence is proved for $0 < p \le 2$. Nevertheless, for $2 \le p < \infty$, it follows from Theorem 3.5 (ii) and from the inequality $S(f) \le Rs(f)$ which holds because of regurality. ∎

The proof of the equivalence between H_p^* and H_p^S is due to Brossard [14] and is based on the next theorem.

Theorem 3.10. *If $\mathcal{F}$ is regular then for a martingale f and for $\lambda > 0$ we have*

$$P(S(f) > \lambda) \le C\Big[P(f^* > \lambda) + \frac{1}{\lambda^2} \int_{\{f^* \le \lambda\}} f^{*2} \, dP\Big].$$

During the proof the following lemmas will be used. One can assume that there exists $(N, M) \in \mathbb{N}^2$ such that $d_{n,m} f = 0$ if $n > N$ or $m > M$. Set

$$F := \{f^* \le \lambda\}, \qquad g_{n,m} := E_{n,m} \chi(F).$$

Moreover, let

$$G := \{(1-g)^* \le \frac{1}{2R^2}\}, \qquad h_{n,m} := E_{n,m} \chi(G)$$

and

$$H := \{(1-h)^* \le \frac{1}{2}\}$$

where R is the constant in the definition of the regularity of $\mathcal{F}$. Obviously, both $g = (g_{n,m})$ and $h = (h_{n,m})$ are martingales.

Lemma 3.11. *If $E_{n,m}\chi(H) \ne 0$ then $1 - h_{n,m} \le 1/2$ and $h_{n-1,m-1} \ge 1/(2R^2)$.*

Proof. We have $1 - h_{n,m} \leq 1/2$ on the set H. Thus

$$\frac{1}{2}E_{n,m}\chi(H) \geq E_{n,m}[(1-h_{n,m})\chi(H)]$$
$$= (1-h_{n,m})E_{n,m}\chi(H).$$

If $E_{n,m}\chi(H) \neq 0$ then $1 - h_{n,m} \leq 1/2$. Hence, by regularity,

$$\frac{1}{2} \leq h_{n,m} \leq R^2 h_{n-1,m-1}$$

which shows the lemma. ■

Lemma 3.12. *If $h_{n-1,m-1} > 0$ then $|f_{n-1,m-1}| \leq \lambda$, $|f_{n,m-1}| \leq \lambda$, $|f_{n-1,m}| \leq \lambda$ and $|f_{n,m}| \leq \lambda$.*

Proof. By the same argument like in the proof of Lemma 3.11 we obtain that $1 - g_{n-1,m-1} \leq 1/(2R^2)$ if $h_{n-1,m-1} > 0$. Therefore

$$1 - g_{n,m-1} \leq \frac{1}{2R},$$

$$1 - g_{n-1,m} \leq \frac{1}{2R}$$

and

$$1 - g_{n,m} \leq \frac{1}{2}.$$

Hence $g_{n-1,m-1} > 0$, $g_{n,m-1} > 0$, $g_{n-1,m} > 0$ and $g_{n,m} > 0$ on the set $\{h_{n-1,m-1} > 0\}$. Again by Lemma 3.11 the lemma is complete. ■

Lemma 3.13. *We have*

$$P(H^c) \leq CP(G^c) \leq CP(f^* > \lambda)$$

where H^c and G^c denote the complementaires of H and G, respectively.

Proof. By Doob's inequality we get

$$P[H^c] = P[(1-h)^* > 1/2]$$
$$\leq 4E[(1-h)^{*2}]$$
$$\leq CE[(1-h_{N,M})^2].$$

Thus $P(H^c) \leq CP(G^c)$. Similarly,

$$P(G^c) \leq CR^4 P(F^c) = CR^4 P(f^* > \lambda)$$

which proves the lemma. ■

For an arbitrary sequence $(u_{n,m})$ of functions the two-parameter differences and the differences taken in one parameter are to be introduced. Let

$$du_{n,m} := u_{n,m} - u_{n-1,m} - u_{n,m-1} + u_{n-1,m-1},$$

$$d_1 u_{n,m} := u_{n,m} - u_{n-1,m}$$

and

$$d_2 u_{n,m} := u_{n,m} - u_{n,m-1}.$$

Lemma 3.14. *For all* $(n,m) \le (N,M)$ *we have*

$$E_{n-1,m-1}\Big[d_1 f^4_{n,m-1}\Big] = E_{n-1,m-1}\Big[(d_1 f_{n,m-1})^2\Big(6f^2_{n-1,m-1} + 4f_{n-1,m-1}d_1 f_{n,m-1} + (d_1 f_{n,m-1})^2\Big)\Big],$$

$$E_{n-1,m-1}\Big[d_2 f^4_{n-1,m}\Big] = E_{n-1,m-1}\Big[(d_2 f_{n-1,m})^2\Big(6f^2_{n-1,m-1} + 4f_{n-1,m-1}d_2 f_{n-1,m} + (d_2 f_{n-1,m})^2\Big)\Big]$$

and

$$\begin{aligned} E_{n-1,m-1}\Big[df^4_{n,m}\Big] = E_{n-1,m-1}\Big[&6(d_1 f_{n,m-1})^2(d_2 f_{n-1,m})^2 \\ &+ 12(df_{n,m})(d_1 f_{n,m-1})(d_2 f_{n-1,m})(f_{n-1,m} + f_{n,m-1}) \\ &+ (df_{n,m})^2\Big[6(f_{n-1,m} + f_{n,m-1} - f_{n-1,m-1})^2 \\ &+ 4(f_{n-1,m} + f_{n,m-1} - f_{n-1,m-1})df_{n,m} + (df_{n,m})^2\Big]\Big]. \end{aligned}$$

The proof of this lemma is omitted because it is a simple but quite a long calculation. For the proof one has to express $f_{n,m}$, $f_{n-1,m}$ and $f_{n,m-1}$ by $f_{n-1,m-1}$, $d_1 f_{n,m-1}$, $d_2 f_{n-1,m}$ and $df_{n,m}$. Moreover, the (F_4) hypothesis, namely, the equation $E_{n,m-1} \circ E_{n-1,m} = E_{n-1,m-1}$ has to be applied several times.

We leave the proof of the next simple algebraic lemma, too.

Lemma 3.15. *Let* α *and* β *be positive numbers. Then it follows from* $x \le \alpha + \beta x^{1/2}$ *that* $x \le 4\alpha + \beta^2$.

In the next lemmas we give some estimates which shall be used in the proof of Theorem 3.10.

Lemma 3.16.

$$E\Big[\sum_{n=1}^{N}\sum_{m=1}^{M}(dh_{n,m})^2\Big] \le CP(f^* > \lambda)$$

and

$$E\Big[\sum_{n=1}^{N}\sum_{m=1}^{M}(d_1 h_{n,m-1})^2(d_2 h_{n-1,m})^2\Big] \le CP(f^* > \lambda).$$

Proof. For every martingale a we have

$$E_{n-1,m-1}(da_{n,m})^2 = E_{n-1,m-1}(da^2_{n,m}). \tag{3.17}$$

Set $a_{n,m} = 1 - h_{n,m}$. Using (3.17) and Lemma 3.13 we obtain

$$\begin{aligned}
(3.18)\qquad E\Big[\sum_{n=1}^{N}\sum_{m=1}^{M}(dh_{n,m})^2\Big] &= E\Big[\sum_{n=1}^{N}\sum_{m=1}^{M}(da_{n,m})^2\Big]\\
&= E\Big[\sum_{n=1}^{N}\sum_{m=1}^{M}da_{n,m}^2\Big]\\
&= E[a_{N,M}^2 - a_{N,0}^2 - a_{0,M}^2 + a_{0,0}^2]\\
&\le 2E[a_{N,M}^2] = 2P(G^c)\\
&\le CP(f^* > \lambda).
\end{aligned}$$

On the other hand, observe that $0 \le a_{n,m} \le 1$ for all $(n,m) \le (N,M)$. Apply this and the third equation of Lemma 3.14 to a to win

$$\begin{aligned}
&6E\Big[\sum_{n=1}^{N}\sum_{m=1}^{M}(d_1h_{n,m-1})^2(d_2h_{n-1,m})^2\Big]\\
&\qquad= 6E\Big[\sum_{n=1}^{N}\sum_{m=1}^{M}(d_1a_{n,m-1})^2(d_2a_{n-1,m})^2\Big]\\
&\qquad\le E\Big[\sum_{n=1}^{N}\sum_{m=1}^{M}da_{n,m}^4\Big] + 24E\Big[\sum_{n=1}^{N}\sum_{m=1}^{M}|da_{n,m}||d_1a_{n,m-1}||d_2a_{n-1,m}|\Big]\\
&\qquad\quad+62E\Big[\sum_{n=1}^{N}\sum_{m=1}^{M}(da_{n,m})^2\Big].
\end{aligned}$$

Similarly to (3.18) one can see

$$E\Big[\sum_{n=1}^{N}\sum_{m=1}^{M}da_{n,m}^4\Big] \le 2E[a_{N,M}^4] \le CP(f^* > \lambda).$$

In (3.18) the third term was also estimated by $CP(f^* > \lambda)$. Applying Hölder's and Cauchy-Schwarz's inequalities for the second term we obtain

$$\begin{aligned}
&E\Big[\sum_{n=1}^{N}\sum_{m=1}^{M}|da_{n,m}||d_1a_{n,m-1}||d_2a_{n-1,m}|\Big]\\
&\qquad\le \Big(E\Big[\sum_{n=1}^{N}\sum_{m=1}^{M}|d_1a_{n,m-1}|^2|d_2a_{n-1,m}|^2\Big]\Big)^{1/2}\Big(E\Big[\sum_{n=1}^{N}\sum_{m=1}^{M}|da_{n,m}|^2\Big]\Big)^{1/2}\\
&\qquad\le C\Big(E\Big[\sum_{n=1}^{N}\sum_{m=1}^{M}|d_1a_{n,m-1}|^2|d_2a_{n-1,m}|^2\Big]\Big)^{1/2}P(f^* > \lambda)^{1/2}.
\end{aligned}$$

Thus

$$E\Big[\sum_{n=1}^{N}\sum_{m=1}^{M}|d_1a_{n,m-1}|^2|d_2a_{n-1,m}|^2\Big]$$
$$\leq CP(f^*>\lambda)+C\Big(E\Big[\sum_{n=1}^{N}\sum_{m=1}^{M}|d_1a_{n,m-1}|^2|d_2a_{n-1,m}|^2\Big]\Big)^{1/2}P(f^*>\lambda)^{1/2}.$$

Now the lemma follows from Lemma 3.15. ∎

Set

$$I:=E\Big[\sum_{n=1}^{N}\sum_{m=1}^{M}(df_{n,m})^2h_{n-1,m-1}^4\Big]$$

and

$$K:=E\Big[\sum_{n=1}^{N}\sum_{m=1}^{M}(d_1f_{n,m-1})^2(d_2f_{n-1,m})^2h_{n-1,m-1}^4\Big].$$

We intend to estimate these expressions by

$$C\Big[\lambda^2P(f^*>\lambda)+\int_{\{f^*\leq\lambda\}}f^{*2}\,dP\Big].$$

Lemma 3.17.

$$K\leq C\Big[\lambda^2I+\lambda^4P(f^*>\lambda)+\lambda^2\int_{\{f^*\leq\lambda\}}f^{*2}\,dP\Big].$$

Proof. Applying the third equation of Lemma 3.14 to the martingale f together with Lemma 3.12 we can conclude that

$$K\leq CE\Big[\sum_{n=1}^{N}\sum_{m=1}^{M}(df_{n,m}^4)h_{n-1,m-1}^4\Big]$$
$$+C\lambda E\Big[\sum_{n=1}^{N}\sum_{m=1}^{M}|df_{n,m}||d_1f_{n,m-1}||d_2f_{n-1,m}|h_{n-1,m-1}^4\Big]$$
$$+C\lambda^2E\Big[\sum_{n=1}^{N}\sum_{m=1}^{M}(df_{n,m})^2h_{n-1,m-1}^4\Big].$$

The third term is equal to $C\lambda^2I$. Based on Hölder's and Cauchy-Schwarz's inequalities the second term can be majorized by $C\lambda I^{1/2}K^{1/2}$. Now the first term is going to be estimated. Let

$$E\Big[\sum_{n=1}^{N}\sum_{m=1}^{M}(df_{n,m}^4)h_{n-1,m-1}^4\Big]=(\alpha)+(\beta)+(\beta')+(\gamma)$$

where

$$(\alpha) := E\Big[\sum_{n=1}^{N}\sum_{m=1}^{M}(df_{n,m}^4)(dh_{n,m}^4)\Big],$$

$$(\beta) := E\Big[\sum_{n=1}^{N}\sum_{m=1}^{M}(df_{n,m}^4)h_{n,m-1}^4\Big],$$

$$(\beta') := E\Big[\sum_{n=1}^{N}\sum_{m=1}^{M}(df_{n,m}^4)h_{n-1,m}^4\Big]$$

and

$$(\gamma) := -E\Big[\sum_{n=1}^{N}\sum_{m=1}^{M}(df_{n,m}^4)h_{n,m}^4\Big].$$

Majorization of (α): Using the algebraic identity

$$\begin{aligned} df_{n,m}^2 &= (f_{n,m} + f_{n,m-1} + f_{n-1,m} - f_{n-1,m-1})df_{n,m} \\ &\quad + 2(d_1 f_{n,m-1})(d_2 f_{n-1,m}) \end{aligned}$$

we can see that

$$\begin{aligned} df_{n,m}^4 &= (f_{n,m}^2 + f_{n,m-1}^2 + f_{n-1,m}^2 - f_{n-1,m-1}^2)df_{n,m}^2 \\ &\quad + 2(f_{n,m-1} + f_{n-1,m-1})(f_{n-1,m} + f_{n-1,m-1}) \\ &\qquad (d_1 f_{n,m-1})(d_2 f_{n-1,m}). \end{aligned} \tag{3.19}$$

By the regularity property of $(\mathcal{F}_n)$ we have

$$h_{n,m} \le R\max(h_{n,m-1}, h_{n-1,m}) \le R^2 h_{n-1,m-1}. \tag{3.20}$$

Applying (3.19) to f and to h and, moreover, (3.20) together with Lemma 3.12 we get that

$$\begin{aligned} &|df_{n,m}^4||dh_{n,m}^4| \\ &\quad \le C\big[\lambda^3|df_{n,m}| + \lambda^2|d_1 f_{n,m-1}||d_2 f_{n-1,m}|\big] \\ &\qquad \big[h_{n-1,m-1}^3|dh_{n,m}| + h_{n-1,m-1}^2|d_1 h_{n,m-1}||d_2 h_{n-1,m}|\big]. \end{aligned} \tag{3.21}$$

By Hölder's and Cauchy-Schwarz's inequalities and by Lemma 3.16 we have

$$\begin{aligned} &\lambda^3 E\Big[\sum_{n=1}^{N}\sum_{m=1}^{M}|df_{n,m}|h_{n-1,m-1}^3|dh_{n,m}|\Big] \\ &\qquad \le \lambda^3 E\Big[\sum_{n=1}^{N}\sum_{m=1}^{M}|df_{n,m}|h_{n-1,m-1}^2|dh_{n,m}|\Big] \\ &\qquad \le \lambda^3 I^{1/2}\Big(E\Big[\sum_{n=1}^{N}\sum_{m=1}^{M}(dh_{n,m})^2\Big]\Big)^{1/2} \\ &\qquad \le C\lambda^3 I^{1/2} P(f^* > \lambda)^{1/2}, \end{aligned} \tag{3.22}$$

$$(3.23)\qquad \lambda^3 E\Big[\sum_{n=1}^{N}\sum_{m=1}^{M} |df_{n,m}|h_{n-1,m-1}^2|d_1 h_{n,m-1}||d_2 h_{n-1,m}|\Big] \le C\lambda^3 I^{1/2} P(f^* > \lambda)^{1/2},$$

$$(3.24)\qquad \lambda^2 E\Big[\sum_{n=1}^{N}\sum_{m=1}^{M} |d_1 f_{n,m-1}||d_2 f_{n-1,m}|h_{n-1,m-1}^3|dh_{n,m}|\Big]$$
$$\le \lambda^2 E\Big[\sum_{n=1}^{N}\sum_{m=1}^{M} |d_1 f_{n,m-1}||d_2 f_{n-1,m}|h_{n-1,m-1}^2|dh_{n,m}|\Big]$$
$$\le C\lambda^2 K^{1/2} P(f^* > \lambda)^{1/2}$$

and

$$(3.25)\qquad \lambda^2 E\Big[\sum_{n=1}^{N}\sum_{m=1}^{M} |d_1 f_{n,m-1}||d_2 f_{n-1,m}|h_{n-1,m-1}^2|d_1 h_{n,m-1}||d_2 h_{n-1,m}|\Big]$$
$$\le C\lambda^2 K^{1/2} P(f^* > \lambda)^{1/2}.$$

Note that $0 \le h_{n-1,m-1} \le 1$. The inequality

$$(\alpha) \le C\big[\lambda^3 I^{1/2} P(f^* > \lambda)^{1/2} + \lambda^2 K^{1/2} P(f^* > \lambda)^{1/2}\big]$$

comes from (3.21)–(3.25). Using the well-known inequality

$$(3.26)\qquad \sqrt{xy} \le x + y \qquad (x, y \ge 0)$$

we have

$$\lambda^3 I^{1/2} P(f^* > \lambda)^{1/2} \le \lambda^2 I + \lambda^4 P(f^* > \lambda)$$

and so

$$(\alpha) \le C\big[\lambda^2 I + \lambda^4 P(f^* > \lambda) + \lambda^2 K^{1/2} P(f^* > \lambda)^{1/2}\big].$$

Majorization of (β) *(and* (β')*)*: From the inequality

$$df_{n,m}^4 = d_2 f_{n,m}^4 - d_2 f_{n-1,m}^4$$

it follows that

$$(df_{n,m}^4)h_{n,m-1}^4$$
$$= (d_2 f_{n,m}^4)h_{n,m-1}^4 - (d_2 f_{n-1,m}^4)h_{n-1,m-1}^4 - (d_2 f_{n-1,m}^4)(d_1 h_{n,m-1}^4).$$

Notice that

$$E_{n-1,m-1}\big[(d_2 f_{n-1,m}^4)(d_1 h_{n,m-1}^4)\big]$$
$$= E_{n-1,m-1}(d_2 f_{n-1,m}^4) E_{n-1,m-1}(d_1 h_{n,m-1}^4) \ge 0$$

holds because f^4 and h^4 are submartingales. Thus

$$(\beta) \leq E\Big[\sum_{n=1}^{N}\sum_{m=1}^{M}\Big((d_2 f_{n,m}^4)h_{n,m-1}^4 - (d_2 f_{n-1,m}^4)h_{n-1,m-1}^4\Big)\Big]$$
$$= E\Big[\sum_{m=1}^{M}(d_2 f_{N,m}^4)h_{N,m-1}^4\Big].$$

From the second equation of Lemma 3.14 we have

$$E_{N,m-1}(d_2 f_{N,m}^4) = E_{N,m-1}\Big[(d_2 f_{N,m})^2\Big(6 f_{N,m-1}^2 + (d_2 f_{N,m})^2 + 4 f_{N,m-1} d_2 f_{N,m}\Big)\Big].$$

Again by Lemma 3.12 we obtain

$$(\beta) \leq C\lambda^2 E\Big[\sum_{m=1}^{M}(f_{N,m} - f_{N,m-1})^2 h_{N,m-1}^4\Big]$$
$$\leq C\lambda^2 E\Big[\sum_{m=1}^{M}(f_{N,m} - f_{N,m-1})^2 h_{N,m-1}^2\Big]$$
$$= C\lambda^2 I'$$

where

$$I' := E\Big[\sum_{m=1}^{M}(f_{N,m} - f_{N,m-1})^2 h_{N,m-1}^2\Big]$$
$$= E\Big[\sum_{m=1}^{M}(f_{N,m}^2 - f_{N,m-1}^2)h_{N,m-1}^2\Big]$$
$$= E\Big[\sum_{m=1}^{M}(f_{N,m}^2 - f_{N,m-1}^2)h_{N,m}^2\Big]$$
$$- E\Big[\sum_{m=1}^{M}(f_{N,m}^2 - f_{N,m-1}^2)(h_{N,m}^2 - h_{N,m-1}^2)\Big].$$

The second term of I' can be majorized by

$$C\lambda E\Big[\sum_{m=1}^{M}|f_{N,m} - f_{N,m-1}||h_{N,m-1}||h_{N,m} - h_{N,m-1}|\Big]$$
$$\leq C\lambda I'^{1/2}\Big(E\Big[\sum_{m=1}^{M}|h_{N,m} - h_{N,m-1}|^2\Big]\Big)^{1/2}.$$

By Lemma 3.16 the second term of I' is less than or equal to

$$C\lambda I'^{1/2} P(f^* > \lambda)^{1/2}.$$

By Abel rearrangement we get for the first term of I' that

$$E\Big[\sum_{m=1}^{M}(f_{N,m}^2-f_{N,m-1}^2)h_{N,m}^2\Big]$$

$$= E[f_{N,M}^2 h_{N,M}^2] - E\Big[\sum_{m=1}^{M-1}(h_{N,m+1}^2 - h_{N,m}^2)f_{N,m}^2\Big]$$

$$\leq E[f_{N,M}^2 h_{N,M}^2]$$

because

$$E_{N,m}(h_{N,m+1}^2 - h_{N,m}^2) \geq 0.$$

For the last expectation one concludes from Lemma 3.11 that

$$\begin{aligned} E[f_{N,M}^2 h_{N,M}^2] &= E[f_{N,M}^2 \chi(G)] \\ &\leq E[f^{*2}\chi(F)] \\ &= \int_{\{f^*\leq\lambda\}} f^{*2}\, dP. \end{aligned}$$

Henceforth

$$I' \leq \int_{\{f^*\leq\lambda\}} f^{*2}\, dP + C\lambda I'^{1/2} P(f^* > \lambda)^{1/2}$$

and, by Lemma 3.15,

$$I' \leq C \int_{\{f^*\leq\lambda\}} f^{*2}\, dP + C\lambda^2 P(f^* > \lambda).$$

Consequently,

$$(\beta) \leq C\lambda^2 \int_{\{f^*\leq\lambda\}} f^{*2}\, dP + C\lambda^4 P(f^* > \lambda).$$

Of course, the same can be proved for (β'), too.

Majorization of (γ): We get by Abel transformation that

$$(\gamma) = (\gamma_1) + (\gamma_2) + (\gamma_2') + (\gamma_3)$$

where

$$(\gamma_1) := -E\Big[\sum_{n=1}^{N}\sum_{m=1}^{M} f_{n-1,m-1}^4 (dh_{n,m}^4)\Big],$$

$$(\gamma_2) := E\Big[\sum_{n=1}^{N-1}(h_{n+1,M}^4 - h_{n,M}^4)f_{n,M}^4\Big],$$

$$(\gamma_2') := E\Big[\sum_{m=1}^{M-1}(h_{N,m+1}^4 - h_{N,m}^4)f_{N,m}^4\Big]$$

and

$$(\gamma_3) := -E[f_{N,M}^4 h_{N,M}^4].$$

Since $(\gamma_3) \le 0$,

$$(\gamma) \le |(\gamma_1)| + |(\gamma_2)| + |(\gamma_2')|.$$

By Lemma 3.14

$$|(\gamma_1)| \le CE\Big[\sum_{n=1}^{N}\sum_{m=1}^{M} f_{n-1,m-1}^4 \Big[(d_1 h_{n,m-1})^2 (d_2 h_{n-1,m})^2 + |dh_{n,m}||d_1 h_{n,m-1}||d_2 h_{n-1,m}| + |dh_{n,m}|^2\Big]\Big].$$

If $h_{n-1,m-1} > 0$ then $|f_{n-1,m-1}| \le \lambda$ (see Lemma 3.12). However, if $h_{n-1,m-1} = 0$ then $h_{n,m-1} = 0$, $h_{n-1,m} = 0$ and $h_{n,m} = 0$ because h is positive. In this case the sum of the right hand side is 0. Hence

$$\begin{aligned}|(\gamma_1)| &\le C\lambda^4 E\Big[\sum_{n=1}^{N}\sum_{m=1}^{M} (d_1 h_{n,m-1})^2 (d_2 h_{n-1,m})^2\Big] \\ &+ C\lambda^4 E\Big[\sum_{n=1}^{N}\sum_{m=1}^{M} |dh_{n,m}||d_1 h_{n,m-1}||d_2 h_{n-1,m}|\Big] \\ &+ C\lambda^4 E\Big[\sum_{n=1}^{N}\sum_{m=1}^{M} (dh_{n,m})^2\Big].\end{aligned}$$

Applying Hölder's and Cauchy-Schwarz's inequalities together with Lemma 3.16 we obtain

$$|(\gamma_1)| \le C\lambda^4 P(f^* > \lambda).$$

Again, by Lemma 3.14, 3.12 and 3.16

$$|(\gamma_2)| \le C\lambda^4 E\Big[\sum_{n=1}^{N-1} (h_{n+1,M} - h_{n,M})^2\Big] \le C\lambda^4 P(f^* > \lambda)$$

is obtained and the same holds for $|(\gamma_2')|$, too. Therefore

$$(\gamma) \le C\lambda^4 P(f^* > \lambda).$$

Summarizing the previous estimates we obtain that

$$K \le C\Big[\lambda^2 I + \lambda I^{1/2} K^{1/2} + \lambda^2 P(f^* > \lambda)^{1/2} K^{1/2} + \lambda^4 P(f^* > \lambda) + \lambda^2 \int_{\{f^* \le \lambda\}} f^{*\,2}\, dP\Big].$$

Finally, Lemma 3.15 completes the lemma. ■

Lemma 3.18.

$$I \le C\Big[\lambda^2 P(f^* > \lambda) + \int_{\{f^* \le \lambda\}} f^{*\,2}\, dP\Big].$$

Proof. From (3.17) it is easy to see that

$$I = E\Big[\sum_{n=1}^{N}\sum_{m=1}^{M}(df_{n,m}^2)h_{n-1,m-1}^4\Big].$$

This expression can be majorized like the term

$$E\Big[\sum_{n=1}^{N}\sum_{m=1}^{M}(df_{n,m}^4)h_{n-1,m-1}^4\Big]$$

in the estimation of K. If, analogously,

$$I = (\tilde{\alpha}) + (\tilde{\beta}) + (\tilde{\beta}') + (\tilde{\gamma})$$

then, in this case,

$$(\tilde{\alpha}) \le C\big[\lambda I^{1/2}P(f^* > \lambda)^{1/2} + K^{1/2}P(f^* > \lambda)^{1/2}\big],$$

$$(\tilde{\beta}), (\tilde{\beta}') \le C\Big[\int_{\{f^*\le\lambda\}} f^{*2}\,dP + \lambda^2 P(f^* > \lambda)\Big]$$

and

$$(\tilde{\gamma}) \le C\lambda^2 P(f^* > \lambda).$$

Using Lemma 3.17 and the equation (3.26) we get that

$$\begin{aligned} K^{1/2}P(f^* > \lambda)^{1/2} &\le C\Big[\lambda I^{1/2}P(f^* > \lambda)^{1/2} + \lambda^2 P(f^* > \lambda) \\ &\quad + (\lambda^2 P(f^* > \lambda))^{1/2}\Big(\int_{\{f^*\le\lambda\}} f^{*2}\,dP\Big)^{1/2}\Big] \\ &\le C\Big[\lambda I^{1/2}P(f^* > \lambda)^{1/2} + \lambda^2 P(f^* > \lambda) \\ &\quad + \lambda^2 P(f^* > \lambda) + \int_{\{f^*\le\lambda\}} f^{*2}\,dP\Big]. \end{aligned}$$

Consequently,

$$I \le C\Big[\lambda I^{1/2}P(f^* > \lambda)^{1/2} + \lambda^2 P(f^* > \lambda) + \int_{\{f^*\le\lambda\}} f^{*2}\,dP\Big].$$

Lemma 3.18 follows from Lemma 3.15. ∎

Proof of Theorem 3.10. It can easily be seen that

$$\begin{aligned} P(S(f) > \lambda) &= P(S(f) > \lambda, H) + P(S(f) > \lambda, H^c) \\ &\le \frac{1}{\lambda^2}\int_H S^2(f)\,dP + P(H^c). \end{aligned}$$

By Lemma 3.13,

$$P(S(f) > \lambda) \le \frac{1}{\lambda^2}\int_H S^2(f)\,dP + CP(f^* > \lambda).$$

On the other hand, using Lemma 3.11 we can conclude that

$$\begin{aligned}\int_H S^2(f)\,dP &= E\Big[\sum_{n=1}^N\sum_{m=1}^M (df_{n,m})^2\chi(H)\Big]\\ &= E\Big[\sum_{n=1}^N\sum_{m=1}^M (df_{n,m})^2 E_{n,m}\chi(H)\Big]\\ &\le (2R^2)^4 E\Big[\sum_{n=1}^N\sum_{m=1}^M (df_{n,m})^2 h_{n-1,m-1}^4\Big]\\ &= CI.\end{aligned}$$

Based on Lemma 3.18 the proof of Theorem 3.10 is complete. ∎

We have reached the point of being able to prove Burkholder-Gundy's inequality for $0 < p \le 1$.

Theorem 3.19. *If $\mathcal{F}$ is regular then for all $0 < p < \infty$*

$$c_p\|f\|_{H_p^S} \le \|f\|_{H_p^s} \le C_p\|f\|_{H_p^S}. \tag{3.27}$$

Proof. By Theorem 3.6 we only need to show this theorem for $0 < p \le 1$. The right hand side of the inequality follows from Theorem 3.5 (i) and from Theorem 3.9.

The other side follows from Theorem 3.10. Multiplying the inequality in Theorem 3.10 by $p\lambda^{p-1}$ and integrating it in λ we get

$$\begin{aligned}&E(S^p(f))\\ &= \int_0^\infty p\lambda^{p-1}P(S(f)>\lambda)\,d\lambda\\ &\le C\int_0^\infty p\lambda^{p-1}P(f^*>\lambda)\,d\lambda + C\int_0^\infty p\lambda^{p-3}\int_\Omega f^{*2}\chi(f^*\le\lambda)\,dP\,d\lambda\\ &= CE(f^{*p}) + C\int_\Omega f^{*2}\int_{f^*}^\infty p\lambda^{p-3}\,d\lambda\,dP.\end{aligned}$$

Since $0 < p \le 1$, the inequality (3.27) follows from the previous inequality. ∎

3.3. DUALITY THEOREMS

In this section some two-parameter duality results are demonstrated. It is proved by Bernard [9] in the two-parameter dyadic case that the dual of H_1^S is BMO_2. This result will be generalized in Theorem 3.20 in which the duality between H_p^s and $\Lambda_2(\alpha)$ is shown ($0 < p \le 1, \alpha = 1/p - 1$) (see Weisz [199]). A convexity inequality due to Stein [178] is generalized for a countable index set with the help of which the duality between H_p^s and H_q^s ($1 < p < \infty, 1/p + 1/q = 1$) is verified in case the stochastic basis is generated by a Vilenkin system. Using $(1,q)$ atoms we introduce the atomic Hardy spaces H_q^{at} and show that the dual of H_q^{at} is $BMO_{q'}$

$(1 < q < \infty, 1/q + 1/q' = 1)$. Note that, by Theorem 3.2, the equivalence $H_1^s \sim H_2^{at}$ holds. Furthermore, in the regular case H_q^{at} is equivalent to H_2^{at} $(1 < q \leq 2)$, hence their duals are equivalent as well: $BMO_{q'} \sim BMO_2$ $(2 \leq q' < \infty)$ (see Weisz [199]). The space VMO_p is defined again as the closure of the step functions in the BMO_p norm and a sufficient and necessary condition for a function to be in VMO_p is given. Like in the one-parameter case, if every σ-algebra is generated by finitely many atoms then the dual of VMO_2 is H_1^s (Weisz [199]). Using simple stopping times, only, we define the so-called simple atomic Hardy spaces H_q^{sat} and the spaces $BMO_p^{\square}$. Duality results between H_p^{sat}, $BMO_q^{\square}$ and $VMO_q^{\square}$ are formulated $(1 < p < \infty, 1/p + 1/q = 1)$. Using an idea due to Carleson [32] and Decamp [56] we give a counterexample which shows that either in the dyadic case BMO_2 is not equivalent to $BMO_2^{\square}$. Hence, opposed to the one-parameter case, H_1^s is not equivalent to H_2^{sat}, in other words, martingales from H_1^s can not be decomposed into simple atoms. This is also a fundamental difference of the one- and the two-parameter cases, since while the support of an atom can be an arbitrary $\mathcal{A}$ measurable set the one of a simple atom is an $\mathcal{F}_n$ measurable set for any n. This is another reason for that the two-parameter case is much more complicated than the one-parameter case.

First the dual of H_p^s is given.

Theorem 3.20. *The dual space of H_p^s is $\Lambda_2(\alpha)$ $(0 < p \leq 1, \alpha = 1/p - 1)$.*

Proof. Taking the same stopping times ν_k, atoms a^k and real numbers μ_k $(k \in \mathbf{Z})$ as in Theorem 3.2 we get that

$$\sum_{k\in\mathbf{Z}} \mu_k a^k = \sum_{k\in\mathbf{Z}} (f^{\nu_{k+1}} - f^{\nu_k}) = f$$

a.e. and also in L_2 norm if $f \in L_2$. By Theorem 3.2 one can establish that L_2 is dense in H_p^s. If

$$l_\phi(f) := E(f\phi) \qquad (f \in L_2),$$

where $\phi \in \Lambda_2(\alpha)$ is arbitrary, then

$$l_\phi(f) = \sum_{k\in\mathbf{Z}} \mu_k E(a^k\phi).$$

By the definition of the atom a^k we have that

$$E(a^k\phi) = E(a^k(\phi - \phi^{\nu_k})).$$

So the inequality

$$|l_\phi(f)| \leq C_p \|f\|_{H_p^s} \|\phi\|_{\Lambda_2(\alpha)} \qquad (0 < p \leq 1, f \in L_2)$$

can be verified like in Theorem 2.24.

The proof of the converse is also similar to the one in Theorem 2.24 with the only difference that the function g given in (2.36) is here a $(p, 2)$ atom. The proof of the theorem is complete. ∎

To characterize the dual of H_p^s $(1 < p < \infty)$ we need the following generalization of an inequality that is analogous to the convexity theorem (Theorem 2.10) and due to Stein ([178], p. 103) in the one-parameter case.

Theorem 3.21. *Let $\mathbf{T}$ be a countable index set and $(\mathcal{A}_t, t \in \mathbf{T})$ be an arbitrary (not necessarily monotone) sequence of σ-algebras with the assumption $\sigma(\cup_{t\in\mathbf{T}}\mathcal{A}_t) = \mathcal{A}$. Suppose that for all $f \in L_p$ Doob's inequality*

$$\|\sup_{t\in\mathbf{T}} |E_t f|\|_p \le C_p \|f\|_p \qquad (p > 1) \tag{3.28}$$

holds. If $(X_t, t \in \mathbf{T})$ is a sequence of measurable functions then for $1 < p < \infty$ we have

$$\|(\sum_{t\in\mathbf{T}} |E_t X_t|^2)^{1/2}\|_p \le c_p \|(\sum_{t\in\mathbf{T}} |X_t|^2)^{1/2}\|_p \qquad (1 < p < \infty) \tag{3.29}$$

where c_p depends only on p.

Proof. The proof follows the original one due to Stein. We shall use the following generalization of Riesz's convexity theorem.

Lemma 3.22. *Let T be a linear operator which maps sequences of functions to sequences of functions. Suppose that, for arbitrary $1 \le p_i, q_i, r_i, s_i \le \infty$ $(i = 0, 1)$, T is a bounded operator from $L_{p_0}(l_{q_0})$ to $L_{r_0}(l_{s_0})$ and from $L_{p_1}(l_{q_1})$ to $L_{r_1}(l_{s_1})$. Then T is also bounded from $L_p(l_q)$ to $L_r(l_s)$ if $p, q < \infty$, $0 < t < 1$,*

$$\frac{1}{p} = \frac{1-t}{p_0} + \frac{t}{p_1}, \qquad \frac{1}{q} = \frac{1-t}{q_0} + \frac{t}{q_1}$$

and

$$\frac{1}{r} = \frac{1-t}{r_0} + \frac{t}{r_1}, \qquad \frac{1}{s} = \frac{1-t}{s_0} + \frac{t}{s_1}.$$

The proof of this lemma is very similar to that of the original Riesz's convexity theorem (see e.g. Benedeck, Panzone [5]).

To prove Theorem 3.21 we consider the operator T defined by

$$T : (X_t, t \in \mathbf{T}) \longrightarrow (E_t X_t, t \in \mathbf{T}).$$

T is linear and bounded on $L_p(l_p)$ $(1 < p < \infty)$ since

$$E(\sum_{t\in\mathbf{T}} |E_t X_t|^p) = \sum_{t\in\mathbf{T}} E|E_t X_t|^p \le \sum_{t\in\mathbf{T}} E|X_t|^p.$$

On the other hand, by the maximal inequality (3.28) we obtain

$$\begin{aligned} E(\sup_{t\in\mathbf{T}} |E_t X_t|)^p &\le E(\sup_{t\in\mathbf{T}} \sup_{u\in\mathbf{T}} E_u |X_t|)^p \\ &\le E(Z^{*p}) \le C_p E(\sup_{t\in\mathbf{T}} |X_t|)^p \end{aligned}$$

where $Z := \sup_{t\in\mathbf{T}} |X_t|$ which shows the boundedness of T on $L_p(l_\infty)$ $(1 < p \le \infty)$. Applying now the generalized Riesz's convexity theorem we can conclude that T is bounded on $L_p(l_q)$ if $1 < p \le q \le \infty$. In particular, if $1 < p \le 2$ then T is bounded on $L_p(l_2)$.

For $2 < p < \infty$ the theorem is verified with a duality argument. By Lemma 2.9,

$$\|(\sum_{t\in\mathbf{T}} |E_t X_t|^2)^{1/2}\|_p = \sup_{\|Y\|_{L_q(l_2)}\leq 1} E[\sum_{t\in\mathbf{T}} (E_t X_t) Y_t]$$

where $1/p + 1/q = 1$ and $Y = (Y_t, t \in \mathbf{T})$. Moreover, by Hölder's inequality and by the inequality (3.29) for $1 < q \leq 2$,

$$\begin{aligned} E[\sum_{t\in\mathbf{T}} (E_t X_t) Y_t] &= E[\sum_{t\in\mathbf{T}} X_t (E_t Y_t)] \\ &\leq \|(X_t)\|_{L_p(l_2)} \|(E_t Y_t)\|_{L_q(l_2)} \\ &\leq C_q \|(X_t)\|_{L_p(l_2)} \end{aligned}$$

which completes the proof of the theorem. ■

Note that Theorem 3.21 does not hold for $p = 1$ even in the one-parameter case (see Lepingle [118]).

If the stochastic basis $\mathcal{F}$ is generated by an unbounded Vilenkin system then H_p^s is not equivalent to H_p^S. So the following theorem includes new result.

Theorem 3.23. *If the σ-algebra sequence $(\mathcal{F}_{n,m})$ is generated by a Vilenkin system then the dual of H_p^s is H_q^s with $1 < p < \infty$ and $1/p + 1/q = 1$.*

Proof. Set $\mathcal{F}_{n,m}^{(i,j)} := \mathcal{F}_{n,m}$ $(i = 1, \ldots, p_n - 1; j = 1, \ldots, q_m - 1)$ where (p_n) and (q_m) are the sequences in the definition of the Vilenkin system. Consider the function sequences of the form

$$X = (X_{n,m}^{(i,j)}; i = 1, \ldots, p_n - 1; j = 1, \ldots, q_m - 1; n, m \in \mathbf{N}).$$

The inequality in Theorem 3.21 can be written as

$$\|(\sum_{n=0}^{\infty} \sum_{m=0}^{\infty} \sum_{i=1}^{p_n-1} \sum_{j=1}^{q_m-1} |E_{n,m} X_{n,m}^{(i,j)}|^2)^{1/2}\|_p \leq C_p \|(\sum_{n=0}^{\infty} \sum_{m=0}^{\infty} \sum_{i=1}^{p_n-1} \sum_{j=1}^{q_m-1} |X_{n,m}^{(i,j)}|^2)^{1/2}\|_p \tag{3.30}$$

if $1 < p < \infty$. Denote by ${}^{\mathrm{a}}L_p(l_2)$ the subspace of $L_p(l_2)$ consisting of the adapted sequences relative to $\mathcal{F}_{n,m}^{(i,j)}$.

We show that the dual of ${}^{\mathrm{a}}L_p(l_2)$ is ${}^{\mathrm{a}}L_q(l_2)$ whenever $1 < p < \infty$ and $1/p + 1/q = 1$. (This result does not hold for $p = 1$ even in the one-parameter case.) If for an arbitrary $Y \in {}^{\mathrm{a}}L_q(l_2)$ one has

$$l_Y(X) := E(\sum_{n=0}^{\infty} \sum_{m=0}^{\infty} \sum_{i=1}^{p_n-1} \sum_{j=1}^{q_m-1} X_{n,m}^{(i,j)} Y_{n,m}^{(i,j)}) \qquad (X \in {}^{\mathrm{a}}L_p(l_2))$$

then l_Y is in the dual of ${}^{\mathrm{a}}L_p(l_2)$ and

$$\|l_Y\| \leq \|Y\|_{L_q(l_2)}.$$

Conversely, if l is in the dual of ${}^{a}L_p(l_2)$ then, by Banach-Hahn's theorem, preserving its norm, it can be extended to the whole $L_p(l_2)$ space. Based on Lemma 2.9 there exists $Y \in L_q(l_2)$ such that $l = l_Y$ and

$$\|Y\|_{L_q(l_2)} \leq \|l\|.$$

As

$$l(X) = E(\sum_{n=0}^{\infty} \sum_{m=0}^{\infty} \sum_{i=1}^{p_n-1} \sum_{j=1}^{q_m-1} X_{n,m}^{(i,j)} E_{n,m} Y_{n,m}^{(i,j)}),$$

we get by (3.30) that

$$\|(E_{n,m} Y_{n,m}^{(i,j)})\|_{L_q(l_2)} \leq \|Y\|_{L_q(l_2)} \leq \|l\|.$$

As a consequence of (1.11) the theorem follows immediately from the isometry between H_p^s and ${}^{a}L_p(l_2)$. ∎

The dual of H_∞^s is considered in Weisz [192] for dyadic martingales. The dual of H_p^* and H_p^S is, of course, L_q whenever $1 < p < \infty$ and $1/p + 1/q = 1$. The duals of the other martingale Hardy spaces are unknown but in the regular case.

Theorem 3.20 will be generalized. For this a so-called *atomic Hardy space* is going to be introduced.

Definition 3.24. *Let us denote by H_q^{at} $(1 < q \leq \infty)$ the space of those martingales $(f_n, n \in \mathbb{N}^2)$ for which there exist a sequence $(a^k, k \in \mathbb{Z})$ of $(1,q)$ atoms and a sequence $\mu = (\mu_k, k \in \mathbb{Z})$ of real numbers such that for all $n \in \mathbb{N}^2$*

$$\sum_{k=-\infty}^{\infty} \mu_k E_n a^k = f_n \tag{3.31}$$

and

$$\sum_{k=-\infty}^{\infty} |\mu_k| < \infty.$$

Endow this space with the norm

$$\|f\|_{H_q^{at}} := \inf \sum_{k=-\infty}^{\infty} |\mu_k|$$

where the infimum is taken over all decompositions of the form (3.31).

A possible definition of the spaces $H_{p,q}^{at}$ with (p,q) atoms $(0 < p < 1)$ can be found in Coifman, Weiss [52] for the classical case. However, we do not deal with these spaces. In the one-parameter case some spaces that are similar to the spaces H_q^{at} were investigated by Herz in [94].

It follows from Hölder's inequality that, if a is a $(1,q)$ atom, then

$$\|a\|_1 \leq 1.$$

Consequently, the series (3.31) converges in L_1 norm, too. Moreover, the series

$$\sum_{k=-\infty}^{\infty} \mu_k a^k$$

is also convergent in L_1 norm and a.e. If we denote the sum of this series by f then $f \in L_1$ and $E_n f = f_n$ for each $n \in \mathbf{N}^2$. Henceforth $H_q^{\mathrm{at}} \subset L_1$ and

$$\|f\|_1 \le \|f\|_{H_q^{\mathrm{at}}} \qquad (1 < q \le \infty). \tag{3.32}$$

The inequalities

$$\|f\|_{H_1^\bullet} \le \|f\|_{H_q^{\mathrm{at}}}, \qquad \|f\|_{H_1^S} \le C_q \|f\|_{H_q^{\mathrm{at}}} \qquad (1 < q \le \infty)$$

can similarly be shown. By $H_2^{\mathrm{at}} \sim H_1^s$, the following theorem is really an extension of Theorem 3.20.

Theorem 3.25. *The dual of H_q^{at} is $BMO_{q'}$ $(1 < q < \infty, 1/q + 1/q' = 1)$.*

Proof. By the definition the linear envelop of the $(1,q)$ atoms is dense in H_q^{at}, therefore L_q is also dense in H_q^{at}. Again, let

$$l_\phi(f) := E(f\phi) \qquad (f \in L_q)$$

where $\phi \in BMO_{q'}$ is arbitrary. First we have to bring the expectation $E(f\phi)$ in the form $\sum_{k\in\mathbf{Z}} \mu_k E(a^k\phi)$ where $\sum_{k\in\mathbf{Z}} \mu_k a^k$ is an atomic decomposition of f. This is not simple so we are going to go in details. Since $\phi_n \to \phi$ in $L_{q'}$ norm as $n \to \infty$, by Hölder's inequality we get that

$$l_\phi(f) = \lim_{n\to\infty} E(f\phi_n) \qquad (f \in L_q). \tag{3.33}$$

Let $n \in \mathbf{N}^2$ be fixed and consider the expectation $E(f\phi_n)$. Create a sequence $(\phi_n^k)_{k\in\mathbf{N}}$ of bounded and $\mathcal{F}_n$ measurable functions for which

$$\|\phi_n^k - \phi_n\|_{q'} \to 0 \quad \text{as} \quad k \to \infty \tag{3.34}$$

and

$$\|\phi_n^k\|_{BMO_{q'}} \le C_q \|\phi\|_{BMO_{q'}}. \tag{3.35}$$

For this choose the $\mathcal{F}_m$ measurable functions $\psi_m^k \in L_\infty$ $(k \in \mathbf{N}, m \le n)$ such that

$$|\psi_m^k| \le |d_m\phi| \qquad (k \in \mathbf{N}, m \le n) \tag{3.36}$$

and

$$\|d_m\phi - \psi_m^k\|_{q'} \le \frac{1}{2^k 2^{m_1} 2^{m_2}}. \tag{3.37}$$

The functions $(\psi_m^k)_{k\in\mathbf{N}}$ with these properties do exist because we have for the functions

$$g^N := (d_m\phi \wedge N) \vee (-N) \qquad (N \in \mathbf{N})$$

that they are $\mathcal{F}_m$ measurable, $g^N \in L_\infty$, $|g^N| \le |d_m\phi|$ and $g^N \to d_m\phi$ in $L_{q'}$ norm as $N \to \infty$. Let

$$d_m\phi^k := \psi_m^k - E_{m_1-1,m_2}\psi_m^k - E_{m_1,m_2-1}\psi_m^k + E_{m-1}\psi_m^k$$

and

$$\phi_n^k := \sum_{m\le n} d_m\phi^k.$$

The functions ϕ_n^k are clearly bounded and $\mathcal{F}_n$ measurable. To prove (3.34) we establish by Burkholder-Gundy's inequality and by the triangle inequality in l_2 that

$$\begin{aligned}
\|\phi_n - \phi_n^k\|_{q'} &\le C_q\|\phi_n - \phi_n^k\|_{H^S_{q'}} \\
&= C_q\|(\sum_{m\le n} |(d_m\phi - E_{m_1-1,m_2}d_m\phi - E_{m_1,m_2-1}d_m\phi + E_{m-1}d_m\phi) \\
&\qquad - (\psi_m^k - E_{m_1-1,m_2}\psi_m^k - E_{m_1,m_2-1}\psi_m^k + E_{m-1}\psi_m^k)|^2)^{1/2}\|_{q'} \\
&\le C_q\|(\sum_{m\le n} |d_m\phi - \psi_m^k|^2)^{1/2}\|_{q'} \\
&\qquad + C_q\|(\sum_{m\le n} |E_{m_1-1,m_2}(d_m\phi - \psi_m^k)|^2)^{1/2}\|_{q'} \\
&\qquad + C_q\|(\sum_{m\le n} |E_{m_1,m_2-1}(d_m\phi - \psi_m^k)|^2)^{1/2}\|_{q'} \\
&\qquad + C_q\|(\sum_{m\le n} |E_{m-1}(d_m\phi - \psi_m^k)|^2)^{1/2}\|_{q'}.
\end{aligned}$$

Based on concavity and Hölder's inequality, each term of the right hand side can be majorized by

$$\begin{aligned}
C_q\|\sum_{m\le n} |E_{\tilde m}(d_m\phi - \psi_m^k)|\|_{q'} &\le C_q \sum_{m\le n} \|E_{\tilde m}(d_m\phi - \psi_m^k)\|_{q'} \\
&\le C_q \sum_{m\le n} \|d_m\phi - \psi_m^k\|_{q'}
\end{aligned}$$

where $\tilde m \in \{m, (m_1-1, m_2), (m_1, m_2-1), m-1\}$. By (3.37),

$$\begin{aligned}
\|\phi_n - \phi_n^k\|_{q'} &\le C_q \sum_{m\le n} \|d_m\phi - \psi_m^k\|_{q'} \\
&\le C_q \sum_{m\le n} \frac{1}{2^k 2^{m_1} 2^{m_2}} \le C_q \frac{1}{2^k}
\end{aligned}$$

which shows (3.34).

For (3.35) we only have to show that

$$\|\phi_n^k - (\phi_n^k)^\nu\|_{q'} \le C_q\|\phi - \phi^\nu\|_{q'}$$

for all stopping times ν. By Burkholder-Gundy's inequality,

$$\begin{aligned}\|\phi_n^k - (\phi_n^k)^\nu\|_{q'} &\leq C_q\|\phi_n^k - (\phi_n^k)^\nu\|_{H^S_{q'}} \\ &= C_q\|(\sum_{m\leq n}\chi(\nu \ll m)|\psi_m^k - E_{m_1-1,m_2}\psi_m^k \\ &\qquad - E_{m_1,m_2-1}\psi_m^k + E_{m-1}\psi_m^k|^2)^{1/2}\|_{q'}.\end{aligned}$$

We remark that $\{\nu \ll m\} \in \mathcal{F}_{m-1}$. So we get by (3.36) that

$$\begin{aligned}\|\phi_n^k - (\phi_n^k)^\nu\|_{q'} &\leq C_q\|(\sum_{m\leq n}\chi(\nu \ll m)|\psi_m^k|^2)^{1/2}\|_{q'} \\ &\quad + C_q\|(\sum_{m\leq n}\chi(\nu \ll m)[E_{m_1-1,m_2}|\psi_m^k|]^2)^{1/2}\|_{q'} \\ &\quad + C_q\|(\sum_{m\leq n}\chi(\nu \ll m)[E_{m_1,m_2-1}|\psi_m^k|]^2)^{1/2}\|_{q'} \\ &\quad + C_q\|(\sum_{m\leq n}\chi(\nu \ll m)[E_{m-1}|\psi_m^k|]^2)^{1/2}\|_{q'} \\ &\leq C_q\|(\sum_{m\leq n}\chi(\nu \ll m)|d_m\phi|^2)^{1/2}\|_{q'} \\ &\quad + C_q\|(\sum_{m\leq n}[E_{m_1-1,m_2}|\chi(\nu \ll m)d_m\phi|]^2)^{1/2}\|_{q'} \\ &\quad + C_q\|(\sum_{m\leq n}[E_{m_1,m_2-1}|\chi(\nu \ll m)d_m\phi|]^2)^{1/2}\|_{q'} \\ &\quad + C_q\|(\sum_{m\leq n}[E_{m-1}|\chi(\nu \ll m)d_m\phi|]^2)^{1/2}\|_{q'}.\end{aligned}$$

Considering Theorem 3.21 we can conclude that

$$\begin{aligned}\|\phi_n^k - (\phi_n^k)^\nu\|_{q'} &\leq C_q\|(\sum_{m\in\mathbf{N}^2}\chi(\nu \ll m)|d_m\phi|^2)^{1/2}\|_{q'} \\ &= C_q\|\phi - \phi^\nu\|_{H^S_{q'}} \\ &\leq C_q\|\phi - \phi^\nu\|_{q'}.\end{aligned}$$

Therefore

$$E(f\phi_n) = \lim_{k\to\infty} E(f\phi_n^k).$$

Since $\phi_n^k \in L_\infty$ and the series (3.31) is convergent in L_1 norm, one has

$$E(f\phi_n^k) = \sum_{l\in\mathbf{Z}}\mu_l E(a^l\phi_n^k).$$

Henceforth

$$|E(f\phi_n^k)| \leq \sum_{l\in\mathbf{Z}}|\mu_l|\|\phi_n^k\|_{BMO_{q'}}.$$

This implies that

$$|E(f\phi_n^k)| \le C_q \|f\|_{H_q^{at}} \|\phi\|_{BMO_{q'}}$$

hence, by (3.33),

$$|l_\phi(f)| \le C_q \|f\|_{H_q^{at}} \|\phi\|_{BMO_{q'}} \qquad (f \in L_q).$$

Namely, it is verified that l_ϕ is a bounded linear functional on H_q^{at}. Since the series (3.31) is also convergent in H_q^{at} norm, the functional l_ϕ can also be written in the following form:

$$l_\phi(f) = \lim_{n\to\infty} E[(\sum_{k=0}^{n} \mu_k a^k)\phi] = \sum_{k=0}^{\infty} \mu_k E(a^k\phi).$$

We prove that L_q can continuously be embedded in H_q^{at}. Indeed, if $f \in L_q$ then $f/\|f^*\|_q$ is surely a $(1,q)$ atom (with the stopping time $\nu(\omega) = (0,0)$ $(\forall\omega \in \Omega)$). Thus

$$\|\frac{f}{\|f^*\|_q}\|_{H_q^{at}} \le 1.$$

Consequently, there exists $\phi \in L_{q'}$ such that

$$l(f) = E(f\phi) \qquad (f \in L_q).$$

The rest of the proof can be worked out as in Theorem 2.24 with the difference that, instead of the test function g defined in (2.36), we take the $(1,q)$ atoms

$$a := C_q^{-1} P(\nu \ne \infty)^{1/q-1}(f - f^\nu)$$

where $f \in L_q$ and $\|f\|_q \le 1$. Since $\|l\| \ge |l(a)| = |E(a\phi)|$, one has

$$P(\nu \ne \infty)^{1/q-1} |E[f(\phi - \phi^\nu)]| \le C_q \|l\|.$$

Let us take the supremum over all functions $f \in L_q$ with $\|f\|_q \le 1$; by Riesz's representation theorem

$$P(\nu \ne \infty)^{-1/q'} \|\phi - \phi^\nu\|_{q'} \le C_q \|l\|$$

which proves Theorem 3.25. ∎

The relations

$$L_\infty \subset BMO_q \subset L_q \qquad (1 \le q < \infty)$$

follow from this and from (3.32).

As the next step it is going to be verified that some of the spaces H_q^{at} are equivalent when $\mathcal{F}$ is regular.

Proposition 3.26. *For a regular sequence of σ-algebras $H_q^{at} \sim H_2^{at}$ $(1 < q \le 2)$.*

Proof. If a is a $(1,q)$ atom then, by Hölder's inequality, it is a $(1,r)$ atom $(1 < r < q)$ as well, thus

$$\|f\|_{H_r^{at}} \le \|f\|_{H_q^{at}} \qquad (1 < r < q).$$

Reversely, suppose that a is a $(1,q)$ atom $(1 < q < 2)$. We have already used several times that $s(a) = s(a)\chi(\nu \neq \infty)$ where ν is the stopping time corresponding to a. Hence

$$\|a\|_{H_2^{at}} \leq C\|s(a)\chi(\nu \neq \infty)\|_1 \leq C\|s(a)\|_q P(\nu \neq \infty)^{1-1/q}. \tag{3.38}$$

Since for a regular sequence of σ-algebras $H_q^s \sim H_q^S \sim L_q$ $(1 < q < \infty)$, the inequality $\|a\|_{H_2^{at}} \leq C_q$ becomes clear from (3.38). Consequently,

$$\|f\|_{H_2^{at}} \leq C_q \|f\|_{H_q^{at}} \qquad (1 < q < 2)$$

which completes the proof. ∎

Keeping the above regularity property, as a simple consequence we obtain that the dual spaces are also equivalent:

Corollary 3.27. *If $\mathcal{F}$ is regular then $BMO_{q'} \sim BMO_2$ $(2 \leq q' < \infty)$.*

Note that, in the one-parameter case, this corollary can be proved with the same method even for $1 \leq q' < \infty$ (see also Corollary 2.51). The reason of the fact that one obtaines more in the one-parameter case is that, opposed to the two-parameter case, a martingale f from H_1^s can be decomposed into $(1,\infty)$ atoms.

Let us consider the dual of VMO for two parameters, too. Of course, the dual of BMO_2 is not H_1^s even in the two-parameter case, however, H_1^s is again equivalent to a subspace of the dual of BMO_2. If $l_f(\phi) := l_\phi(f)$ $(\phi \in BMO_2$ and $f \in H_1^s)$ then l_f is a bounded linear functional on BMO_2, moreover, the inequality (2.49) holds also.

Let the definition of L' be the same as in (2.51) (of course, we write now $\mathbf{N}^2$ instead of $\mathbf{N}$) and denote by L the vectorspace

$$\{\phi \in L : E_{0,k}\phi = E_{k,0}\phi = 0 \quad (k \in \mathbf{N})\}.$$

Again, let VMO_p be the closure of L in BMO_p norm. In order to give a characterization of VMO_p by limit, let us introduce the following concept of the limit of the number sequences $(\alpha_\nu, \nu \in \mathcal{T})$ indexed by stopping times. For $N \in \mathbf{N}$ let $\mathcal{T}_N'$ denote the set of the stopping times $\nu \in \mathcal{T}$ such that $\nu(\omega) \not\leq (N,N)$ for all $\omega \in \Omega$. We say that the limit of the sequence $(\alpha_\nu, \nu \in \mathcal{T})$ is 0 as $\nu \to \infty$ if for every $\epsilon > 0$ there exists $N \in \mathbf{N}$ such that $|\alpha_\nu| < \epsilon$ for all $\nu \in \mathcal{T}_N'$.

The next result corresponds to Proposition 2.38.

Proposition 3.28. *If every σ-algebra $\mathcal{F}_n$ is generated by finitely many atoms then f is in VMO_p if and only if $f \in BMO_p$ and*

$$\lim_{\nu \to \infty} |\nu \neq \infty|^{-1/p} \|f - f^\nu\|_p = 0.$$

Proof. Let $f \in VMO_p$. Then there exists a step function $g \in L$ for all $\epsilon > 0$ such that

$$\|f - g\|_{BMO_p} < \epsilon.$$

If $\nu \in \mathcal{T}$ is arbitrary then

$$|\nu \neq \infty|^{-1/p}\|f - f^\nu\|_p \leq |\nu \neq \infty|^{-1/p}\|(f-g) - (f-g)^\nu\|_p + |\nu \neq \infty|^{-1/p}\|g - g^\nu\|_p.$$

The first term is less then ϵ and the second term clearly tends to 0 if $\nu \to \infty$.

To prove the converse let $f \in BMO_p$ and

$$\lim_{\nu\to\infty} |\nu \neq \infty|^{-1/p}\|f - f^\nu\|_p = 0.$$

We show that for all $\epsilon > 0$ there exists $N \in \mathbf{N}$ (the same as the one in the definition of the limit) such that

$$\|f - f_{N,N}\|_{BMO_p} < \epsilon.$$

Let $\nu \in \mathcal{T}'_N$ and $g = f - f_{N,N}$. In this case we have $g - g^\nu = f - f^\nu$, so, by the definition of the limit, we have

$$|\nu \neq \infty|^{-1/p}\|g - g^\nu\|_p < \epsilon.$$

Let $\nu \notin \mathcal{T}'_N$. Then another stopping time ν' should be defined such that

$$\{\nu \neq \infty\} = \{\nu' \neq \infty\}, \ \nu' \in \mathcal{T}'_N \quad \text{and} \quad g - g^\nu = f - f^{\nu'}.$$

In this case there exist $\omega \in \Omega$ and $(n,m) \in \nu(\omega)$ such that $(n,m) \ll (N,N)$. Let us fix this ω and denote by B the set of these pairs. Take the minimum both in the first and in the second coordinates of the pairs in B:

$$n_0 := \min\{n : (n,m) \in B\} \qquad \text{and} \qquad m_0 := \min\{m : (n,m) \in B\}$$

and let

$$(N, m_0) \in \nu'(\omega),\ (n_0, N) \in \nu'(\omega) \quad \text{while} \quad (n,m) \notin \nu'(\omega) \quad \text{if} \quad (n,m) \in B.$$

Proceed the same way for all $\omega \in \Omega$. If, for an ω, one has $(n,m) \in \nu(\omega)$ and $(n,m) \not\ll (N,N)$ then let $(n,m) \in \nu'(\omega)$. It is easy to see that ν' is a stopping time and satisfies the property above, indeed. Therefore

$$|\nu \neq \infty|^{-1/p}\|g - g^\nu\|_p = |\{\nu' \neq \infty\}|^{-1/p}\|f - f^{\nu'}\|_p < \epsilon.$$

The proof of Proposition 3.28 is complete. ∎

The proof of the next theorem is similar to the one of Theorem 2.39, however, it needs to be worked out more carefully.

Theorem 3.29. *If every σ-algebra $\mathcal{F}_n$ $(n \in \mathbf{N}^2)$ is generated by finitely many atoms then the dual of VMO_2 is H_1^s.*

Proof. By Theorem 3.20 the functional

$$l_f(\phi) := E(f\phi) \qquad (\phi \in L)$$

is bounded linear on VMO_2.

Conversely, we can show that if $l \in VMO_2'$ then there exists $f \in H_1^s$ such that

$$l(\phi) = E(f\phi) \qquad (\phi \in L) \tag{3.39}$$

and

$$\|f\|_{H_1^s} \leq 4\|l\|. \tag{3.40}$$

For this, we embed again the normed vector space $(L, \|\cdot\|_{VMO_2})$ isometrically in a space the dual of which can be identified. For $\nu \in \mathcal{T}$, let X_ν be the subspace of L_2 of which elements ξ satisfy the condition $\{\xi \neq 0\} \subset \{\nu \neq \infty\}$ and let

$$\|\xi\|_{X_\nu} := P(\nu \neq \infty)^{-1/2}\|\xi\|_2.$$

By Riesz's representation theorem $\Psi \in X_\nu'$ if and only if there exists an only $f \in L_2$ such that $\{f \neq 0\} \subset \{\nu \neq \infty\}$,

$$\Psi(\xi) = E(f\xi) \qquad (\xi \in X_\nu)$$

and

$$\|\Psi\| = P(\nu \neq \infty)^{1/2}\|f\|_2.$$

Let

$$X := \times_{\nu \in \mathcal{T}} X_\nu$$

and let us endow it again with the supremum norm:

$$\|\xi\|_X := \sup_{\nu \in \mathcal{T}} \|\xi_\nu\|_{X_\nu} \qquad (\xi = (\xi_\nu, \nu \in \mathcal{T}) \in X).$$

For a pair $n \in \mathbb{N}^2$, denote by $\mathcal{T}_n$ the subset of $\mathcal{T}$ for an element ν of which $m \leq n$ holds if $m \in \nu$ ($m \in \mathbb{N}^2$). (Of course, ν can take ∞ as well if $\nu \in \mathcal{T}_n$.) Since every σ-algebra $\mathcal{F}_n$ is generated by finitely many atoms, $\mathcal{T}_n$ is a finite set. The dual of the subspace X_n of X is easy to be formulated: let $\xi \in X_n \subset X$ if $\xi_\nu = 0$ for each $\nu \notin \mathcal{T}_n$ ($n \in \mathbb{N}^2$). As a simple conclusion we get that $\Lambda \in X_n'$ if and only if there exist functions $f_\nu \in X_\nu$ ($\nu \in \mathcal{T}_n$) such that

$$\Lambda(\xi) = \sum_{\nu \in \mathcal{T}_n} E(f_\nu \xi_\nu) \qquad (\xi \in X_n)$$

and

$$\|\Lambda\| = \sum_{\nu \in \mathcal{T}_n} P(\nu \neq \infty)^{1/2}\|f_\nu\|_2 < \infty.$$

Denote by $L(\mathcal{F}_n)$ the set of the $\mathcal{F}_n$ measurable functions from L. It is easy to see that, for $\phi \in L(\mathcal{F}_n)$,

$$\|\phi\|_{BMO_2} = \sup_{\nu \in \mathcal{T}_n} P(\nu \neq \infty)^{-1/2}\|\phi - \phi^\nu\|_2.$$

Thus the mapping

$$R_n : (L(\mathcal{F}_n), \|\cdot\|_{VMO_2}) \longrightarrow X_n, \qquad R_n\phi := (\phi_\nu, \nu \in \mathcal{T})$$

is an isometric isomorphism where $\phi_\nu = \phi - \phi^\nu$ if $\nu \in \mathcal{T}_n$ and $\phi_\nu = 0$ if $\nu \notin \mathcal{T}_n$. Hence, if $l \in VMO_2'$ then $l \circ R_n^{-1}$ is a bounded linear functional on a subspace of X_n and it can be extended onto X_n preserving its norm. Consequently, there exist functions $f_{\nu,n} \in X_\nu$ ($\nu \in \mathcal{T}_n$) for which

$$\|l\| \geq \|l \circ R_n^{-1}\| = \sum_{\nu \in \mathcal{T}_n} P(\nu \neq \infty)^{1/2} \|f_{\nu,n}\|_2 \tag{3.41}$$

and

$$l(\phi) = \sum_{\nu \in \mathcal{T}_n} E[f_{\nu,n}(\phi - \phi^\nu)] \qquad (\phi \in L(\mathcal{F}_n)).$$

Obviously, the lattest inequality can be written as

$$l(\phi) = \sum_{\nu \in \mathcal{T}_n} E[\phi E_n(f_{\nu,n} - f_{\nu,n}^\nu)] \qquad (\phi \in L(\mathcal{F}_n)). \tag{3.42}$$

If

$$f_n = \sum_{\nu \in \mathcal{T}_n} E_n(f_{\nu,n} - f_{\nu,n}^\nu)$$

then it can immediately be seen from (3.42) that

$$E(f_n\phi) = E(f_m\phi) = E(\phi E_n f_m) \qquad (\phi \in L(\mathcal{F}_n), m \geq n).$$

As $f_n \in L_2$ and it is $\mathcal{F}_n$ measurable, by the previous equation we obtain

$$f_n = E_n f_m \qquad (m \geq n),$$

therefore $f = (f_n, n \in \mathbb{N}^2)$ is a martingale. We are to show that $f \in H_1^s$ and that (3.40) holds. On the one hand,

$$\|f_n\|_{H_1^s} \leq \sum_{\nu \in \mathcal{T}_n} \|E_n(f_{\nu,n} - f_{\nu,n}^\nu)\|_{H_1^s}.$$

On the other hand,

$$a_{\nu,n} := \frac{f_{\nu,n} - f_{\nu,n}^\nu}{4\|f_{\nu,n} - f_{\nu,n}^\nu\|_2} P(\nu \neq \infty)^{-1/2}$$

is evidently a $(1,2)$ atom, thus

$$\|f_{\nu,n} - f_{\nu,n}^\nu\|_{H_1^s} \leq 4\|f_{\nu,n} - f_{\nu,n}^\nu\|_2 P(\nu \neq \infty)^{1/2}.$$

Since $H_2^S \sim L_2$, we get immediately that

$$\|f_{\nu,n} - f_{\nu,n}^\nu\|_{H_1^s} \leq 4\|f_{\nu,n}\|_2 P(\nu \neq \infty)^{1/2}.$$

Hence, by (3.41)

$$\|f_n\|_{H_1^s} \leq 4\|l\|$$

can be obtained. Now we can conclude that $f \in H_1^s$ and $\|f\|_{H_1^s} \leq 4\|l\|$. Since $f_n \to f$ ($n \to \infty$) a.e. and also in L_1 (as well as in H_1^s) norm, therefore (3.39) follows from (3.42). The proof of Theorem 3.29 is complete. ∎

Denoting the closure of L in $\Lambda_q(\alpha)$ norm by $VMO_q(\alpha)$ $(VMO_q = VMO_q(0))$, analogously to the previous proof we can characterize the dual space of $VMO_2(\alpha)$ which is similar to the envelop Banach space of H_p^s $(\alpha = 1/p - 1, 0 < p \leq 1)$. It can also be shown that the dual space of VMO_q $(1 < q < \infty)$ contains the martingales $f = (f_n, n \in \mathbf{N}^2)$ for which

$$\sup_{n \in \mathbf{N}^2} \|f_n\|_{H_{q'}^{\bullet s}} < \infty \qquad (1/q + 1/q' = 1).$$

It is natural to put the question how to generalize the one-parameter BMO_p spaces for two parameters. One of the most evident ways of the generalization is to consider the space BMO_p that was investigated above. Another one is to define a space with the norm

$$\|\phi\|_{\mathcal{BMO}_p} := \sup_{t \in \mathbf{N}^2} \|(E_t|\phi - E_t\phi|^p)^{1/p}\|_\infty \qquad (1 \leq p < \infty).$$

Studies of these spaces can be found in the papers Weisz [189], [192] and [196]. A third possible way is to use another norm

$$\|\phi\|_{BMO_p^\square} := \sup_{n,m \in \mathbf{N}} \|(E_{n,m}|\phi - E_{n,\infty}\phi - E_{\infty,m}\phi + E_{n,m}\phi|^p)^{1/p}\|_\infty$$

where $1 \leq p < \infty$. Notice that this norm can be got from the definition of the BMO_p norm by restricting the supremum. A stopping time ν is said to be ***simple*** if there exist $n, m \in \mathbf{N}$ and $A \in \mathcal{F}_{n,m}$ such that

$$\nu(\omega) = \nu_A(\omega) := \begin{cases} (n,m) & \text{if } \omega \in A \\ \infty & \text{if } \omega \notin A. \end{cases}$$

On the other hand, a stopping time is *elementaire* if for all $\omega \in \Omega$ there exists an only pair (n, m) with the property $(n, m) \in \nu(\omega)$. The sets of simple and elementaire stopping times are denoted by $\mathcal{T}_s$ and $\mathcal{T}_e$, respectively. It is easy to prove that

$$\|\phi\|_{BMO_p^\square} = \sup_{\nu \in \mathcal{T}_s} P(\nu \neq \infty)^{-1/p} \|\phi - \phi^\nu\|_p \leq \|\phi\|_{BMO_p}$$

and

$$\|\phi\|_{BMO_p^\square} = \sup_{\nu \in \mathcal{T}_e} P(\nu \neq \infty)^{-1/p} \|\phi - \phi^\nu\|_p \leq \|\phi\|_{BMO_p}.$$

We are going to show that, as it is expected, the spaces BMO_p and $BMO_p^\square$ are not equivalent, in general. The idea of the next counterexample is due to Carleson [32] and Decamp [56] (see also Fefferman [70]).

Theorem 3.30. *Let $(\mathcal{F}_{n,m})$ be the sequence of the dyadic σ-algebras. Then there exist a sequence (f^N) of martingales and a sequence (ν_N) of stopping times such that for all $N \in \mathbf{N}$*

$$\|f^N\|_{BMO_2^\square} = 1, \tag{3.43}$$

$$E[f^N - (f^N)^{\nu_N}]^2 = 1 \tag{3.44}$$

and

$$P(\nu_N \neq \infty) \to 0 \quad \text{as} \quad N \to \infty. \tag{3.45}$$

To prove this theorem some definitions and lemmas are needed. The enlargener transformations θ_n $(n \in \mathbf{N})$ are defined on $[0,1)$ by

$$\theta_n(x) := \begin{cases} 2^n(x - k2^{-n}) & \text{if } x \in [k2^{-n}, (k+1)2^{-n}) \\ \infty & \text{else} \end{cases}$$

where k is even and less than 2^n. The transformation θ_n maps an interval $[k2^{-n}, (k+1)2^{-n})$ onto $[0,1)$ or onto ∞. The two-parameter transformations $\theta_{n,m}$ of $\Omega = [0,1) \times [0,1)$ can be defined with

$$\theta_{n,m}(x,y) := \theta_n(x)\theta_m(y) \qquad (n, m \in \mathbf{N}).$$

We shall use the convention that, for a function $f \in L_1(\Omega)$, $f(x,\infty) = f(\infty, y) = 0$.

Lemma 3.31. *For $A \in \mathcal{F}_{p,q}$ the sets $\theta_{n,0}^{-1}(A)$ and $\theta_{0,m}^{-1}(A)$ are independent if $(p,q) \leq (n,m)$.*

Proof. Let

$$A = \bigcup_{(i,j) \in H} [i2^{-p}, (i+1)2^{-p}) \times [j2^{-q}, (j+1)2^{-q})$$

where H is an arbitrary index set. If $\theta_{n,0}(x,y) \in A$ then for some i, j and even k

$$i2^{-p} \leq 2^n(x - k2^{-n}) < (i+1)2^{-p}$$

and

$$j2^{-q} \leq y < (j+1)2^{-q}. \tag{3.46}$$

Thus

$$2^{-n}(i2^{-p} + k) \leq x < 2^{-n}((i+1)2^{-p} + k). \tag{3.47}$$

Similarly, if $\theta_{0,m}(x,y) \in A$ then

$$i2^{-p} \leq x < (i+1)2^{-p} \tag{3.48}$$

and

$$2^{-m}(j2^{-q} + l) \leq y < 2^{-m}((j+1)2^{-q} + l). \tag{3.49}$$

By an easy calculation we get from (3.46) and from (3.47) that

$$P[\theta_{n,0}^{-1}(A)] = |H| \sum_k 2^{-n}2^{-p}2^{-q} = |H|2^{-p-q-1}$$

where $|H|$ denotes the cardinality of H. Analogously,

$$P[\theta_{0,m}^{-1}(A)] = |H|2^{-p-q-1}.$$

If $(x,y) \in \theta_{n,0}^{-1}(A) \cap \theta_{0,m}^{-1}(A)$ then, by (3.46)–(3.49),

$$i2^{-p} \le 2^{-n}(i2^{-p} + k),$$
$$2^{-n}((i+1)2^{-p} + k) \le (i+1)2^{-p}$$

and

$$j2^{-q} \le 2^{-m}(j2^{-q} + l),$$
$$2^{-m}((j+1)2^{-q} + l) \le (j+1)2^{-q}.$$

Henceforth

$$i2^{n-p} \le k < (i+1)2^{n-p}$$

and

$$j2^{m-q} \le l < (j+1)2^{m-q}.$$

The numbers of such even k and l are 2^{n-p-1} and 2^{m-q-1}, respectively. Therefore

$$\begin{aligned} P[\theta_{n,0}^{-1}(A) \cap \theta_{0,m}^{-1}(A)] &= |H|^2 \sum_k \sum_l 2^{-n} 2^{-p} 2^{-m} 2^{-q} \\ &= |H|^2 2^{-2p-1} 2^{-2q-1}. \end{aligned}$$

Consequently,

$$P[\theta_{n,0}^{-1}(A) \cap \theta_{0,m}^{-1}(A)] = P[\theta_{n,0}^{-1}(A)] P[\theta_{0,m}^{-1}(A)]$$

which proves the lemma. ∎

Lemma 3.32. *For a function $f \in L_1[0,1)$ we have*

$$\int_0^1 f \circ \theta_n \, dP = \frac{1}{2} \int_0^1 f \, dP \qquad (n \in \mathbf{N}).$$

Moreover, if I is an atom of $\mathcal{F}_{n_0}$ and $n_0 < n$ $(n_0, n \in \mathbf{N})$ then

$$\int_I f \circ \theta_n \, dP = 2^{-n_0} \int_0^1 f \circ \theta_n \, dP.$$

Proof. By a substitution we obtain

$$\begin{aligned} \int_0^1 f \circ \theta_n \, dP &= \sum_{\substack{k=0 \\ 2|k}}^{2^n - 1} \int_{k2^{-n}}^{(k+1)2^{-n}} f \circ \theta_n \, dP \\ &= 2^{-n} \sum_{\substack{k=0 \\ 2|k}}^{2^n - 1} \int_0^1 f \, dP \\ &= \frac{1}{2} \int_0^1 f \, dP. \end{aligned}$$

On the other hand,

$$\int_I f \circ \theta_n \, dP = \sum_k \int_{k2^{-n}}^{(k+1)2^{-n}} f \circ \theta_n \, dP = 2^{-n} \sum_k \int_0^1 f \, dP$$

where the sum is taken over all even k for which $[k2^{-n}, (k+1)2^{-n}) \subset I$. So, by the first statement,

$$\int_I f \circ \theta_n \, dP = 2^{n-n_0-1} 2^{-n} \int_0^1 f \, dP = 2^{-n_0} \int_0^1 f \circ \theta_n \, dP.$$

The proof of the lemma is finished. ■

Proof of Theorem 3.30. The stopping times ν_N are created in a recursive way. Let

$$\nu_0(\omega) := (0,0) \quad \text{for all} \quad \omega \in [0,1) \times [0,1)$$

and let

$$\nu_N := \tau_{2^{N-1},0} \circ \nu_{N-1} \circ \theta_{2^{N-1},0} \wedge \tau_{0,2^{N-1}} \circ \nu_{N-1} \circ \theta_{0,2^{N-1}}$$

where $\tau_{i,j}$ is the shift operator, more precisely,

$$\tau_{i,j}(k,l) := (i+k, j+l).$$

Let us verify that if $(k,l) \in \nu_N$ then $(k,l) \ll (2^N, 2^N)$. This holds evidently for ν_0. Suppose that ν_{N-1} satisfies this condition and for a set $A \in \mathcal{F}_{k,l}$ one has $(k,l) \in \nu_{N-1}(\omega)$ for all $\omega \in A$ where $(k,l) \ll (2^{N-1}, 2^{N-1})$. Let $B = \theta^{-1}_{2^{N-1},0}(A)$; then $B \in \mathcal{F}_{k+2^{N-1},l}$ and

$$(k + 2^{N-1}, l) \in \tau_{2^{N-1},0} \circ \nu_{N-1} \circ \theta_{2^{N-1},0}(\omega)$$

for all $\omega \in B$. Since $k + 2^{N-1} < 2^N$, this proves our statement. Hence

$$\{\nu_N \neq \infty\} \in \mathcal{F}_{2^N-1,2^N-1}. \tag{3.50}$$

With the previous method it can also be derived that $\tau_{2^{N-1},0} \circ \nu_{N-1} \circ \theta_{2^{N-1},0}$ is a stopping time. As the minimum of two stopping times is again a stopping time, we get that ν_N is really a stopping time.

Notice that

$$\begin{aligned} P(\nu_N \neq \infty) &= P\big(\{\tau_{2^{N-1},0} \circ \nu_{N-1} \circ \theta_{2^{N-1},0} \neq \infty\} \\ &\qquad \cup \{\tau_{0,2^{N-1}} \circ \nu_{N-1} \circ \theta_{0,2^{N-1}} \neq \infty\}\big) \\ &= P[\theta^{-1}_{2^{N-1},0}(\nu_{N-1} \neq \infty)] + P[\theta^{-1}_{0,2^{N-1}}(\nu_{N-1} \neq \infty)] \\ &\qquad - P\big[\theta^{-1}_{2^{N-1},0}(\nu_{N-1} \neq \infty) \cap \theta^{-1}_{0,2^{N-1}}(\nu_{N-1} \neq \infty)\big]. \end{aligned}$$

Applying (3.50) together with Lemmas 3.31 and 3.32 we get

$$\begin{aligned} P(\nu_N \neq \infty) &= P(\nu_{N-1} \neq \infty) - P[\theta^{-1}_{2^{N-1},0}(\nu_{N-1} \neq \infty)] P[\theta^{-1}_{0,2^{N-1}}(\nu_{N-1} \neq \infty)] \\ &= P(\nu_{N-1} \neq \infty) - \frac{1}{4} P(\nu_{N-1} \neq \infty)^2. \end{aligned}$$

From this it follows that $P(\nu_N \neq \infty)$ is decreasing in N and

$$P(\nu_N \neq \infty) \to 0 \quad \text{as} \quad N \to \infty.$$

The martingales are going to be created as follows. Let

$$f^0(x,y) := r_0(x)r_0(y)$$

and

$$f^N_{n,m} := f^{N-1}_{n-2^{N-1},m} \circ \theta_{2^{N-1},0} + f^{N-1}_{n,m-2^{N-1}} \circ \theta_{0,2^{N-1}}$$

where r_0 is the Rademacher function, i.e.

$$r_0(x) := \begin{cases} 1 & \text{if } x \in [0,1/2) \\ -1 & \text{if } x \in [1/2,1). \end{cases}$$

f^0 is obviously a martingale. To verify that f^N is a martingale it is enough to show that

$$\left(g_{n,m} := f^{N-1}_{n-2^{N-1},m} \circ \theta_{2^{N-1},0}\right)_{n,m\in\mathbf{N}}$$

is a martingale as well. One can establish that

$$E_{n,m-1}(f^{N-1}_{n-2^{N-1},m} \circ \theta_{2^{N-1},0}) = f^{N-1}_{n-2^{N-1},m-1} \circ \theta_{2^{N-1},0}.$$

On the other hand, let $n > 2^{N-1}$ and $I \in \mathcal{F}_{n-1,m}$. Then, by a substitution,

$$\int_I f^{N-1}_{n-2^{N-1},m} \circ \theta_{2^{N-1},0}\, dP = 2^{-2^{N-1}} \int_{\theta_{2^{N-1},0}(I)} f^{N-1}_{n-2^{N-1},m}\, dP$$

where the convention $\int_\infty f\,dP = 0$ is used. Using the martingale property of f^{N-1} and the fact that $\theta_{2^{N-1},0}(I)$ is $\mathcal{F}_{n-1-2^{N-1},m}$ measurable we get

$$\begin{aligned}\int_I f^{N-1}_{n-2^{N-1},m} \circ \theta_{2^{N-1},0}\, dP &= 2^{-2^{N-1}} \int_{\theta_{2^{N-1},0}(I)} f^{N-1}_{n-1-2^{N-1},m}\, dP \\ &= \int_I f^{N-1}_{n-1-2^{N-1},m} \circ \theta_{2^{N-1},0}\, dP.\end{aligned}$$

Thus

$$E_{n-1,m}(f^{N-1}_{n-2^{N-1},m} \circ \theta_{2^{N-1},0}) = f^{N-1}_{n-1-2^{N-1},m} \circ \theta_{2^{N-1},0} \tag{3.51}$$

because $f^{N-1}_{n-1-2^{N-1},m} \circ \theta_{2^{N-1},0}$ is $\mathcal{F}_{n-1,m}$ measurable. The equation (3.51) holds evidently if $n \le 2^{N-1}$ since in this case $f^{N-1}_{n-2^{N-1},m} = 0$. So f^N is really a martingale.

Lemma 3.33. *For the above constructed martingales and for all $N \in \mathbf{N}$ we have that*

$$d_{n,m}f^N = 0 \quad \text{if} \quad (n,m) \gg (2^{N-1},2^{N-1}) \quad \text{or} \quad n > 2^N \quad \text{or} \quad m > 2^N.$$

Proof. The statement is trivially valid for $N = 0$ (of course, the condition $(n,m) \gg (2^{N-1},2^{N-1})$ is omitted in this case). The lemma will be proved by induction. Suppose that it is true for $N-1$. Since

$$d_{n,m}f^N = d_{n-2^{N-1},m}f^{N-1} \circ \theta_{2^{N-1},0} + d_{n,m-2^{N-1}}f^{N-1} \circ \theta_{0,2^{N-1}}, \tag{3.52}$$

we get that $d_{n,m}f^N = 0$ if one of the conditions $n - 2^{N-1} > 2^{N-1}$ and $m > 2^{N-1}$ together with one of $n > 2^{N-1}$ and $m - 2^{N-1} > 2^{N-1}$ hold. In this case $m > 2^{N-1}$ with $n > 2^{N-1}$ or $n > 2^N$ or $m > 2^N$. The lemma is proved. ■

Considering the square norm of $f^N - (f^N)^{\nu_N}$ we get by the definition of the stopped martingale that

$$E[f^N - (f^N)^{\nu_N}]^2 = E\Big[\sum_{n,m\in\mathbf{N}} \chi(\nu_N \ll (n,m))|d_{n,m}f^N|^2\Big].$$

Obviously,

$$\begin{aligned}\chi(\nu_N \ll (n,m)) = {}& \chi\big(\tau_{2^{N-1},0} \circ \nu_{N-1} \circ \theta_{2^{N-1},0} \ll (n,m)\big) \\ & + \chi\big(\tau_{0,2^{N-1}} \circ \nu_{N-1} \circ \theta_{0,2^{N-1}} \ll (n,m)\big) \\ & - \chi\big(\tau_{2^{N-1},0} \circ \nu_{N-1} \circ \theta_{2^{N-1},0} \ll (n,m) \\ & \qquad \cap \tau_{0,2^{N-1}} \circ \nu_{N-1} \circ \theta_{0,2^{N-1}} \ll (n,m)\big).\end{aligned}$$

In case the set

$$\{\tau_{2^{N-1},0} \circ \nu_{N-1} \circ \theta_{2^{N-1},0} \ll (n,m)\} \cap \{\tau_{0,2^{N-1}} \circ \nu_{N-1} \circ \theta_{0,2^{N-1}} \ll (n,m)\}$$

is non-empty we have $n > 2^{N-1}$ and $m > 2^{N-1}$. By Lemma 3.33, $d_{n,m}f^N = 0$ on the previous set. Henceforth

$$\begin{aligned}& E[f^N - (f^N)^{\nu_N}]^2 \\ & \quad = E\Big[\sum_{n,m\in\mathbf{N}} \chi(\tau_{2^{N-1},0} \circ \nu_{N-1} \circ \theta_{2^{N-1},0} \ll (n,m))|d_{n,m}f^N|^2\Big] \\ & \qquad + E\Big[\sum_{n,m\in\mathbf{N}} \chi(\tau_{0,2^{N-1}} \circ \nu_{N-1} \circ \theta_{0,2^{N-1}} \ll (n,m))|d_{n,m}f^N|^2\Big].\end{aligned}$$

If $\tau_{2^{N-1},0} \circ \nu_{N-1} \circ \theta_{2^{N-1},0} \ll (n,m)$ then $n > 2^{N-1}$ and, by Lemma 3.33, $d_{n,m-2^{N-1}}f^{N-1} = 0$. An analogous statement is valid for the second index. Consequently, by (3.52),

$$\begin{aligned}E[f^N - (f^N)^{\nu_N}]^2 = {}& E\Big[\sum_{n,m\in\mathbf{N}} \chi(\tau_{2^{N-1},0} \circ \nu_{N-1} \circ \theta_{2^{N-1},0} \ll (n,m)) \\ & \qquad |d_{n-2^{N-1},m}f^{N-1}|^2 \circ \theta_{2^{N-1},0}\Big] \\ & + E\Big[\sum_{n,m\in\mathbf{N}} \chi(\tau_{0,2^{N-1}} \circ \nu_{N-1} \circ \theta_{0,2^{N-1}} \ll (n,m)) \\ & \qquad |d_{n,m-2^{N-1}}f^{N-1}|^2 \circ \theta_{0,2^{N-1}}\Big] \\ =:{}& (A) + (B).\end{aligned}$$

Using Lemma 3.32 we get for the first term that

$$(A) = E\Big[\sum_{n,m\in\mathbf{N}} \chi(\nu_{N-1}\circ\theta_{2^{N-1},0} \ll (n-2^{N-1},m))$$
$$|d_{n-2^{N-1},m}f^{N-1}|^2\circ\theta_{2^{N-1},0}\Big]$$
$$= \frac{1}{2}E\Big[\sum_{n,m\in\mathbf{N}} \chi(\nu_{N-1}\ll(n,m))|d_{n,m}f^{N-1}|^2\Big]$$
$$= \frac{1}{2}E[f^{N-1}-(f^{N-1})^{\nu_{N-1}}]^2.$$

The same holds for (B) as well. So

$$E[f^N-(f^N)^{\nu_N}]^2 = E[f^{N-1}-(f^{N-1})^{\nu_{N-1}}]^2.$$

As

$$E[f^0-(f^0)^{\nu_0}]^2 = 1,$$

(3.44) is proved.

To verify (3.43) we only have to show that

$$\sup_{k,l\in\mathbf{N}}\sup_{I\in\mathcal{F}_{k,l}} P(I)^{-1}\int_I |f^N-E_{k,\infty}f^N-E_{\infty,l}f^N+E_{k,l}f^N|^2\,dP = 1 \tag{3.53}$$

where I is a dyadic rectangle. Notice that

$$P(I)^{-1}\int_I |f^N-E_{k,\infty}f^N-E_{\infty,l}f^N+E_{k,l}f^N|^2\,dP$$
$$= P(I)^{-1}\int_I \sum_{(n,m)\gg(k,l)} |d_{n,m}f^N|^2\,dP \qquad (I\in\mathcal{F}_{k,l}).$$

We prove (3.53) by induction. Suppose that it is true for $N-1$. Let $I\in\mathcal{F}_{n_0,m_0}$ be a dyadic rectangle.

Case 1. If $n_0\ge 2^{N-1}$ and $m_0\ge 2^{N-1}$ then, by Lemma 3.33, $d_{n,m}f^N=0$ holds for $(n,m)\gg(n_0,m_0)$. Thus

$$\int_I \sum_{(n,m)\gg(n_0,m_0)} |d_{n,m}f^N|^2\,dP = 0.$$

Case 2. If $n_0\ge 2^{N-1}$ and $m_0<2^{N-1}$ then, again by Lemma 3.33,

$$d_{n,m-2^{N-1}}f^{N-1}\circ\theta_{0,2^{N-1}} = 0$$

for $(n,m) \gg (n_0,m_0)$. Hence, applying (3.52),

$$\int_I \sum_{(n,m)\gg(n_0,m_0)} |d_{n,m}f^N|^2\,dP$$

$$= \int_I \sum_{(n,m)\gg(n_0,m_0)} |d_{n-2^{N-1},m}f^{N-1}|^2 \circ \theta_{2^{N-1},0}\,dP$$

$$= \int_I \sum_{(n,m)\gg(n_0-2^{N-1},m_0)} |d_{n,m}f^{N-1}|^2 \circ \theta_{2^{N-1},0}\,dP.$$

Again, by a substitution,

$$\int_I \sum_{(n,m)\gg(n_0,m_0)} |d_{n,m}f^N|^2\,dP$$

$$= 2^{-2^{N-1}} \int_{\theta_{2^{N-1},0}(I)} \sum_{(n,m)\gg(n_0-2^{N-1},m_0)} |d_{n,m}f^{N-1}|^2\,dP.$$

Since $\theta_{2^{N-1},0}(I)$ is an atom of $\mathcal{F}_{n_0-2^{N-1},m_0}$, we can apply (3.53) to $N-1$. Thus

$$\int_I \sum_{(n,m)\gg(n_0,m_0)} |d_{n,m}f^N|^2\,dP \le 2^{-2^{N-1}} 2^{-n_0+2^{N-1}} 2^{-m_0}$$

$$= 2^{-n_0-m_0} = P(I).$$

Case 3. With $n_0 < 2^{N-1}$ and $m_0 \ge 2^{N-1}$ the way of proving is similar to case 2.

Case 4. Let $n_0 < 2^{N-1}$ and $m_0 < 2^{N-1}$. It follows from Lemma 3.33 that at least one of $d_{n-2^{N-1},m}f^{N-1}$ and $d_{n,m-2^{N-1}}f^{N-1}$ is equal to 0. Therefore

$$\int_I \sum_{(n,m)\gg(n_0,m_0)} |d_{n,m}f^N|^2\,dP$$

$$= \int_I \sum_{(n,m)\gg(n_0,m_0)} |d_{n-2^{N-1},m}f^{N-1}|^2 \circ \theta_{2^{N-1},0}\,dP$$

$$+ \int_I \sum_{(n,m)\gg(n_0,m_0)} |d_{n,m-2^{N-1}}f^{N-1}|^2 \circ \theta_{0,2^{N-1}}\,dP$$

$$=: (C) + (D).$$

Setting $I = I_1 \times I_2$ and using Lemma 3.32 we obtain

$$(C) = \int_I \sum_{(n,m)\gg(0,m_0)} |d_{n,m}f^{N-1}|^2 \circ \theta_{2^{N-1},0}\,dP$$

$$= 2^{-n_0-1} \int_{[0,1)\times I_2} \sum_{(n,m)\gg(0,m_0)} |d_{n,m}f^{N-1}|^2\,dP.$$

As $[0,1) \times I_2 \in \mathcal{F}_{0,m_0}$, one has

$$(C) \leq 2^{-n_0-1}2^{-m_0} = \frac{1}{2}P(I).$$

The second term can be estimated by $P(I)/2$ in the same way. Thus

$$\int_I \sum_{(n,m)\gg(n_0,m_0)} |d_{n,m}f^N|^2 \, dP \leq P(I)$$

which proves that the supremum in (3.53) is not greater than 1. Note that, if $n_0 = m_0 = 0$ and $I = \Omega$ then we get that

$$\begin{aligned} E|f^N|^2 &= (C) + (D) = \frac{1}{2}E|f^{N-1}|^2 + \frac{1}{2}E|f^{N-1}|^2 \\ &= E|f^{N-1}|^2 = \ldots = E|f^0|^2 = 1 \end{aligned}$$

which shows the equality in (3.53). The proof of Theorem 3.30 is complete. ■

As a consequence of this theorem we obtain

Corollary 3.34. *In the dyadic case BMO_2 is not equivalent to $BMO_2^\square$, more exactly, there is no constant c such that for all $f \in BMO_2^\square$*

$$\|f\|_{BMO_2} \leq c\|f\|_{BMO_2^\square}. \tag{3.54}$$

Proof. If (3.54) is valid then for the martingales f^N constructed in Theorem 3.30

$$\begin{aligned} P(\nu_N \neq \infty)^{-1/2} &= P(\nu_N \neq \infty)^{-1/2}\|f^N - (f^N)^{\nu_N}\|_2 \\ &\leq \|f^N\|_{BMO_2} \leq c\|f^N\|_{BMO_2^\square} = c \end{aligned}$$

would hold. However, this is a contradiction to (3.45). ■

Note that the converse of (3.54) holds obviously. It is unknown even in the dyadic case, whether BMO_r is equivalent to $BMO_r^\square$ $(r \neq 2)$. The proofs of Theorem 3.30 and Corollary 3.34 can easily be extended to (p,q) martingales, i.e., for martingales generated by a Vilenkin system having the basic sequences $p_n = p$ and $q_n = q$ $(n \in \mathbb{N})$. It is an open problem to find a condition for the stochastic basis $\mathcal{F}$ such that BMO_r is not equivalent to $BMO_r^\square$ $(1 \leq r < \infty)$. This does not hold for each stochastic basis. For example, let $\mathcal{F}_{k,0} := \mathcal{F}_{0,k} := \{\emptyset, \Omega\}$ for all $k \in \mathbb{N}$ and $\mathcal{F}_{k,l} := \mathcal{A}$ if $k \neq 0$ and $l \neq 0$. In this case $BMO_r = BMO_r^\square = L_r$ $(1 \leq r < \infty)$.

Some atomic Hardy spaces the dual of which are the $BMO_p^\square$ spaces are to be defined. For this let us introduce the concept of a simple atom.

Definition 3.35. *A function $a \in L_q$ is called a $(1,q)$ simple atom if there exist $(n,m) \in \mathbb{N}^2$ and $A \in \mathcal{F}_{n,m}$ such that*

$$\begin{aligned} &(i) \qquad E_{n,m}a = 0 \\ &(ii) \qquad \|a^*\|_q \leq P(A)^{1/q-1} \\ &(iii) \qquad \{a \neq 0\} \subset A. \end{aligned}$$

If we allow in Definition 3.24 simple atoms, only, then we get the definition of the $H_q^{\rm sat}$ spaces. Denote by $VMO_p^{\square}$ the closure of L in the $BMO_p^{\square}$ norm. Analogously to Proposition 2.38 the spaces $VMO_p^{\square}$ can also be characterized by limit.

Proposition 3.36. *If every σ-algebra $\mathcal{F}_{n,m}$ is generated by finitely many atoms then $\phi \in VMO_p^{\square}$ if and only if $\phi \in BMO_p^{\square}$ and*

$$\lim_{\max(n,m)\to\infty} \|(E_{n,m}|\phi - E_{n,\infty}\phi - E_{\infty,m}\phi + E_{n,m}\phi|^p)^{1/p}\|_\infty = 0.$$

The proof is similar to the one of Proposition 2.38, so we leave it. The following results can be proved in the same way as we did in Theorems 3.25, 3.29, 2.39 and 2.41.

Theorem 3.37. *The dual of $H_q^{\rm sat}$ is $BMO_{q'}^{\square}$ $(1 < q < \infty, 1/q + 1/q' = 1)$.*

Theorem 3.38. *If every σ-algebra $\mathcal{F}_{n,m}$ is generated by countably many atoms then the dual of $VMO_q^{\square}$ is $H_{q'}^{\rm sat}$ $(1 < q < \infty, 1/q + 1/q' = 1)$.*

Corollary 3.39. *Let every σ-algebra $\mathcal{F}_{n,m}$ be generated by countably many atoms and $f \in H_{q'}^{\rm sat}$ $(1 < q' < \infty)$. Then there exist functions $f_{k,l} \in L_{q'}$ $(k, l \in \mathbf{N})$ such that*

$$f = \sum_{k,l\in\mathbf{N}} (f_{k,l} - E_{k,\infty} f_{k,l} - E_{\infty,l} f_{k,l} + E_{k,l} f_{k,l})$$

a.e. and also in L_1 norm. Moreover,

$$C_{q'} \sum_{k,l\in\mathbf{N}} \|(E_{k,l}|f_{k,l}|^{q'})^{1/q'}\|_1 \le \|f\|_{H_{q'}^{\rm sat}} \le 4q^2 \sum_{k,l\in\mathbf{N}} \|(E_{k,l}|f_{k,l}|^{q'})^{1/q'}\|_1.$$

Corollary 3.34 and Theorem 3.37 show that Theorem 2.5 does not hold in general for two parameters.

Corollary 3.40. *In the two-parameter dyadic case the martingales from H_1^s can not be decomposed into $(1,2)$ simple atoms, more precisely, the spaces H_1^s and $H_2^{\rm sat}$ are not equivalent.*

This corollary was proved by Carleson [32] for the two-parameter classical case. If H_1^s were equivalent to $H_2^{\rm sat}$ then the proofs of some theorems later could be simplified very much. However, by the lattest corollary, this can not be done.

3.4. MARTINGALE TRANSFORMS

In this section some theorems of Section 2.5 are extended to the two-parameter case. Amongst others we show that the martingale transform operator T_v is bounded from H_q^s to H_r^s and from H_q^S to H_r^S if $v \in V_p$ as well as from L_p to L_p if v is uniformly bounded. The lattest result is due to Metraux [125]. Moreover, it is proved that the martingales from H_p^s $(0 < p < \infty)$ can be obtained as a transform of a martingale from BMO_2 with a multiplier sequence $v \in V_p$. In special cases the converse is also verified. The results of this section can be found in Weisz [201].

The classes V_p of processes $v = (v_n, n \in \mathbb{N}^2)$ adapted to $\mathcal{F}$ are defined as in Section 2.5:

$$V_p := \{v : \|v\|_{V_p} := \|v^*\|_p < \infty\}, \qquad 0 < p \leq \infty.$$

The martingale transform T_v for a given $v \in V_p$ is defined again by the martingale $T_v f = (T_v f_n, n \in \mathbb{N}^2)$ where

$$T_v f_n := \sum_{m \leq n} v_{m-1} d_m f.$$

Similarly to the proof of Theorem 2.55, the pointwise estimations

$$s(T_v f) \leq v^* s(f)$$

and

$$S(T_v f) \leq v^* S(f)$$

imply the following result.

Theorem 3.41. *If $0 < p, q \leq \infty$, $v \in V_p$ and $1/r = 1/p + 1/q$ then T_v is of types (H_q^s, H_r^s) and (H_q^S, H_r^S) with $\|T_v\| \leq \|v\|_{V_p}$.*

Since $H_q^S \sim L_q$ $(1 < q < \infty)$ in the two-parameter case, too, it follows from the preceding theorem that T_v is bounded from L_q to L_q for $1 < q < \infty$ if $v \in V_\infty$.

By the duality argument one can prove

Theorem 3.42. *Let $0 \leq \alpha < \infty$, $1/(1+\alpha) < p \leq \infty$ and $v \in V_p$. Then*

(i) T_v is of type $(\Lambda_2(\alpha), \Lambda_2(\beta))$ where $\beta = \alpha - 1/p \geq 0$ (i.e. $1/\alpha \leq p \leq \infty$)

(ii) T_v is of type $(\Lambda_2(\alpha), H_r^s)$ whenever $\mathcal{F}$ is regular or it is generated by a Vilenkin system where $0 < 1/r = 1/p - \alpha < 1$ (i.e. $1/(1+\alpha) < p < 1/\alpha$).

In both cases $\|T_v\| \leq C_p \|v\|_{V_p}$.

For the proof we note that, if $\mathcal{F}$ is generated by a Vilenkin system then, by Theorem 3.23, the dual of H_q^s is H_r^s whenever $1 < q < \infty$ and $1/q + 1/r = 1$. This result holds also if $\mathcal{F}$ is regular because in this case $H_q^s \sim L_q$ $(1 < q < \infty)$ (see Proposition 3.4, Theorem 3.6 and 3.9).

Analogously to Theorem 2.63 we give a characterization of H_p^s by martingale transforms.

Theorem 3.43. *If $0 < p < \infty$ and $f \in H_p^s$ then there exist $g \in BMO_2$ with $\|g\|_{BMO_2} \leq 1$ and a (non-negative, non-decreasing) $v \in V_p$ with $\|v\|_{V_p} \leq C_p \|f\|_{H_p^s}$ such that $f = T_v g$.*

Proof. Let $f \in H_p^s$. Choose p_0 such that $0 < p_0 < p$ with $p_0 < 2$. Define $v_0 = 1$ and, for $n \geq 1$,

$$v_n := \sup_{m \leq n} (E_m[s(f)^{p_0}])^{1/p_0}$$

and

$$g_n := \sum_{m \leq n} v_{m-1}^{-1} d_m f \qquad (n \in \mathbb{N}^2).$$

Thus $f = T_v g$. Furthermore,

$$\begin{aligned}\|g\|_{BMO_2} &= \sup_{\nu\in\mathcal{T}_2} P(\nu \neq \infty)^{-1/2}\|g - g^\nu\|_2 \\ &= \sup_{\nu\in\mathcal{T}_2} P(\nu \neq \infty)^{-1/2}\Big(\sum_{m\in\mathbf{N}^2} E[\chi(\nu \ll m)|d_m g|^2]\Big)^{1/2} \\ &= \sup_{\nu\in\mathcal{T}_2} P(\nu \neq \infty)^{-1/2}\Big(\sum_{m\in\mathbf{N}^2} E[\chi(\nu \ll m)v_{m-1}^{-2}E_{m-1}|d_m f|^2]\Big)^{1/2}.\end{aligned}$$

Since

$$1 \leq E_{m-1}(s(f)^{p_0})E_{m-1}(s(f)^{-p_0}) \leq v_{m-1}^{p_0}E_{m-1}(s(f)^{-p_0}),$$

we have

$$(v_{m-1})^{-2} \leq (E_{m-1}[s(f)^{-p_0}])^{2/p_0} \leq E_{m-1}[s(f)^{-2}].$$

Therefore, we get

$$\begin{aligned}\|g\|_{BMO_2} &\leq \sup_{\nu\in\mathcal{T}_2} P(\nu \neq \infty)^{-1/2}\Big(\sum_{m\in\mathbf{N}^2} E[\chi(\nu \ll m)s(f)^{-2}E_{m-1}|d_m f|^2]\Big)^{1/2} \\ &\leq \sup_{\nu\in\mathcal{T}_2} P(\nu \neq \infty)^{-1/2}\Big(\sum_{m\in\mathbf{N}^2}\int_{\{\nu\neq\infty\}} s(f)^{-2}E_{m-1}|d_m f|^2\,dP\Big)^{1/2} \\ &\lesssim 1.\end{aligned}$$

By the equivalence $H_q^* \sim L_q$ $(q > 1)$ we obtain

$$\|v\|_{V_p} = \big(E\big[(s(f)^{p_0})^{*\,p/p_0}\big]\big)^{1/p} \leq C_p\big(E\big[s(f)^{p_0 p/p_0}\big]\big)^{1/p} = C_p\|f\|_{H_p^s}$$

which completes the proof. ∎

It follows from Theorem 3.42 (ii) that the converse of Theorem 3.43 holds in special cases, too.

Corollary 3.44. *Let $1 < p < \infty$. If $\mathcal{F}$ is regular or it is generated by a Vilenkin system and $f = T_v g$ with $v \in V_p$ and with $g \in BMO_2$ then $f \in H_p^s$ and $\|f\|_{H_p^s} \leq C_p\|v\|_{V_p}\|g\|_{BMO_2}$.*

3.5. TWO-PARAMETER STRONG MARTINGALES

In this section two-parameter strong martingales are investigated. Some of these martingales can be obtained as a sum of independent random variables with zero mean. The analogues of several foregoing theorems are given. Some of the theorems below can be reduced to the analogous one-parameter results. Atomic decompositions, martingale inequalities and duality theorems are given. Contrary to the two-parameter martingale case, Davis's inequality can be proved for strong martingales in the same way as for the one-parameter martingales. The dual of H_1^* are characterized, moreover, the equivalence of the BMO_q^- spaces is verified.

First of all let us see some new definitions. Let

$$\mathcal{F}_n^- := \mathcal{F}_{n_1-1,n_2} \vee \mathcal{F}_{n_1,n_2-1}, \qquad \mathcal{F}_n^+ := \mathcal{F}_{n_1+1,n_2} \vee \mathcal{F}_{n_1,n_2+1}$$

and

$$\mathcal{F}_n^* := \mathcal{F}_{n_1,\infty} \vee \mathcal{F}_{\infty,n_2} \qquad (n \in \mathbb{N}^2).$$

The corresponding conditional expectations are denoted by E_n^-, E_n^+ and E_n^*, respectively. A martingale f is said to be a *strong martingale* if

$$E_n^- d_n f = 0 \qquad (n \in \mathbb{N}^2).$$

It follows from the (F_4) hypothesis that $\mathcal{F}_{n-1}^*$ and $\mathcal{F}_n$ are conditionally independent with respect to $\mathcal{F}_n^-$, namely, $E_{n-1}^* \circ E_n = E_n^-$. So the above condition for the strong martingales is equivalent to

$$E_{n-1}^* d_n f = 0 \qquad (n \in \mathbb{N}^2).$$

To give an example for strong martingales let $\left(u_n^{(i)}; 0 \le i \le N \in \mathbb{N}, n \in \mathbb{N}^2\right)$ be a sequence of random variables with zero expectation such that, for each $n \in \mathbb{N}^2$, $u_n^{(i)}$ is independent of $\left(u_m^{(j)}; 0 \le j \le N, m \not\ge n\right)$ for all $0 \le i \le N$. Let $\mathcal{F}_n$ be generated by the variables $\left(u_m^{(i)}; 0 \le i \le N, m \le n\right)$. If

$$f_n := \sum_{m \le n} \sum_{i=0}^{N} u_m^{(i)}$$

then $(f_n, n \in \mathbb{N}^2)$ is obviously a strong martingale. More generally, if $v_m^{(i)}$ is $\mathcal{F}_m^-$ measurable for all $0 \le i \le N$ $(m \in \mathbb{N}^2)$ then

$$\left(f_n := \sum_{m \le n} \sum_{i=0}^{N} v_m^{(i)} u_m^{(i)}\right)_{n \in \mathbb{N}^2} \tag{3.55}$$

is also a strong martingale.

For strong martingales we have to modify the definitions of regularity and previsibility. A strong martingale f is called *previsible* if there exists a real number $R > 0$ such that

$$|d_n f|^2 \le R E_n^- |d_n f|^2 \qquad (n \in \mathbb{N}^2). \tag{3.56}$$

The class of previsible strong martingales having the same constant R in (3.56) is denoted again by $\mathcal{V}_R$. Note that Brossard considered a little bit more general condition than (3.56) (see [15] p. 105). Similarly to Lemma 2.18, for all $0 < p < \infty$, the condition (3.56) is equivalent to

$$|d_n f|^p \le R_p E_n^- |d_n f|^p \qquad (n \in \mathbb{N}^2)$$

where $R_p = R$ if $0 < p \leq 2$ and $R_p = R^{p/2}$ if $2 \leq p < \infty$. In this section the stochastic basis $\mathcal{F}$ is said to be *regular* if every strong martingale is previsible with the same previsibility constant R.

An example for previsible strong martingales is demonstrated. Together with the above mentioned conditions let

$$E(u_m^{(i)} u_m^{(j)}) = 0 \qquad (0 \leq i < j \leq N, m \in \mathbf{N}^2),$$

$$E|u_m^{(i)}|^2 = 1 \qquad (0 \leq i \leq N, m \in \mathbf{N}^2)$$

and

$$|u_m^{(i)}| \leq C \qquad (0 \leq i \leq N, m \in \mathbf{N}^2)$$

be required to be satisfied. In this case

$$|d_n f|^2 \leq C^2 (\sum_{i=0}^{N} |v_n^{(i)}|)^2 \leq C^2 N \sum_{i=0}^{N} |v_n^{(i)}|^2 = C^2 N E_n^- |d_n f|^2,$$

so the martingale $(f_n, n \in \mathbf{N}^2)$ is previsible, indeed. These martingales are similar on the one hand, to the ones considered by Gundy [86] and by Ledoux [114] and, on the other hand, to the Vilenkin martingales. An example for a regular stochastic basis is delayed to later.

We suppose again that a martingale vanishes on the axes. The subspaces of L_p, H_p^*, H_p^s and H_p^S containing strong martingales are denoted by sL_p, sH_p^*, sH_p^s and sH_p^S $(0 < p \leq \infty)$, respectively. Now it is more useful to consider a new conditional square function instead of s. For a strong martingale f let

$$\sigma(f) := (\sum_{n \in \mathbf{N}^2} E_n^- |d_n f|^2)^{1/2}.$$

Let us denote by sH_p^σ $(0 < p \leq \infty)$ the space of strong martingales f for which

$$\|f\|_{sH_p^\sigma} := \|\sigma(f)\|_p < \infty.$$

Of course, $sH_2^\sigma = sL_2$ and the norms are also equal. The next theorem describes the relation between sH_p^s and sH_p^σ.

Theorem 3.45. *For a strong martingale f we have*

(i)

$$\|f\|_{sH_p^\sigma} \leq C_p \|f\|_{sH_p^s} \qquad (0 < p \leq 2)$$

(ii)

$$\|f\|_{sH_p^s} \leq C_p \|f\|_{sH_p^\sigma} \qquad (2 \leq p < \infty).$$

Proof. The proof is similar to the one of Theorem 3.5. From (3.13) we get that

$$\sigma^2(f) = \sum_{k \in \mathbf{Z}} \sigma^2(\mu_k a^k)$$

and

$$E(\sigma^p(f)) \le \sum_{k\in\mathbf{Z}} |\mu_k|^p E(\sigma^p(a^k)) \qquad (0 < p \le 2)$$

where μ_k and a^k are the real numbers and atoms used in Theorem 3.2. If a is a $(p,2)$ atom then it easy to see that $\sigma(a) = 0$ on the set $\{\nu = \infty\}$ where ν is the stopping time in the definition of a. So we have

$$E(\sigma^p(a)) \le 1 \qquad (0 < p \le 2)$$

which proves (i).

(ii) comes easily from Theorem 2.10. ■

Note that (i) was proved by Brossard in [15] with another argument.

Of course, $sH_p^* \sim sH_p^S \sim sL_p$ for $1 < p < \infty$. The discussion of the connection between the other spaces comes later.

A new maximal theorem that is similar to Proposition 3.4 and to Corollary 3.8 is established.

Proposition 3.46. *For an arbitrary function $f \in L_1$, $\lambda > 0$ and $n \in \mathbf{N}^2$ we have*

$$\lambda P(\sup_{m\le n} |E_m^- f| > \lambda) \le \int_{\{\sup_{m\le n} |E_m^- f| > \lambda\}} f_n^* \, dP \tag{3.57}$$

and if $f \in L_p$ $(1 < p \le \infty)$ then

$$\| \sup_{m\in\mathbf{N}^2} |E_m^- f| \|_p \le \left(\frac{p}{p-1}\right)^2 \|f\|_p. \tag{3.58}$$

Proof. For $\lambda > 0$ let

$$\nu_1 := \inf\{k \le n_1 : \sup_{l\le n_2} |E_{k,l}^- f| > \lambda\},$$

$$\nu_2 := \inf\{l \le n_2 : |E_{\nu_1,l}^- f| > \lambda\}$$

and $\nu = (\nu_1, \nu_2)$. Then ν is not a stopping time, however, it is easy to see that the set $\{\nu = m\}$ is in $\mathcal{F}_{m-1}^*$. Using the stochastic independence of $\mathcal{F}_{m-1}^*$ and $\mathcal{F}_m$ with respect to $\mathcal{F}_m^-$ we have

$$\begin{aligned}
\lambda P(\sup_{m\le n} |E_m^- f| > \lambda) &= \lambda \sum_{m\le n} P(\nu = m) \le \sum_{m\le n} \int_{\{\nu=m\}} |E_m^- f| \, dP \\
&= \sum_{m\le n} \int_{\{\nu=m\}} |E_{m-1}^* E_m f| \, dP \\
&\le \sum_{m\le n} \int_{\{\nu=m\}} E_{m-1}^* |E_m f| \, dP \\
&= \sum_{m\le n} \int_{\{\nu=m\}} |E_m f| \, dP \le \int_{\{\nu\le n\}} f_n^* \, dP
\end{aligned}$$

which shows (3.57). With Proposition 2.6 one can prove the inequality (3.58) by integration. ■

We give the atomic decomposition of the space sH_p^σ, too. In this section under a stopping time we mean always a stopping time relative to $(\mathcal{F}_n^+)$ and we write briefly $\nu \in \mathcal{T}(\mathcal{F}_n^+)$.

Definition 3.47. *A strong martingale $a \in sL_q$ is called a strong (p,q) atom if there exists a stopping time $\nu \in \mathcal{T}(\mathcal{F}_n^+)$ such that*

$$(i) \qquad a_n := E_n a = 0 \qquad \text{if} \qquad \nu \not\leq n$$

$$(ii) \qquad \|a^*\|_q \leq P(\nu \neq \infty)^{1/q-1/p} \qquad (0 < p \leq q, 1 < q \leq \infty).$$

Since the following theorem has high importance in the proof of duality theorems, it will precisely be given, though it is similar to Theorem 3.2.

Theorem 3.48. *If the strong martingale $f = (f_n; n \in \mathrm{N}^2)$ is in sH_p^σ $(0 < p \leq 2)$ then there exist a sequence $(a^k, k \in \mathrm{Z})$ of strong $(p,2)$ atoms and a sequence $\mu = (\mu_k, k \in \mathrm{Z}) \in l_p$ of real numbers such that for all $n \in \mathrm{N}^2$*

$$\sum_{k=-\infty}^{\infty} \mu_k E_n a^k = f_n \tag{3.59}$$

and

$$\Big(\sum_{k=-\infty}^{\infty} |\mu_k|^p\Big)^{1/p} \leq C_p \|f\|_{sH_p^\sigma}. \tag{3.60}$$

Moreover, the sum $\sum_{k=l}^{m} \mu_k a^k$ converges to f in sH_p^σ norm as $m \to \infty$, $l \to -\infty$, too. Conversely, if $0 < p \leq 1$ and the martingale f has a decomposition of type (3.59) then $f \in sH_p^\sigma$ and

$$\|f\|_{sH_p^\sigma} \sim \inf\Big(\sum_{k=-\infty}^{\infty} |\mu_k|^p\Big)^{1/p} \tag{3.61}$$

where the infimum is taken over all decompositions of f of the form (3.59).

Proof. Assume that $f \in sH_p^\sigma$. For every $k \in \mathrm{Z}$ introduce the sets

$$F_k := \{\sigma(f) > 2^k\}$$

and the stopping times

$$\nu_k := \inf\{n \in \mathrm{N}^2 : E_n^+ \chi(F_k) > 1/2\}.$$

Obviously, ν_k is a stopping time relative to $(\mathcal{F}_n^+)$. The atoms and the real numbers are given in a similar way:

$$\mu_k := 2^{k+3}\sqrt{2}P(\nu_k \neq \infty)^{1/p}$$

and

$$a_n^k := \frac{1}{\mu_k}(f_n^{\nu_{k+1}} - f_n^{\nu_k}).$$

It is easy to show that a^k is a strong martingale for all $k \in \mathbf{Z}$. The condition (i) in Definition 3.47 can be proved like in Theorem 3.2. For the proof of (ii) of Definition 3.47 we only note that the set $\{\nu_k \ll n\}$ is $\mathcal{F}_n^-$ measurable ($k \in \mathbf{Z}, n \in \mathbf{N}^2$).

The inequality (3.60) can be proved as (3.2). Applying Proposition 3.46 we obtain

$$\begin{aligned}\sum_{k\in\mathbf{Z}} |\mu_k|^p &\le C_p \sum_{k\in\mathbf{Z}} 4(2^k)^p E\Big[\sup_{n\in\mathbf{N}^2} \big(E_n^+ \chi(F_k)\big)^2\Big] \\ &\le C_p \sum_{k\in\mathbf{Z}} (2^k)^p P(F_k) \\ &= C_p \sum_{k\in\mathbf{Z}} (2^k)^p P(\sigma(f) > 2^k)\end{aligned}$$

which implies (3.60).

The rest of the proof is the same as in Theorem 3.2. ∎

Because of the next theorem, the analogue of Theorem 3.3 is not worthy to be given.

From a two-parameter strong martingale we intend to create a one-parameter martingale that inherits the Hardy norms. This idea is belonging to L. Chevalier and can be found in Brossard [15] and in Frangos, Imkeller [76]. Let us introduce the following bijective map $\rho : \mathbf{N} \longrightarrow \mathbf{N}^2$:

$$\begin{aligned}\rho(k^2+i) &= (k,i) \quad \text{if} \quad 0 \le i \le k-1 \\ \rho(k^2+i) &= (i-k,k) \quad \text{if} \quad k \le i \le 2k.\end{aligned}$$

Set

$$\tilde{\mathcal{F}}_k := \bigvee_{l\le k} \mathcal{F}_{\rho(l)} \quad \text{and} \quad \tilde{\mathcal{F}} := (\tilde{\mathcal{F}}_k, k \in \mathbf{N}).$$

Note that $\tilde{\mathcal{F}}_k \subset \mathcal{F}^*_{\rho(k)}$. Moreover, let

$$\tilde{f}_k := \sum_{l=0}^{k} d_{\rho(l)} f \quad \text{and} \quad \tilde{f} := (\tilde{f}_k, k \in \mathbf{N}).$$

Denote by $\tilde{E}_k$ the conditional expectation operator relative to $\tilde{\mathcal{F}}_k$ ($k \in \mathbf{N}$). If f is a strong martingale then $\tilde{f}$ is a one-parameter martingale relative to $\tilde{\mathcal{F}}$ because

$$\tilde{E}_{k-1}(d_k\tilde{f}) = \tilde{E}_{k-1} E^*_{\rho(k)-1}(d_{\rho(k)} f) = 0.$$

Note that $\tilde{f}_0 = 0$.

Reversely, if $\tilde{f}$ is a martingale relative to $\tilde{\mathcal{F}}$ and if $d_k\tilde{f} = 0$ in case $k = \rho^{-1}(i,0)$ or $k = \rho^{-1}(0,i)$ for an $i \in \mathbf{N}$ then

$$f_n = \sum_{m\le n} E_m d_{\rho^{-1}(m)} \tilde{f}$$

is a strong martingale. In order to avoid misunderstandig, during this section let us add an extra tilde above the letters denoting the one-parameter martingale Hardy spaces relative to $\tilde{\mathcal{F}}$, that is to say, we write $\tilde{X}$ for the space $X \in \{H_p^*, H_p^s, H_p^S, \mathcal{Q}_p, \mathcal{P}_p, \mathcal{G}_p\}$ relative to $\tilde{\mathcal{F}}$. Since $d_k\tilde{f} = d_{\rho(k)}f$, one has $S(\tilde{f}) = S(f)$. Moreover, $\|\cdot\|_{sH_p^S} = \|\cdot\|_{\tilde{H}_p^S}$ and $\|\cdot\|_{s\mathcal{G}_p} = \|\cdot\|_{\tilde{\mathcal{G}}_p}$. We recall that the $\mathcal{G}_p$ norm is defined by

$$\|f\|_{\mathcal{G}_p} := \| \sum_{n\in\mathbf{N}^2} |d_n f| \|_p \qquad (0 < p < \infty).$$

As

$$\begin{aligned}\tilde{E}_{k-1}|d_k\tilde{f}|^2 &= \tilde{E}_{k-1}E^*_{\rho(k)-1}|d_{\rho(k)}f|^2 \\ &= \tilde{E}_{k-1}E^-_{\rho(k)}|d_{\rho(k)}f|^2 \\ &= E^-_{\rho(k)}|d_{\rho(k)}f|^2,\end{aligned}$$

one can see that $\sigma(f) = s(\tilde{f})$. Thus $\|\cdot\|_{sH_p^\sigma} = \|\cdot\|_{\tilde{H}_p^s}$. Furthermore, if f is previsible then so is $\tilde{f}$ and if $\tilde{\mathcal{F}}$ is regular then so is $\mathcal{F}$.

An example for a regular $\mathcal{F}$ is demonstrated. Let $\tilde{\mathcal{F}}_k = \mathcal{F}_{\rho(k)} := \sigma(r_0, \ldots, r_{k-1})$ where r_l are the Rademacher functions. Thus $(\tilde{\mathcal{F}}_k)$ is the sequence of dyadic σ-algebras and so it is regular. Consequently, the corresponding stochastic basis $\mathcal{F}$ is also regular.

It is easy to check that

$$\tilde{f}_{k^2+i} = \begin{cases} f_{k,i} + f_{k-1,k-1} - f_{k-1,i} & \text{if } 0 \le i \le k-1 \\ f_{i-k,k} + f_{k,k-1} - f_{i-k,k-1} & \text{if } k \le i \le 2k. \end{cases}$$

Hence $\tilde{f}^* \le 3f^*$ and $\|\cdot\|_{\tilde{H}_p^*} \le 3\|\cdot\|_{sH_p^*}$. The converse inequality can not easily be seen. However, for $1 \le p < \infty$, it follows from Theorem 3.52 and from Burkholder-Gundy's inequality (see Theorem 3.6).

So the following theorem is proved.

Theorem 3.49. *If f is a strong martingale then $\tilde{f}$ is a one-parameter martingale relative to $\tilde{\mathcal{F}}$. Moreover, $\|\cdot\|_{sH_p^S} = \|\cdot\|_{\tilde{H}_p^S}$, $\|\cdot\|_{s\mathcal{G}_p} = \|\cdot\|_{\tilde{\mathcal{G}}_p}$, $\|\cdot\|_{sH_p^\sigma} = \|\cdot\|_{\tilde{H}_p^s}$ and $\|\cdot\|_{\tilde{H}_p^*} \le 3\|\cdot\|_{sH_p^*}$ $(0 < p \le \infty)$. If f is previsible then so is $\tilde{f}$.*

Using the atomic decomposition of sH_p^σ (see Theorem 3.48) and Burkholder-Gundy's inequality we can prove the next result similarly to Theorem 3.5.

Proposition 3.50.

(i)

$$\|f\|_{sH_p^*} \le C_p\|f\|_{sH_p^\sigma}, \quad \|f\|_{sH_p^S} \le C_p\|f\|_{sH_p^\sigma} \qquad (0 < p \le 2)$$

(ii)

$$\|f\|_{sH_p^\sigma} \le C_p\|f\|_{sH_p^*}, \quad \|f\|_{sH_p^\sigma} \le C_p\|f\|_{sH_p^S} \qquad (2 \le p < \infty).$$

Note that (ii) and the second inequality of (i) can also be proved with the help of Theorem 3.49 and the one-parameter results.

For strong martingales Davis's inequality can be proved with the same method as we did for one-parameter martingales. First we describe Davis decomposition for strong martingales.

Lemma 3.51. *Let $f \in X$ where $X \in \{sH_1^*, sH_1^S\}$. Then there exist $h \in s\mathcal{G}_1$ and $g \in sH_1^\sigma$ such that $f_n = h_n + g_n$ for all $n \in \mathbb{N}^2$ and*

$$\|h\|_{s\mathcal{G}_1} \leq C\|f\|_X, \qquad \|g\|_{sH_1^\sigma} \leq C\|f\|_X.$$

Proof. The result is going to be verified and later utilized for sH_1^S, only. Nevertheless, the proof for sH_1^* is similar. Let $\tilde{f}$ be the corresponding one-parameter martingale to f. One can not simply apply Lemma 2.15 because if the one-parameter martingales $\tilde{h}$ and $\tilde{g}$ are created as in Lemma 2.13, the martingale differences $d_k\tilde{h}$ and $d_k\tilde{g}$ are not necessarily $\mathcal{F}_{\rho(k)}$ measurable and so, h and g are not necessarily strong martingales, either. Therefore in the construction of $d_k\tilde{h}$ and $d_k\tilde{g}$ we have to take the conditional expectation operator $E_{\rho(k)}$. Suppose that $\tilde{\lambda}_0 \leq \tilde{\lambda}_1 \leq \ldots$ is an adapted sequence of functions with respect to $\tilde{\mathcal{F}}$ such that

$$S_k(\tilde{f}) \leq \tilde{\lambda}_k, \qquad \tilde{\lambda}_\infty := \sup_{k\in\mathbb{N}} \tilde{\lambda}_k \in L_p.$$

Let

$$d_k\tilde{h} := E_{\rho(k)}[d_k\tilde{f}\chi(\tilde{\lambda}_k > 2\tilde{\lambda}_{k-1})] - E_{\rho(k)}\tilde{E}_{k-1}[d_k\tilde{f}\chi(\tilde{\lambda}_k > 2\tilde{\lambda}_{k-1})]$$

and

$$d_k\tilde{g} := E_{\rho(k)}[d_k\tilde{f}\chi(\tilde{\lambda}_k \leq 2\tilde{\lambda}_{k-1})] - E_{\rho(k)}\tilde{E}_{k-1}[d_k\tilde{f}\chi(\tilde{\lambda}_k \leq 2\tilde{\lambda}_{k-1})].$$

Since $d_k\tilde{f} = d_{\rho(k)}\tilde{f}$ is $\mathcal{F}_{\rho(k)}$ measurable, we get immediately that

$$d_k\tilde{f} = d_k\tilde{h} + d_k\tilde{g} \qquad (k \in \mathbb{N}).$$

Let

$$d_nh := d_{\rho^{-1}(n)}\tilde{h} \quad \text{and} \quad d_ng := d_{\rho^{-1}(n)}\tilde{g} \qquad (n \in \mathbb{N}^2).$$

As d_nh and d_ng are $\mathcal{F}_n$ measurable, it follows that

$$\Big(h_n := \sum_{m\leq n} d_nh\Big)_{n\in\mathbb{N}^2} \quad \text{and} \quad \Big(g_n := \sum_{m\leq n} d_ng\Big)_{n\in\mathbb{N}^2}$$

are two-parameter strong martingales with respect to $\mathcal{F}$. Moreover,

$$d_nf = d_nh + d_ng \qquad (n \in \mathbb{N}^2).$$

Similarly to the proof of Lemma 2.13 we can show that

$$|d_k\tilde{h}| \leq 2E_{\rho(k)}(\tilde{\lambda}_k - \tilde{\lambda}_{k-1}) + 2E_{\rho(k)}\tilde{E}_{k-1}(\tilde{\lambda}_k - \tilde{\lambda}_{k-1}) \qquad (k \in \mathbb{N}) \tag{3.62}$$

and

$$|d_k\tilde{g}| \leq 2E_{\rho(k)}\tilde{\lambda}_{k-1} + 2E_{\rho(k)}\tilde{E}_{k-1}\tilde{\lambda}_{k-1} \qquad (k \in \mathbb{N}). \tag{3.63}$$

By (3.62) we have

$$\|\tilde{h}\|_{\tilde{\mathcal{Q}}_1} = E(\sum_{k=0}^{\infty} |d_k \tilde{h}|) \le 4E(\tilde{\lambda}_\infty). \tag{3.64}$$

On the other hand,

$$E_{\rho(k)} \circ \tilde{E}_{k-1} = E_{\rho(k)} \circ E^*_{\rho(k)-1} \circ \tilde{E}_{k-1} = E^-_{\rho(k)} \circ \tilde{E}_{k-1} = E^-_{\rho(k)}. \tag{3.65}$$

So (3.63) implies that

$$|d_k \tilde{g}| \le 4E^-_{\rho(k)}(\tilde{\lambda}_{k-1}).$$

Clearly, $E^-_{\rho(k)}(\tilde{\lambda}_{k-1})$ is $\tilde{\mathcal{F}}_{k-1}$ measurable. Furthermore,

$$S_k(\tilde{g}) \le S_{k-1}(\tilde{g}) + |d_k \tilde{g}| \le S_{k-1}(\tilde{f}) + S_{k-1}(\tilde{h}) + 4E^-_{\rho(k)}(\tilde{\lambda}_{k-1}).$$

Using (3.62) and (3.65) we can conclude that

$$S_k(\tilde{g}) \le \tilde{\lambda}_{k-1} + 2\sum_{j=1}^{k-1} [E_{\rho(j)}(\tilde{\lambda}_j - \tilde{\lambda}_{j-1}) + E^-_{\rho(j)}(\tilde{\lambda}_j - \tilde{\lambda}_{j-1})] + 4E^-_{\rho(k)}(\tilde{\lambda}_{k-1}).$$

The right hand side is obviously $\tilde{\mathcal{F}}_{k-1}$ measurable, so we got a predictable process of $S_k(\tilde{g})$. Hence

$$\|\tilde{g}\|_{\tilde{\mathcal{Q}}_1} \le 9E(\tilde{\lambda}_\infty). \tag{3.66}$$

Setting $\tilde{\lambda}_k := S_k(\tilde{f})$ and applying (3.64), (3.66) and Theorem 2.11 (v) we obtain

$$\|h\|_{s\mathcal{G}_1} = \|\tilde{h}\|_{\tilde{\mathcal{G}}_1} \le 4\|S(\tilde{f})\|_1 = 4\|S(f)\|_1$$

and

$$\|g\|_{sH_1^\sigma} = \|\tilde{g}\|_{\tilde{H}_1^s} \le C\|\tilde{g}\|_{\tilde{\mathcal{Q}}_1} \le 9C\|S(\tilde{f})\|_1 = 9C\|S(f)\|_1$$

which proves the lemma. ∎

Note that, by the convexity theorem (see Theorem 2.10) and by Proposition 3.46, Lemma 3.51 can also be proved for all $1 < p < \infty$.

We have reached the point of being able to verify Davis's inequality for two-parameter strong martingales. This inequality was first proved by Frangos and Imkeller [76] with another method. The left hand side of the following inequality can be found in Brossard [15] as well.

Theorem 3.52. **(Davis's inequality)** *There exist real numbers $c > 0$ and $C > 0$ such that for all strong martingales*

$$c\|f\|_{sH_1^S} \le \|f\|_{sH_1^*} \le C\|f\|_{sH_1^S}.$$

Proof. The proof of the left hand side is simple. Let again $\tilde{f}$ be the corresponding one-parameter martingale to f. Then, by the one-parameter Davis's inequality and by Theorem 3.49,

$$\|f\|_{sH_1^s} = \|\tilde{f}\|_{\tilde{H}_1^s} \le C\|\tilde{f}\|_{\tilde{H}_1^*} \le 3C\|f\|_{sH_1^*}.$$

To prove the right hand side, let $f \in sH_1^S$. Then there exist two strong martingales $h \in s\mathcal{G}_1$ and $g \in sH_1^\sigma$ such that Lemma 3.51 holds. In Proposition 3.50 it was proved that

$$\|f\|_{sH_1^*} \le C\|f\|_{sH_1^\sigma} \qquad (f \in sH_1^\sigma).$$

Using this together with Lemma 3.51 and the well-known inequality

$$\|f\|_{sH_1^*} \le \|f\|_{s\mathcal{G}_1} \qquad (f \in s\mathcal{G}_1)$$

we obtain that

$$\|f\|_{sH_1^*} \le \|h\|_{sH_1^*} + \|g\|_{sH_1^*} \le \|h\|_{s\mathcal{G}_1} + C\|g\|_{sH_1^\sigma} \le C\|f\|_{sH_1^S}.$$

The proof of Davis's inequality is complete. ∎

It follows from Theorem 3.49 and from Corollary 2.36 that, for a strong martingale f,

$$\|f^*\|_q \le C_q\|\sigma(f)\|_q + C_q\|\sup_{n\in\mathbb{N}^2}|d_nf|\|_q \qquad (2 \le q < \infty).$$

Some inequalities for previsible strong martingales are going to be proved. During the proofs the one-parameter results will be used.

Theorem 3.53. *For a previsible martingale $f \in V_R$ one has for every $0 < p < \infty$ that*

$$\|f\|_{sH_p^\sigma} \le C_p\|f\|_{sH_p^S} \le C_p\|f\|_{sH_p^*} \le C_p\|f\|_{sH_p^\sigma}$$

where the constant C_p is depending only on the previsibility constant R and on p.

Proof. Since previsibility is inherited from the two-parameter strong martingales to the corresponding one-parameter martingales, the first and the second inequalities follow from Theorem 3.49 and from the one-parameter results (see Theorem 2.22). The third inequality for $0 < p \le 2$ comes from Proposition 3.50.

For $2 \le p < \infty$ it follows from Burkholder-Gundy's inequality (Theorem 3.6) and from Theorem 2.22 that

$$\|f\|_{sH_p^*} \le C_p\|f\|_{sH_p^S} = C_p\|\tilde{f}\|_{\tilde{H}_p^S} \le C_p\|\tilde{f}\|_{\tilde{H}_p^\sigma} = C_p\|f\|_{sH_p^\sigma}$$

which completes the proof. ∎

The inequalities between $\|\cdot\|_{sH_p^*}$ and $\|\cdot\|_{sH_p^S}$ can be found in Brossard [15].

If every martingale is previsible then $\mathcal{F}$ is regular. So, as a consequence of Theorem 3.53, we get immediately the next result.

Corollary 3.54. *If $\mathcal{F}$ is regular then sH_p^σ, sH_p^S and sH_p^* are all equivalent* ($0 < p < \infty$).

We are going to prove some duality theorems. The inequality

$$E[(\sum_{n\in\mathbf{N}^2} E_n^- g_n)^r] \le r^{2r} E[(\sum_{n\in\mathbf{N}^2} g_n)^r] \qquad (1 \le r < \infty) \tag{3.67}$$

follows easily from Theorem 2.10 where $(g_n, n \in \mathbf{N}^2)$ is a sequence of non-negative functions. Indeed, we only have to check whether there exists a maximal inequality with respect to $(\mathcal{F}_n^-)_{n\in\mathbf{N}^2}$. For this maximal inequality see Proposition 3.46.

First we prove a duality theorem for the sL_p spaces.

Theorem 3.55. *For $1 < p < \infty$ the dual of sL_p is sL_q where $1/p + 1/q = 1$.*

Proof. It is easy to show that sL_p is a closed subspace of L_p. Since L_p is reflexive, so is the space sL_p. Therefore we need to prove the theorem for $1 < p \le 2$, only. If $g \in sL_q$ and

$$l_g(f) := E(fg) \qquad (f \in sL_p)$$

then l_g is in the dual of sL_p and

$$\|l_g\| \le \|g\|_q.$$

Reversely, if l is in the dual of sL_p then, by Banach-Hahn's theorem, it can be extended preserving its norm onto the space L_p. Therefore there exists $g \in L_q$ such that $l = l_g$ and

$$\|g\|_q \le \|l\|.$$

Clearly,

$$\begin{aligned} l(f) &= E(fg) \\ &= \sum_{n\in\mathbf{N}^2} E(d_n f d_n g) \\ &= \sum_{n\in\mathbf{N}^2} E[(d_n f)(d_n g - E_n^- d_n g)]. \end{aligned}$$

As

$$g' := \Big(g'_m := \sum_{n\le m} (d_n g - E_n^- d_n g)\Big)_{m\in\mathbf{N}^2}$$

is a strong martingale, by the equivalence between sH_q^S and sL_q we get that

$$\begin{aligned} \|g'\|_q &\le C_q \|(\sum_{n\in\mathbf{N}^2} |d_n g - E_n^- d_n g|^2)^{1/2}\|_q \\ &\le C_q \|(\sum_{n\in\mathbf{N}^2} |d_n g|^2)^{1/2}\|_q + C_q \|(\sum_{n\in\mathbf{N}^2} E_n^- |d_n g|^2)^{1/2}\|_q. \end{aligned}$$

Since $2 \le q < \infty$, we can apply (3.67) to $r = q/2$. Hence, by Burkholder-Gundy's inequality, we can conclude that

$$\begin{aligned} \|g'\|_q &\le C_q \|g\|_q + C_q \|(\sum_{n\in\mathbf{N}^2} |d_n g|^2)^{1/2}\|_q \\ &\le C_q \|g\|_q \le C_q \|l\|. \end{aligned}$$

This completes the proof of Theorem 3.55. ∎

We have proved that the space sH_1^σ can isometrically be embedded in $\tilde{H}_1^s$. However, the argument of the previous proof can not be used to characterize the dual space of sH_1^σ because $\tilde{\mathcal{F}}_j \not\subset \mathcal{F}_{\rho(k)}$ for all $k \geq j$. Though the dual of sH_1^σ is known as a factor space of the dual of $\tilde{H}_1^s$, it is worthy to characterize it directly.

In the definition of the strong BMO and $\Lambda_q(\alpha)$ norms we use again the stopping times relative to $(\mathcal{F}_n^+)$. Let us write $s\Lambda_q(\alpha)$ to denote the set of those strong martingales $f \in sL_q$ for which

$$\|f\|_{s\Lambda_q(\alpha)} = \sup_{\nu \in \mathcal{T}(\mathcal{F}_n^+)} P(\nu \neq \infty)^{-1/q-\alpha} \|f - f^\nu\|_q < \infty \qquad (1 \leq q < \infty, \alpha \geq 0).$$

The $s\Lambda_q(0)$ space will be denoted by $sBMO_q$. Of course,

$$\|f\|_{\Lambda_q(\alpha)} \leq \|f\|_{s\Lambda_q(\alpha)} \qquad (1 \leq q < \infty, \alpha \geq 0).$$

The proof of the next theorem is similar to the one of Theorem 2.24. One has to notice only that the dual of sL_2 is sL_2, itself (see Theorem 3.55).

Theorem 3.56. *The dual space of sH_p^σ is $s\Lambda_2(\alpha)$ $(0 < p \leq 1, \alpha = 1/p - 1)$.*

With the argument used in Theorem 3.55 one can prove the next theorem.

Theorem 3.57. *If $1 < p < \infty$ then the dual of sH_p^σ is sH_q^σ where $1/p + 1/q = 1$.*

Proof. Since the space sH_p^σ can isometrically be embedded into $\tilde{H}_p^s$, it can be considered as a closed subspace of $\tilde{H}_p^s$. By Theorem 2.26 the space $\tilde{H}_p^s$ is reflexive. Hence so is sH_p^σ which means that the theorem is enough to be shown for $1 < p \leq 2$. Let $g \in sH_q^\sigma$ and

$$l_g(f) := E\Big(\sum_{n \in \mathbf{N}^2} d_n f d_n g\Big) = E\Big(\sum_{k=0}^{\infty} d_{\rho(k)} f d_{\rho(k)} g\Big) \qquad (f \in sH_p^\sigma).$$

Then, by Hölder's inequality,

$$|l_g(f)| \leq E\Big[\sum_{n \in \mathbf{N}^2} (E_n^- |d_n f|^2)^{1/2} (E_n^- |d_n g|^2)^{1/2}\Big]$$
$$\leq \|f\|_{sH_p^\sigma} \|g\|_{sH_q^\sigma}.$$

Thus

$$\|l_g\| \leq \|g\|_{sH_q^\sigma}.$$

Reversely, if l is in the dual of $sH_p^\sigma \subset \tilde{H}_p^s$ then, by Theorem 2.26, there exists $\tilde{g} \in \tilde{H}_q^s$ such that

$$l(f) = l(\tilde{f}) = E(\sum_{k=0}^{\infty} d_k \tilde{f} d_k \tilde{g}) \qquad (f \in sH_p^\sigma)$$

and

$$\|\tilde{g}\|_{\tilde{H}_q^s} \leq C_q \|l\|.$$

Since $d_k\tilde{f} = d_{\rho(k)}f$ is $\mathcal{F}_{\rho(k)}$ measurable,

$$l(f) = E\Big[\sum_{k=0}^{\infty}(d_k\tilde{f})E_{\rho(k)}(d_k\tilde{g})\Big].$$

If

$$d_k\tilde{h} := E_{\rho(k)}(d_k\tilde{g})$$

then the corresponding martingale h is a strong one. By Theorem 3.49

$$\begin{aligned}\|h\|_{sH_q^\sigma} = \|\tilde{h}\|_{\tilde{H}_q^s} &= \|(\sum_{k=0}^{\infty}\tilde{E}_{k-1}|E_{\rho(k)}(d_k\tilde{g})|^2)^{1/2}\|_q \\ &\le \|(\sum_{k=0}^{\infty}\tilde{E}_{k-1}E_{\rho(k)}|d_k\tilde{g}|^2)^{1/2}\|_q \\ &= \|(\sum_{k=0}^{\infty}E^-_{\rho(k)}\tilde{E}_{k-1}|d_k\tilde{g}|^2)^{1/2}\|_q.\end{aligned}$$

Applying (3.67) to $q/2 \ge 1$ we obtain

$$\begin{aligned}\|h\|_{sH_q^\sigma} &\le C_q\|(\sum_{k=0}^{\infty}\tilde{E}_{k-1}|d_k\tilde{g}|^2)^{1/2}\|_q \\ &= C_q\|\tilde{g}\|_{\tilde{H}_q^s} \le C_q\|l\|.\end{aligned}$$

The proof of Theorem 3.57 is complete. ∎

The following atomic Hardy spaces will be used to prove the equivalence of the $sBMO_q$ and the $sBMO_q^-$ norms. If we take only strong atoms in Definition 3.24 then we get the definition of the space sH_q^{at} $(1 < q < \infty)$.

Theorem 3.58. *The dual of sH_q^{at} is $sBMO_{q'}$ $(1 < q < \infty, 1/q + 1/q' = 1)$.*

This theorem can be shown in the same way as Theorem 3.25. In the proof we use the fact that the dual of sL_q is $sL_{q'}$ and the inequality

$$\|(\sum_{n=0}^{\infty}\sum_{m=0}^{\infty}|E^-_{k_n,l_m}X_{n,m}|^2)^{1/2}\|_p \le C_p\|(\sum_{n=0}^{\infty}\sum_{m=0}^{\infty}|X_{n,m}|^2)^{1/2}\|_p \qquad (1 < p < \infty)$$

which follows from Theorem 3.21 and Proposition 3.46 where (k_n) and (l_m) are non-decreasing sequences.

The next two results can easily be verified (see e.g. Proposition 3.26).

Proposition 3.59. *For a regular sequence of σ-algebras one has $sH_q^{at} \sim sH_2^{at}$ $(1 < q \le 2)$.*

Corollary 3.60. *If $\mathcal{F}$ is regular then $sBMO_{q'} \sim sBMO_2$ $(2 \le q' < \infty)$.*

Now the BMO^- spaces of strong martingales are introduced. The space $sBMO_q^-$ $(1 \leq q < \infty)$ consists of all strong martingales $f \in sL_q$ for which

$$\|f\|_{sBMO_q^-} = \sup_{\nu \in \mathcal{T}(\mathcal{F}_n^+)} \sup_{m \in \mathbf{N}^2} P(\nu \neq \infty)^{-1/q} \|f - f^\nu + d_m f \chi(m \in \nu)\|_q < \infty.$$

We are going to show that the dual of sH_1^* is $sBMO_2^-$ and that the $sBMO_q^-$ spaces $(2 \leq q < \infty)$ are all equivalent.

Analogously to Definition 2.31 one can define the sBD_q space for strong martingales with the norm

$$\|f\|_{sBD_q} := \| \sup_{n \in \mathbf{N}^2} |d_n f| \|_q \qquad (1 \leq q \leq \infty).$$

An analogue of Theorem 2.32 can be stated for two-parameter strong martingales as well:

Theorem 3.61. *The dual space of $s\mathcal{G}_p$ is sBD_q where $1 \leq p < \infty$ and $1/p + 1/q = 1$.*

During the proof of the duality result mentioned above the next proposition will be used.

Proposition 3.62. *For a strong martingale f we have*

$$\frac{1}{2} \sup\{\|f\|_{sBMO_q}, \|f\|_{sBD_\infty}\} \leq \|f\|_{sBMO_q^-}$$
$$\leq \|f\|_{sBMO_q} + \|f\|_{sBD_\infty} \qquad (1 \leq q < \infty).$$

Thus, as in the one-parameter case, $sBMO_q^- = sBMO_q \cap sBD_\infty$.

Proof. On the one hand,

$$\begin{aligned} \|f\|_{sBMO_q^-} &\leq \sup_{\nu \in \mathcal{T}(\mathcal{F}_n^+)} P(\nu \neq \infty)^{-1/q} \|f - f^\nu\|_q \\ &\quad + \sup_{\nu \in \mathcal{T}(\mathcal{F}_n^+)} \sup_{m \in \mathbf{N}^2} P(\nu \neq \infty)^{-1/q} \|d_m f \chi(m \in \nu)\|_q \\ &\leq \|f\|_{sBMO_q} + \sup_{m \in \mathbf{N}^2} \|d_m f\|_\infty \\ &= \|f\|_{sBMO_q} + \|f\|_{sBD_\infty}. \end{aligned}$$

Since for a fixed stopping time $\nu \in \mathcal{T}(\mathcal{F}_n^+)$ and $m \in \mathbf{N}^2$ the sequence

$$(f_n - f_n^\nu + d_m f \chi(m \in \nu))_{n \geq m}$$

is a strong martingale relative to $(\mathcal{F}_n)_{n \geq m}$, we obtain that

$$\|f_n\|_{sBMO_q^-} \leq \|f\|_{sBMO_q^-} \qquad (n \in \mathbf{N}^2)$$

where $f = (f_n)$. Let ν be a simple stopping time, namely, let $A \in \mathcal{F}_n^+$ and

$$\nu(\omega) = \nu_A(\omega) := \begin{cases} n & \text{if } \omega \in A \\ \infty & \text{if } \omega \notin A. \end{cases}$$

Then

$$\|d_n f\|_\infty = \sup_{A\in\mathcal{F}_n^+} P(A)^{-1/q}\|d_n f\chi(A)\|_q$$
$$= \sup_{A\in\mathcal{F}_n^+} P(\nu_A \neq \infty)^{-1/q}\|f_n - f_n^{\nu_A} + d_n f\chi(n\in\nu_A)\|_q$$
$$\leq \|f_n\|_{sBMO_q^-} \leq \|f\|_{sBMO_q^-}.$$

Note that $f_n - f_n^{\nu_A} = 0$. Consequently,

$$\|f\|_{sBD_\infty} \leq \|f\|_{sBMO_q^-}.$$

On the other hand, the left hand side of the inequality in Proposition 3.62 follows from

$$\|f\|_{sBMO_q} \leq \|f\|_{sBMO_q^-} + \|f\|_{sBD_\infty}.$$

The proof of the proposition is finished. ■

Theorem 3.63. *The dual of sH_1^* is $sBMO_2^-$.*

Proof. Let $\phi \in sBMO_2^-$ and

$$l_\phi(f) := E(f\phi) \qquad (f\in sL_2).$$

Since sL_2 is dense in sH_1^*, the functional l_ϕ is well defined. As $f_n \to f$ in sL_2 norm $(n\to\infty)$, we have again

$$l_\phi(f) = \lim_{n\to\infty} E(f_n\phi) \qquad (f\in sL_2).$$

There exist two strong martingales h and g such that $f_n = h_n + g_n$ for all $n\in\mathbb{N}^2$ and Lemma 3.51 holds. If $f\in sL_2$ then the functions h_n and g_n are finite sums of square integrable differences, so they are in sL_2, too. Hence

$$|E(f_n\phi)| \leq |E(g_n\phi)| + |E(h_n\phi)|.$$

By Proposition 3.62 we have $\phi\in sBMO_2$ and $\phi\in sBD_\infty$. Applying Theorems 3.56 and 3.61 we obtain

$$|E(f_n\phi)| \leq C\|g_n\|_{sH_1^\sigma}\|\phi\|_{sBMO_2} + \|h_n\|_{s\mathcal{G}_1}\|\phi\|_{sBD_\infty}.$$

From Lemma 3.51 and Proposition 3.62 one concludes that

$$|E(f\phi)| \leq C\|f\|_{sH_1^*}\|\phi\|_{sBMO_2^-}.$$

Thus l_ϕ is really a bounded linear functional on sH_1^*.

Conversely, assume that l is an arbitrary bounded linear functional on sH_1^*. Since $\|f\|_{sH_1^*} \leq 4\|f\|_{sL_2}$, the functional l is also bounded on sL_2. Consequently, there exists $\phi\in sL_2$ such that

$$l(f) = l_\phi(f) = E(f\phi) \qquad (f\in sL_2).$$

On the other hand,

$$\|f\|_{sH_1^*} \leq C\|f\|_{sH_1^\sigma}$$

(see Proposition 3.50 (i)) and

$$\|f\|_{sH_1^*} \leq \|f\|_{s\mathcal{G}_1}.$$

Henceforth l is bounded on sH_1^σ and on $s\mathcal{G}_1$ as well. Therefore Theorems 3.56 and 3.61 imply that

$$\|\phi\|_{sBMO_2} \leq \|l\|$$

and

$$\|\phi\|_{sBD_\infty} \leq 2\|l\|.$$

It follows from Proposition 3.62 that

$$\|\phi\|_{sBMO_2^-} \leq 3\|l\|$$

which completes the proof of the theorem. ∎

John-Nirenberg's equivalence theorem for two-parameter strong martingales is going to be proved.

Theorem 3.64. *The $sBMO_q^-$ spaces are all equivalent if $2 \leq q < \infty$.*

Proof. The inequality

$$\|\phi\|_{sBMO_2^-} \leq \|\phi\|_{sBMO_q^-} \qquad (2 \leq q < \infty)$$

comes from Hölder's inequality.

By the often used atomic decomposition method we can prove that

$$\|f\|_{sH_1^*} \leq \|f\|_{sH_p^{at}} \qquad (1 < p \leq \infty).$$

Therefore, if $l = l_\phi$ is a bounded linear functional on sH_1^* then it is bounded on sH_p^{at} $(1 < p \leq \infty)$ as well. Theorem 3.58 implies that $\phi \in sBMO_q$ $(1/p + 1/q = 1)$ and

$$\|\phi\|_{sBMO_q} \leq C_q\|l\|.$$

On the other hand, by Theorem 3.63,

$$\|l\| \leq C\|\phi\|_{sBMO_2^-}.$$

Thus

$$\|\phi\|_{sBMO_q} \leq C_q\|\phi\|_{sBMO_2^-} \qquad (1 < q < \infty).$$

Notice that in Theorem 3.63 it is proved that $\phi \in sBD_\infty$ and

$$\|\phi\|_{sBD_\infty} \leq 2\|l\| \leq C\|\phi\|_{sBMO_2^-}.$$

Applying Proposition 3.62 we have

$$\|\phi\|_{sBMO_q^-} \leq C_q\|\phi\|_{sBMO_2^-} \qquad (1 \leq q < \infty).$$

The proof of the theorem is complete. ∎

Since $sL_p \subset sH_1^* \subset sL_1$ $(1 < p < \infty)$, from Theorems 3.63 and 3.64 it follows that

$$sL_\infty \subset sBMO_p^- \subset sL_q \qquad (2 \leq p < \infty, 1 \leq q < \infty).$$

Moreover,

$$\begin{aligned} \|f\|_{sBMO_2} &= \sup_{\nu \in \mathcal{T}(\mathcal{F}_n^+)} P(\nu \neq \infty)^{-1/2} \|f - f^\nu\|_2 \\ &= \sup_{\nu \in \mathcal{T}(\mathcal{F}_n^+)} P(\nu \neq \infty)^{-1/2} \|\sigma(f - f^\nu)\|_2 \\ &\leq \sup_{\nu \in \mathcal{T}(\mathcal{F}_n^+)} P(\nu \neq \infty)^{-1/2} \|\sigma(f)\|_2 \leq \|\sigma(f)\|_\infty. \end{aligned}$$

It was verified in this section that the dual of sH_1^σ is $sBMO_2$ and the dual of sH_p^s is sH_q^s $(1 < p < \infty, 1/p + 1/q = 1)$. Obviously, $sH_p^\sigma \subset sH_1^\sigma$ $(1 < p < \infty)$. Hence

$$sH_\infty^\sigma \subset sBMO_2 \subset sH_q^\sigma \qquad (1 \leq q < \infty).$$

By Proposition 3.62 the following relation holds as well:

$$sL_\infty \subset sBMO_q^- \subset sBMO_q \subset sL_q \qquad (2 \leq q < \infty).$$

We proved in Corollary 3.54 that sH_1^* is equivalent to sH_1^σ in case $\mathcal{F}$ is regular. Henceforth, in this case, $sBMO_2$ is also equivalent to $sBMO_2^-$. From Corollary 3.60 and Theorem 3.64 we obtain the analogue of Corollary 2.51.

Corollary 3.65. *If $\mathcal{F}$ is regular then the spaces $sBMO_q$ and $sBMO_p^-$ are all equivalent for every $2 \leq p, q < \infty$.*

With the following theorem the completeness of this section is intended to be served. Denote by sL the set of the strong martingales in L and let $sVMO_p$ be the closure of sL in the $sBMO_p$ norm.

Theorem 3.66. *If every σ-algebra $\mathcal{F}_n$ is generated by finitely many atoms then the dual of $sVMO_2$ is sH_1^σ.*

CHAPTER 4

TREE MARTINGALES

After the study of the one- and two-parameter martingales a discussion on martingales having a so-called tree stochastic basis is coming. A few previous results are going to be generalized for tree martingales. In Section 4.1 some inequalities relative to tree martingales are verified. A maximal inequality as well as Burkholder-Gundy's inequality for $2 < p < \infty$ are proved. For a regular stochastic basis the latter inequality is extended to all $1 < p < \infty$. Moreover, it will be shown that the L_p norm of the conditional quadratic variation and of the maximal function of a martingale transform can be estimated by the L_p norm of the martingale in case the stochastic basis is regular. The partial sums of the Vilenkin-Fourier series of an integrable function can be majorized by the maximal function of a suitable martingale transform. As a consequence, we obtain in Section 4.2 that the one-parameter Vilenkin-Fourier series of a function $f \in L_p$ $(1 < p < \infty)$ converges a.e. to the function f whenever the Vilenkin system is bounded. This result for the Walsh system was proved by Billard [12] for $p = 2$ and by Sjölin [174] for other parameter p while for bounded Vilenkin systems by Gosselin [84] (see also Schipp [164]; $p = 2$) and, moreover, for the trigonometric Fourier series by Carleson [33] $(p = 2)$ and by Hunt [96] $(1 < p < \infty)$. The idea of the proof presented below is due to Schipp [166]. In his paper this result was proved for the Walsh system. In case the Vilenkin system is unbounded the same convergence result is obtained for lacunary functions. Finally, we verify that, for an arbitrary Vilenkin system and for $f \in L_p$ $(1 < p < \infty)$, the Vilenkin-Fourier series of f converges in L_p norm to itself. This result was shown by Schipp [163] and by Young [203]. For a tree stochastic basis duality theorems between atomic Hardy, BMO and VMO spaces can be found in the paper Weisz [189].

4.1. INEQUALITIES

Let $\mathbf{T}$ be a countable, upward directed index set with respect to the partial ordering $\leq$ satisfying the following two conditions: for every $t \in \mathbf{T}$

$$\mathbf{T}^t := \{u \in \mathbf{T} : u \leq t\}$$

is finite and the set

$$\mathbf{T}_t := \{u \in \mathbf{T} : t \leq u\}$$

is linearly ordered.

Thus $\mathbf{T}$ is a tree and every non-empty subset of $\mathbf{T}$ has at least one minimum element. Denote by $\mathbf{T}_0$ the set of the minimum elements of $\mathbf{T}$.

Let us fix a non-decreasing sequence $\mathcal{F} = (\mathcal{F}_t, t \in \mathbf{T})$ of sub-σ-algebras of $\mathcal{A}$ with respect to the partial ordering such that $(\mathcal{F}_t, t \in \mathbf{T})$ can be ordered linearly and

$$\mathcal{A} = \sigma(\bigcup_{t\in\mathbf{T}} \mathcal{F}_t).$$

Denote again by E_t the conditional expectation operator with respect to $\mathcal{F}_t$.

In the tree case it is more useful to work with projections instead of the conditional expectation operators. We consider the *projections*

$$P_t f := \phi_t E_t(f\overline{\phi}_t) \qquad (f \in L_1, t \in \mathbf{T}) \tag{4.1}$$

where $|\phi_t| = 1$ for every $t \in \mathbf{T}$. P_t is a projection, indeed, since $\|P_t\| \leq 1$ and $P_t \circ P_t = P_t$ $(t \in \mathbf{T})$. It is obvious that the conditional expectation operators are projections of the form (4.1) with $\phi = 1$.

Definition 4.1. *The sequence $(\mathcal{F}_t, P_t; t \in \mathbf{T})$ is called a tree basis if for the projections defined in (4.1) we have*
(i) $P_t f = \phi_u E_t(f\overline{\phi}_u)$ for every $u \leq t$ and $f \in L_1$ $(t \in \mathbf{T})$
(ii) $P_t P_u = 0$ for each incomparable u and t from $\mathbf{T}$.

Note that the equality $P_t P_u = P_u P_t = P_u$ for $u \leq t$ follows from (i). We say that a sequence $(f_t, t \in \mathbf{T})$ of integrable functions is a *tree martingale* if $u \leq t$ implies $P_u f_t = f_u$ $(u, t \in \mathbf{T})$. Of course, the sequence $(P_t f, t \in \mathbf{T})$ is a tree martingale for all $f \in L_1$. Suppose that there is a distinguished minimal element $v_0 \in \mathbf{T}_0$ for which $f_{v_0} = 0$ for all tree martingales.

The succeeding element of $t \in \mathbf{T}$, namely, the minimum element of $\mathbf{T}_t \setminus \{t\}$ is denoted by t^+. For simplicity we suppose that if $t = u^+ = r^+$ then $\mathcal{F}_u = \mathcal{F}_r$. This common σ-algebra will be denoted by $\mathcal{F}_t^-$. If t is a minimal element of $\mathbf{T}$ then set $\mathcal{F}_t^- = \mathcal{F}_t$.

The *martingale difference sequence* is going to be defined as in the one-parameter case. Let

$$d_t f := f_{t^+} - f_t \qquad (t \in \mathbf{T}).$$

Using the notations

$$S_t(f) := \Big(\sum_{u\in\mathbf{T}_t} |d_u f|^2\Big)^{1/2}, \qquad s_t(f) := \Big(\sum_{u\in\mathbf{T}_t} E_u |d_u f|^2\Big)^{1/2} \qquad (t \in \mathbf{T})$$

we define the *quadratic variation*, the *conditional quadratic variation* and the *maximal function* for tree martingales by

$$S(f) := \sup_{t\in\mathbf{T}} S_t(f), \qquad s(f) := \sup_{t\in\mathbf{T}} s_t(f)$$

and

$$f^* := \sup_{t\in\mathbf{T}} |f_t|,$$

respectively.

In order to be able to prove the a.e. convergence of a Vilenkin-Fourier series we introduce the following tree martingale transforms.

Definition 4.2. *Suppose that the sequence $T = (T^t, t \in \mathbf{T})$ of linear operators satisfies the following conditions for all $f \in L_1$ and $t \in \mathbf{T}$:*

(i) $P_{t+}(T^t f) = T^t f$

(ii) $P_t(T^t f) = 0$

(iii) for every $\mathcal{F}_t$ measurable function ξ one has $T^t(\xi f) = \xi T^t f$

(iv) $|T^t f|^2 \le R^2 E_t |f|^2$ where the constant R is independent of t and of f.

The maximal function of a tree martingale transform is defined by

$$T_t^* f := \sup_{r \in \mathbf{T}_t} \Big| \sum_{t \le u < r} T^u(d_u f) \Big|, \qquad T^* f := \sup_{t \in \mathbf{T}} T_t^* f.$$

Because of the fact that a tree martingale transform can not be defined as a martingale, the stopped martingale can not be introduced unfortunately. These martingale transforms are more general than the ones investigated in Sections 2.5 and 2.6. However, working with them does not need new ideas. In the linear case one can see that T^* is of type (L_p, L_p) for $2 \le p < \infty$ (cf. the proof of Lemma 4.9).

Martingales with respect to a linearly ordered stochastic basis are special tree martingales. In this case the functions f^*, $S(f)$, $s(f)$ and $T^* f$ correspond to the one-parameter maximal function, the quadratic variation, the conditional quadratic variation and the maximal function of the martingale transform, respectively. It is important to remark that for every $t \in \mathbf{T}$ the sequence

$$(\overline{\phi}_t f_u, u \in \mathbf{T}_t) \tag{4.2}$$

is a linear martingale with respect to $(\mathcal{F}_u, u \in \mathbf{T}_t)$. Indeed, $\overline{\phi}_t f_u = E_u(f_u \overline{\phi}_t)$, so it is $\mathcal{F}_u$ measurable and $E_r(\overline{\phi}_t f_u) = E_r(f_u \overline{\phi}_t) = \overline{\phi}_t f_r$ for all $t \le r \le u$. In particular, $S_t(f)$ resp. $s_t(f)$ is the quadratic variation resp. the conditional quadratic variation of the linear martingale (4.2) and $T_t^* f$ is the maximal function of its martingale transform.

For tree martingales we are going to introduce the quasi-norm $\|\cdot\|_{\mathbf{M}^{p,q}}$. Let $f = (f_t, t \in \mathbf{T})$ be a sequence of $\mathcal{A}$ measurable functions defined on Ω and indexed by a tree. Introducing the map

$$\nu_y^f := \inf\{t \in \mathbf{T} : |f_t| > y\}$$

for every $y > 0$ we have

$$\{t \in \nu_y^f\} = \{\omega \in \Omega : |f_t(\omega)| > y, |f_u(\omega)| \le y \ (\forall u < t)\}$$

where $u < t$ means that $u \le t$ but $u \ne t$. For $0 < p, q < \infty$ set

$$\|f\|_{\mathbf{M}^{pq}} := \sup_{y > 0} y \Big(\int_\Omega \big(\sum_{t \in \mathbf{T}} \chi(t \in \nu_y^f) \big)^{p/q} dP \Big)^{1/p}. \tag{4.3}$$

Denote by $\mathbf{M}^{pq}$ the set of such sequences f which satisfy $\|f\|_{\mathbf{M}^{pq}} < \infty$. Notice that, for each fixed sequence f, the function $q \mapsto \|f\|_{\mathbf{M}^{pq}}$ decreases, $p \mapsto \|f\|_{\mathbf{M}^{pq}}$ increases and, consequently, the limit does exist as $q \to \infty$ satisfying

$$\|f\|_{\mathbf{M}^{p\infty}} := \lim_{q \to \infty} \|f\|_{\mathbf{M}^{pq}} = \sup_{y > 0} y P(f^* > y)^{1/p} \qquad (0 < p < \infty).$$

Note that the lattest expression is the weak L_p norm of f^* (for the exact definition see the next chapter). To distinguish the spaces of this chapter from the ones of the next chapter we write here the double indexes up and there down.

First of all it is checked that $\|\cdot\|_{\mathrm{M}^{pq}}$ is really a quasi-norm.

Proposition 4.3. *The map $f \mapsto \|f\|_{\mathrm{M}^{pq}}$ is a quasi-norm, namely, for any two sequences $f = (f_t, t \in \mathbf{T})$ and $g = (g_t, t \in \mathbf{T})$ of functions and for any $\lambda \in \mathbf{C}$*

$$\text{(4.4)} \qquad \begin{aligned} &\text{(i)} \quad \|\lambda f\|_{\mathrm{M}^{pq}} = |\lambda| \|f\|_{\mathrm{M}^{pq}}, \\ &\text{(ii)} \quad \|f+g\|_{\mathrm{M}^{pq}} \leq \kappa_{pq} (\|f\|_{\mathrm{M}^{pq}} + \|g\|_{\mathrm{M}^{pq}}) \end{aligned}$$

where $0 < p < \infty$, $0 < q \leq \infty$ and κ_{pq} depends only on p and q. Moreover, the map $f \mapsto \|f\|_{\mathrm{M}^{pq}}$ is non-decreasing in the following sense: if $|f_t| \leq |g_t|$ for all $t \in \mathbf{T}$ then

$$\|f\|_{\mathrm{M}^{pq}} \leq \|g\|_{\mathrm{M}^{pq}} \qquad (0 < p < \infty, 0 < q \leq \infty).$$

Proof. To prove the monotony of the map in question, fix $y > 0$ and $\omega \in \Omega$. It is easy to see that

$$\text{(4.5)} \qquad \sum_{t \in \mathbf{T}} \chi(t \in \nu_y^f)(\omega) = |\nu_y^f(\omega)|$$

where $|\cdot|$ denotes cardinality. Since

$$\{t \in \mathbf{T} : |f_t(\omega)| > y\} \subset \{t \in \mathbf{T} : |g_t(\omega)| > y\},$$

it is clear that for any $t \in \nu_y^f(\omega)$ there exists at least one $t' \in \nu_y^g$ such that $t' \leq t$. If $t, u \in \nu_y^f$ and $t \neq u$ then t and u are incomparable. This and the fact that $\mathbf{T}_{t'}$ is linearly ordered imply $t' \neq u'$. Consequently, $|\nu_y^f(\omega)| \leq |\nu_y^g(\omega)|$ and monotony follow from (4.3) and (4.5).

The property (4.4) (i) is an immediate consequence of the definition. To prove (4.4) (ii) set

$$h_t := \max\{|f_t|, |g_t|\} \qquad (t \in \mathbf{T}).$$

Then $|f_t + g_t| \leq 2h_t$ $(t \in \mathbf{T})$ and by monotony and (4.4) (i) we get

$$\|f+g\|_{\mathrm{M}^{pq}} \leq 2\|h\|_{\mathrm{M}^{pq}}.$$

As

$$\nu_y^h = (\{|f_t| > y\} \cup \{|g_t| > y\}) \bigcap_{u<t} (\{|f_u| \leq y\} \cap \{|g_u| \leq y\}) \subset \nu_y^f \cup \nu_y^g,$$

the rest of the proof follows directly from (4.3).

Note that $\kappa_{pq} = 2$ if $1 \leq p, q$. ∎

Since for an integrable function f

$$|P_t f| \leq E_t |f|$$

and $(\mathcal{F}_t, t \in \mathbf{T})$ is linearly oredered, by the one-parameter result we get the following

Proposition 4.4. *For a function $f \in L_p$ we have*

$$\|f\|_p \le \|f^*\|_p \le \frac{p}{p-1}\|f\|_p \qquad (p>1).$$

Notice that if $\mathbf{T}$ is linearly ordered then the sets on the right hand side in (4.3) are pairwise disjoint and

$$\|f\|_{\mathbf{M}^{pq}} = \sup_{y>0} yP(f^* > y)^{1/p}$$

for any $0 < p < \infty$ and $0 < q \le \infty$. Thus, in the linear case, by using the $\mathbf{M}^{pq}$ quasi-norms, the maximal inequalities (2.14) and (2.15) can be reformulated as follows:

$$\|(f_t, t \in \mathbf{T})\|_{\mathbf{M}^{pq}} \le \|f\|_p \qquad (1 \le p < \infty, 0 < q \le \infty).$$

This form of the maximal inequality can be transfered to the non-linear case. Let Δ denote the closure of the triangle in $\mathbf{R}^2$ with vertices $(0,0)$, $(1/2,1/2)$ and $(1,0)$ except the points $(x, 1-x)$, $0 \le x < 1/2$.

Theorem 4.5. *Let $f = (f_t, t \in \mathbf{T})$ be a tree martingale and suppose that $1 < p, q < \infty$ satisfy $(1/p, 1/q) \in \Delta$. Then, depending only on p and q, there exists a constant $C_{p,q}$ such that*

$$\|(f_t, t \in \mathbf{T})\|_{\mathbf{M}^{pq}} \le C_{p,q}\|f\|_p$$

for all $f \in L_p$.

This maximal theorem is a special case of the next statement. Let $\delta : \mathbf{T} \longrightarrow \mathbf{T}$ be a map such that $\delta(t) \le t$ and $\mathcal{F}_{\delta(u)} \subset \mathcal{F}_{\delta(t)}$ for each $u \le t$. Set

$$f_t^\delta := E_{\delta(t)}|f_t| \qquad (t \in \mathbf{T}).$$

Since $|f_t|$ is $\mathcal{F}_t$ measurable for all $t \in \mathbf{T}$, in the special case $\delta(t) = t$ we derive Theorem 4.5 from the next result.

Theorem 4.6. *Let $f = (f_t, t \in \mathbf{T})$ be a tree martingale and suppose that $1 < p, q < \infty$ satisfy $(1/p, 1/q) \in \Delta$. Then, depending only on p and q, there exists a constant $C_{p,q}$ such that*

$$\|(f_t^\delta, t \in \mathbf{T})\|_{\mathbf{M}^{pq}} \le C_{p,q}\|f\|_p \tag{4.6}$$

for all $f \in L^p$.

Proof. For fixed $y > 0$ and $f \in L_p$ let

$$\alpha_t := \chi(t \in \nu_y^{f^\delta}) = \chi(f_t^\delta > y, f_u^\delta \le y \ (\forall u < t)) \qquad (t \in \mathbf{T}).$$

Then α_t is $\mathcal{F}_{\delta(t)}$ measurable and, obviously,

$$\alpha_t \alpha_u = 0 \quad \text{if} \quad t < u \quad \text{or} \quad u < t.$$

For all $t \in \mathbf{T}$ introduce the projections $F_t := \alpha_t P_t$ and observe that by the martingale property we have $P_t \circ P_u = 0$ for every incomparable t and u. Therefore, we get for every $g \in L_1$ and $t, u \in \mathbf{T}$ that

$$F_t(F_u g) = \alpha_t P_t(P_u(\alpha_u g)) = P_t(\alpha_t \alpha_u P_u g) = \delta_{t,u} F_t g$$

where $\delta_{t,u}$ is the Kronecker symbol. Thus the projections F_t are orthogonal and Bessel's inequality implies for any $g \in L_2$ that

$$\|(F_t g, t \in \mathbf{T})\|^2_{L_2(l_2)} = \sum_{t \in \mathbf{T}} \|F_t g\|_2^2 \le \|g\|_2^2. \tag{4.7}$$

To prove (4.6) introduce the operators

$$G_t g := E_{\delta(t)}(\eta_t F_t g) \qquad (g \in L_1, t \in \mathbf{T})$$

where $(\eta_t, t \in \mathbf{T})$ is a fixed sequence of functions satisfying $\|\eta_t\|_\infty \le 1$ for each $t \in \mathbf{T}$. By (4.7),

$$\begin{aligned} \|(G_t g, t \in \mathbf{T})\|^2_{L_2(l_2)} &\le E(\sum_{t \in \mathbf{T}} E_{\delta(t)} |F_t g|^2) \\ &= E(\sum_{t \in \mathbf{T}} |F_t g|^2) \le \|g\|_2^2. \end{aligned}$$

Furthermore, by Doob's inequality,

$$\|(G_t g, t \in \mathbf{T})\|_{L_s(l_\infty)} \le \|\sup_{t \in \mathbf{T}} E_{\delta(t)} |g|\|_s \le \frac{s}{s-1} \|g\|_s$$

for any $1 < s \le \infty$ and $g \in L_s$. It follows by interpolation (see Lemma 3.22) that

$$\|(G_t g, t \in \mathbf{T})\|_{L_p(l_q)} \le C_{p,q} \|g\|_p \qquad (g \in L_p)$$

where $1/p = (1-t)/2 + t/s$ and $1/q = (1-t)/2$ for any $0 \le t \le 1$. Setting $g := f$ and $\eta_t := \operatorname{sign} P_t f$ we have

$$\Big(\int_\Omega [\sum_{t \in \mathbf{T}} (\alpha_t E_{\delta(t)} |f_t|)^q]^{p/q} \, dP\Big)^{1/p} \le C_{p,q} \|f\|_p.$$

Using the fact that

$$\alpha_t E_{\delta(t)} |f_t| = \alpha_t f_t^\delta > y \alpha_t$$

we can see that (4.6) holds. ∎

The predictable martingales can be defined in the tree case, too, as in Section 1.1. A martingale $f = (f_t, t \in \mathbf{T})$ is said to be *predictable* by a sequence $\lambda = (\lambda_t, t \in \mathbf{T})$ if λ is increasing and each λ_t is $\mathcal{F}_t^-$ measurable with

$$|f_t| \le \lambda_t \qquad (t \in \mathbf{T}).$$

Denote by $\mathbf{P}^{pq}$ $(0 < p < \infty, 0 < q \le \infty)$ the collection of such sequences f for which $\lambda \in \mathbf{M}^{pq}$ and set

$$\|f\|_{\mathbf{P}^{pq}} := \inf \|\lambda\|_{\mathbf{M}^{pq}}$$

where the infimum is taken over all predictions $\lambda \in \mathrm{M}^{pq}$ belonging to f.

If f is a tree martingale then the $\mathrm{M}^{p\infty}$ quasi-norm of T^*f, $S(f)$ and $s(f)$ can be estimated by the M^{pq} quasi-norm of λ. A version of the next result can be found in Schipp [166] for the dyadic tree (see Section 4.2).

Theorem 4.7. *If $f = (f_t, t \in \mathrm{T})$ is a tree martingale, $1 \le p$ and $2, p \le q < \infty$ then*

$$\|(T_t^* f, t \in \mathrm{T})\|_{\mathrm{M}^{p\infty}} \le C_{p,q} \|f\|_{\mathrm{P}^{pq}}, \tag{4.8}$$

$$\|(S_t(f), t \in \mathrm{T})\|_{\mathrm{M}^{p\infty}} \le C_{p,q} \|f\|_{\mathrm{P}^{pq}} \tag{4.9}$$

and

$$\|(s_t(f), t \in \mathrm{T})\|_{\mathrm{M}^{p\infty}} \le C_{p,q} \|f\|_{\mathrm{P}^{pq}} \tag{4.10}$$

where $C_{p,q}$ depends only on p and q.

During the proof of Theorem 4.7 the following lemmas and notations will be used. Assume that $\epsilon = (\epsilon_t, t \in \mathrm{T})$ is a sequence of functions such that ϵ_t is $\mathcal{F}_t$ measurable. Set

$$T_{\epsilon;t,u} f := \sum_{t \le r < u} \epsilon_r T^r(d_r f),$$

$$S_{\epsilon;t,u}(f) := \Big(\sum_{t \le r < u} |\epsilon_r d_r f|^2\Big)^{1/2}$$

and

$$s_{\epsilon;t,u}(f) := \Big(\sum_{t \le r < u} E_r |\epsilon_r d_r f|^2\Big)^{1/2}.$$

The operator $T_{\epsilon;t}^* f$ resp. $S_{\epsilon;t}(f)$ resp. $s_{\epsilon;t}(f)$ is defined by the supremum of $|T_{\epsilon;t,u} f|$ resp. $S_{\epsilon;t,u}(f)$ resp. $s_{\epsilon;t,u}(f)$ taken over all $u \ge t$ and, moreover, the one $T_\epsilon^* f$ resp. $S_\epsilon(f)$ resp. $s_\epsilon(f)$ is defined by the supremum of $T_{\epsilon;t}^* f$ resp. $S_{\epsilon;t}(f)$ resp. $s_{\epsilon;t}(f)$ taken over all $t \in \mathrm{T}$.

Lemma 4.8. *Suppose that $\lambda = (\lambda_v, v \in \mathrm{T})$ is a prediction of the tree martingale $f = (f_v, v \in \mathrm{T})$ and*

$$\epsilon_v := \chi(x < \lambda_{v+} \le 2x)$$

for any real number $x > 0$. Then

$$T_\epsilon^* f \le 2 \sup_{v \in \mathrm{T}} \alpha_v T_{\epsilon;v}^* f + 4Rx\chi(\lambda^* > x), \tag{4.11}$$

$$S_\epsilon(f) \le \sup_{v \in \mathrm{T}} \alpha_v S_{\epsilon;v}(f) + 4x\chi(\lambda^* > x) \tag{4.12}$$

and

$$s_\epsilon(f) \le \sup_{v \in \mathrm{T}} \alpha_v s_{\epsilon;v}(f) + 4x\chi(\lambda^* > x) \tag{4.13}$$

where

$$\alpha_v := \chi\big(\lambda_v > x, \lambda_u \le x \ (\forall u < v)\big) \qquad (v \in \mathrm{T})$$

and $\lambda^* = \sup_{v\in\mathbf{T}} \lambda_v$ *is the maximal function of* λ.

Proof. Let us fix $t < u$ in $\mathbf{T}$, ω in Ω and set

$$\tau_v := \chi(\lambda_v > x) \qquad (v \in \mathbf{T}).$$

Therefore $\epsilon_v = \tau_{v^+}\epsilon_v$. Consequently, if the set

$$\{v \in \mathbf{T}_t : v < u, \tau_{v^+}(\omega) = 1\}$$

is empty then $T_{\epsilon;t,u}f(\omega) = 0$ or else let t_1 be its minimum element. Moreover, denote by t_0 a minimum element of the set

$$\{v \in \mathbf{T} : v \leq t_1^+, \tau_v(\omega) = 1\}.$$

Thus $\alpha_{t_0}(\omega) = 1$ and $T_{\epsilon;t,u}$ can be written in the form

$$\begin{aligned} T_{\epsilon;t,u}f(\omega) &= T_{\epsilon;t_1,u}f(\omega) = \epsilon_{t_1} T^{t_1}(d_{t_1}f)(\omega) + T_{\epsilon;t_1^+,u}f(\omega) \\ &= T^{t_1}(\epsilon_{t_1} d_{t_1} f)(\omega) + \alpha_{t_0}(\omega)\big(T_{\epsilon;t_0,u}f(\omega) - T_{\epsilon;t_0,t_1^+}f(\omega)\big). \end{aligned}$$

By predictability and by (iv) of Definition 4.2 we have

$$\begin{aligned} |T^{t_1}(\epsilon_{t_1} d_{t_1} f)| &\leq R[E_{t_1}(\epsilon_{t_1}|d_{t_1}f|^2)]^{1/2} \\ &\leq R[E_{t_1}(4\epsilon_{t_1}\lambda^2_{t_1^+})]^{1/2} \leq 4Rx\chi(\lambda^* > x). \end{aligned}$$

On the other hand,

$$|T_{\epsilon;t_0,u}f(\omega) - T_{\epsilon;t_0,t_1^+}f(\omega)| \leq 2T^*_{\epsilon;t_0}f(\omega).$$

Taking the supremum over all $u \in \mathbf{T}_t$ and $t \in \mathbf{T}$ we get (4.11). The inequalities (4.12) and (4.13) can be shown in the same way. ■

The one-parameter martingale inequalities can be in use during the estimation of the maximal function Y where Y denotes one of the functions

$$\sup_{v\in\mathbf{T}} \alpha_v T^*_{\epsilon;v} f, \qquad \sup_{v\in\mathbf{T}} \alpha_v S_{\epsilon;v}(f), \qquad \sup_{v\in\mathbf{T}} \alpha_v s_{\epsilon;v}(f).$$

Lemma 4.9. *If the conditions of Lemma 4.8 are satisfied then for* $2, p \leq q < \infty$, $1 \leq p$ *and* $z > 0$

$$P(Y > zx) \leq C_{p,q} z^{-q} x^{-p} \|\lambda\|^p_{\mathbf{M}^{pq}}. \tag{4.14}$$

Proof. Suppose that

$$Y = \sup_{v\in\mathbf{T}} \alpha_v T^*_{\epsilon;v} f.$$

By Definition 4.2 we have

$$T^*_{\epsilon;v}(\xi f) = \xi T^*_{\epsilon;v} f \tag{4.15}$$

for any non-negative $\mathcal{F}_v$ measurable function ξ. By (i) of Definitions 4.1 and 4.2 one can see that

$$T^u(d_u f) = P_{u^+}(T^u(d_u f)) = \phi_v E_{u^+}[T^u(d_u f)\overline{\phi}_v] \qquad (u \geq v),$$

hence $\overline{\phi}_v T^u(d_u f)$ is $\mathcal{F}_{u^+}$ measurable. On the other hand, by (ii) of Definition 4.2 we have

$$P_u(T^u(d_u f)) = \phi_v E_u[T^u(d_u f)\overline{\phi}_v] = 0 \qquad (u \geq v),$$

consequently, $\left(T^u(d_u f)\overline{\phi}_v\right)_{u\geq v}$ is a one-parameter martingale difference sequence relative to $(\mathcal{F}_{u^+})_{u\geq v}$. So the one-parameter martingale inequalities can be applied. By $|\phi_v| = 1$, one has for each $v \in \mathbf{T}$

$$T^*_{\epsilon;v} f := \sup_{r\geq v} \Big| \sum_{v\leq u<r} \epsilon_u T^u(d_u f)\overline{\phi}_v \Big|$$

Using (iv) of Definition 4.2 together with Theorem 2.12 and (4.15) we obtain

$$\begin{aligned} E_v[(T^*_{\epsilon;v} f)^{p_0}] &\leq C_{p_0} E_v\left[\Big(\sum_{u\geq v} \epsilon_u |T^u(d_u f)\overline{\phi}_v|^2\Big)^{p_0/2}\right] \\ &\leq C_{p_0} E_v\left[\Big(\sum_{u\geq v} \epsilon_u E_u|d_u f|^2\Big)^{p_0/2}\right] \end{aligned}$$

for $p_0 \geq 1$. Moreover, for $p_0 \geq 2$, the convexity theorem (Theorem 2.10) implies

$$\begin{aligned} E_v[(T^*_{\epsilon;v} f)^{p_0}] &\leq C_{p_0} E_v\left[\Big(\sum_{u\geq v} \epsilon_u |d_u f|^2\Big)^{p_0/2}\right] \\ &= C_{p_0} E_v\left[\Big(\sum_{u\geq v} \epsilon_u |E_{u^+}(f\overline{\phi}_v) - E_u(f\overline{\phi}_v)|^2\Big)^{p_0/2}\right]. \end{aligned}$$

Applying Burkholder-Gundy's inequality (see Theorem 2.12) to the martingale $\left(E_u(f\overline{\phi}_v)\right)_{u\geq v}$ one can establish that

$$\begin{aligned} E_v[(T^*_{\epsilon;v} f)^{p_0}] &\leq C_{p_0} E_v\left[\Big|\sum_{u\geq v} \epsilon_u (E_{u^+}(f\overline{\phi}_v) - E_u(f\overline{\phi}_v))\Big|^{p_0}\right] \\ &= C_{p_0} E_v\left[\Big|\sum_{u\geq v} \epsilon_u (f_{u^+} - f_u)\Big|^{p_0}\right]. \end{aligned}$$

It follows from the definition of ϵ_u that

$$E_v[(T^*_{\epsilon;v} f)^{p_0}] \leq C_{p_0} x^{p_0}.$$

By Tsebisev's inequality and the concavity theorem (see Theorem 2.10), for $p_0 \geq q \geq 2$, one can see that

$$\begin{aligned} P(Y > zx) &\leq (zx)^{-q} E\Big[(\sup_{v\in\mathbf{T}} \alpha_v (T^*_{\epsilon;v} f)^q\Big] \\ &\leq (zx)^{-q} E\Big[\Big(\sum_{v\in\mathbf{T}} \alpha_v (T^*_{\epsilon;v} f)^{p_0}\Big)^{q/p_0}\Big] \\ &\leq C_q (zx)^{-q} E\Big[\Big(\sum_{v\in\mathbf{T}} \alpha_v E_v[(T^*_{\epsilon;v} f)^{p_0}]\Big)^{q/p_0}\Big] \\ &\leq C_q z^{-q} E\Big[\Big(\sum_{v\in\mathbf{T}} \alpha_v\Big)^{q/p_0}\Big]. \end{aligned}$$

Set $p_0 := q^2/p \geq q \geq 2$ and observe that

$$\begin{aligned} E\Big[\Big(\sum_{v\in\mathbf{T}} \alpha_v\Big)^{q/p_0}\Big] &= E\Big[\Big(\sum_{v\in\mathbf{T}} \chi(\lambda_v > x, \lambda_u \leq x \ (\forall u < v))\Big)^{p/q}\Big] \\ &\leq x^{-p} \|\lambda\|^p_{\mathbf{M}^{pq}} \end{aligned} \tag{4.16}$$

which proves (4.14). For $Y = \sup_{v\in\mathbf{T}} \alpha_v S_{\epsilon;v}(f)$ or for $Y = \sup_{v\in\mathbf{T}} \alpha_v s_{\epsilon;v}(f)$ the inequality (4.14) can be shown similarly. ■

Proof of Theorem 4.7. We are going to deal with the inequality (4.8), only. The proofs of the inequalities (4.9) and (4.10) are similar. We shall show that for any $y > 0$, $1 \leq p$ and $2, p \leq q < \infty$

$$y^p P(T^* f > (2 + 8R)y) \leq C_{p,q} \|\lambda\|^p_{\mathbf{M}^{pq}} \tag{4.17}$$

where R appeared in Definition 4.2. The proof proceeds in three steps.

First a decomposition generated by the sequences

$$\epsilon^k = (\epsilon^k_t, t \in \mathbf{T}) := \big(\chi(2^k < \lambda_{t+} \leq 2^{k+1}), t \in \mathbf{T}\big) \qquad (k \in \mathbf{Z})$$

is used. Notice that

$$\chi(\lambda^* = 0) T^t(d_t f) = \chi(\lambda^* = 0)\chi(\lambda_{t+} = 0) T^t(d_t f) = 0$$

comes from Definition 4.2. Henceforth

$$T^t(d_t f) = \chi(\lambda^* > 0) T^t(d_t f) = \sum_{k\in\mathbf{Z}} \epsilon^k_t T^t(d_t f). \tag{4.18}$$

Note that, in the linear case, this decomposition was in use while the atomic decomposition of the spaces $\mathcal{P}_p$ was being proved.

The equation (4.18) implies

$$T^* f \leq \sum_{k\in\mathbf{Z}} T^*_{\epsilon^k} f. \tag{4.19}$$

Secondly, apply Lemma 4.8 to ϵ^k and $x = 2^k$ to write

$$T^*_{\epsilon^k} f \leq 2 \sup_{v\in\mathrm{T}} \alpha_v^k T^*_{\epsilon^k;v} f + 2^{k+2} R\chi(\lambda^* > 2^k) \tag{4.20}$$

where

$$\alpha_v^k := \chi\big(\lambda_v > 2^k, \lambda_u \leq 2^k \ (\forall u < v)\big) \qquad (v \in \mathbf{T}).$$

In the third step set

$$Y_k := \sup_{v\in\mathrm{T}} \alpha_v^k T^*_{\epsilon^k;v} f \qquad (k \in \mathbf{Z}). \tag{4.21}$$

Then, by Lemma 4.9, for any $k \in \mathbf{Z}$ and $z_k > 0$ we have

$$P(Y_k > z_k 2^k) \leq C_{p,q} z_k^{-q} 2^{-pk} \|\lambda\|^p_{\mathrm{M}^{pq}} \tag{4.22}$$

if $1 \leq p$ and $2, p \leq q < \infty$. Choose $j \in \mathbf{Z}$ such that $2^j < y \leq 2^{j+1}$ and combine (4.19), (4.20) and (4.21) to write

$$\begin{aligned}\chi(\lambda^* \leq y) T^* f &\leq 2 \sum_{k\leq j} \sup_{v\in\mathrm{T}} \alpha_v^k T^*_{\epsilon^k;v} f + 8Ry \\ &\leq 2 \sum_{k\leq j} Y_k + 8Ry.\end{aligned}$$

Consequently,

$$\begin{aligned} y^p P\big(T^* f > (2+8R)y\big) &\leq y^p P(\lambda^* > y) + y^p P\big(T^* f > (2+8R)y, \lambda^* \leq y\big) \\ &\leq \|\lambda\|^p_{\mathrm{M}^{p\infty}} + y^p P\Big(\sum_{k\leq j} Y_k > y\Big) \\ &\leq \|\lambda\|^p_{\mathrm{M}^{p\infty}} + y^p P\Big(\sum_{k\leq j} Y_k > 2^j\Big).\end{aligned}$$

To use (4.22) observe that $c_\beta \sum_{k\leq j} 2^{\beta(k-j)} = 1$ if $\beta > 0$ and $c_\beta = 1 - 2^{-\beta}$. Set

$$2^j c_\beta 2^{\beta(k-j)} = c_\beta 2^{(\beta-1)(k-j)} \cdot 2^k =: z_k \cdot 2^k. \tag{4.23}$$

Then for $\beta = (q-p)/(2q)$ we get

$$z_k^{-q} 2^{-pk} \leq C_{p,q} 2^{-pj} 2^{p(j-k)+q(\beta-1)(j-k)} \leq C_{p,q} y^{-p} 2^{(q-p)(k-j)/2}.$$

Thus, by (4.22),

$$\begin{aligned} P\Big(\sum_{k\leq j} Y_k > 2^j\Big) &\leq \sum_{k\leq j} P(Y_k > z_k 2^k) \\ &\leq C_{p,q} y^{-p} \|\lambda\|^p_{\mathrm{M}^{pq}} \sum_{k\leq j} 2^{(q-p)(k-j)/2} \\ &\leq C_{p,q} y^{-p} \|\lambda\|^p_{\mathrm{M}^{pq}},\end{aligned}$$

so (4.17) is proved.

To verify (4.9) and (4.10) we just note that

$$d_t f = \sum_{k\in\mathbf{Z}} \epsilon_t^k (d_t f).$$

Since, for a fixed $t \in \mathbf{T}$, the sets $\{2^k < \lambda_{t^+} \le 2^{k+1}\}$ $(k \in \mathbf{Z})$ are disjoint, we have

$$|d_t f|^2 = \sum_{k\in\mathbf{Z}} \epsilon_t^k |d_t f|^2.$$

The theorem is proved. ∎

The concept of regularity and previsibility are going to be introduced as well. A tree stochastic basis $\mathcal{F}$ is said to be *regular* if there exists a constant $R > 0$ such that for all $f \in L_1$

(4.24) $$|E_t f| \le R E_t^- |E_t f| \qquad (t \in \mathbf{T})$$

where E_t^- denotes the conditional expectation operator with respect to $\mathcal{F}_t^-$. A tree martingale $f = (f_t, t \in \mathbf{T})$ is *previsible* if there exists an $N \in \mathbf{N}$ such that for every $t \in \mathbf{T}$ there are indices $t_1, \dots, t_N \in \mathbf{T}$ satisfying $t_i^+ = t$ $(i = 1, \dots, N)$ and

(4.25) $$|f_t| \le C \sum_{i=1}^{N} |f_{t_i}| \qquad (t \in \mathbf{T}).$$

The following result is analogous to Corollary 2.23.

Corollary 4.10. *Suppose that $1 < p, q < \infty$ with $(1/p, 1/q) \in \Delta$ and $f = (f_t, t \in \mathbf{T}) \in L_p$ is a previsible tree martingale. Then there exists a constant $C_{p,q}$ depending only on p and q such that*

$$\|f\|_{\mathbf{P}^{pq}} \le C_{p,q} \|f\|_p.$$

If $\mathcal{F}$ is regular then the previous inequality holds for all $f \in L_p$.

Proof. Assume that f is previsible. By (4.25), the sequence

$$\lambda_t := C \sup_{u \le t} \sum_{i=1}^{N} |f_{u_i}| \qquad (t \in \mathbf{T})$$

is a prediction of f. Obviously,

$$\lambda_t \le C \sum_{i=1}^{N} \sup_{u \le t} |f_{u_i}| \qquad (t \in \mathbf{T}).$$

By the definition of the $\mathbf{M}^{pq}$ quasi-norm,

$$\|f\|_{\mathbf{P}^{pq}} \le \|\lambda\|_{\mathbf{M}^{pq}} \le C \sum_{i=1}^{N} \|(f_{u_i}, u \in \mathbf{T})\|_{\mathbf{M}^{pq}}.$$

If u and t are incomparable then so are u_i and t_j $(u, t \in \mathbf{T}; i, j = 1, \dots, N)$, hence Theorem 4.5 can be applied. The desired inequality follows easily from this theorem.

Assume that $\mathcal{F}$ is regular and let $\delta(u) := u_i$ for any $1 \le i \le N$ $(u \in \mathbf{T})$. By (4.24), $|f_t| \le RE_t^-|f_t|$ $(t \in \mathbf{T})$ and so the sequence

$$\lambda_t := R \sup_{u \le t} E_t^- |f_t| \qquad (t \in \mathbf{T})$$

is a prediction of f. The corollary can be derived from Theorem 4.6. ∎

The next consequence, that is a weak version of Theorems 2.12 and 2.55, comes immediately from Theorem 4.7 and Corollary 4.10.

Corollary 4.11. *If $1 < p < \infty$ and $f = (f_t, t \in \mathbf{T}) \in L_p$ is a previsible tree martingale then*

$$\|(T_t^* f, t \in \mathbf{T})\|_{\mathrm{M}^{p\infty}} \le C_p \|f\|_p,$$

$$\|(S_t(f), t \in \mathbf{T})\|_{\mathrm{M}^{p\infty}} \le C_p \|f\|_p$$

and

$$\|(s_t(f), t \in \mathbf{T})\|_{\mathrm{M}^{p\infty}} \le C_p \|f\|_p.$$

In case $\mathcal{F}$ is regular the previous inequalities hold for every $f \in L_p$ $(1 < p < \infty)$.

Applying Theorem 5.1 and 5.5 the strong version of Corollary 4.11 can also be obtained which includes Burkholder-Gundy's inequality.

Corollary 4.12. *If $1 < p < \infty$ and $f = (f_t, t \in \mathbf{T}) \in L_p$ is a previsible tree martingale then*

$$\|T^* f\|_p \le C_p \|f\|_p,$$

$$c_p \|f\|_p \le \|S(f)\|_p \le C_p \|f\|_p$$

and

$$c_p \|f\|_p \le \|s(f)\|_p \le C_p \|f\|_p.$$

In case $\mathcal{F}$ is regular the previous inequalities hold for every $f \in L_p$ $(1 < p < \infty)$.

Proof. The right hand sides of the inequalities come by interpolation and from Corollary 4.11. To prove the left hand side of the second inequality observe that, by the one-parameter Burkholder-Gundy's inequality (Theorem 2.12),

$$\|f\|_p = \|f \overline{\phi}_{v_0}\|_p \le C_p \|(\sum_{t \ge v_0} |d_t f|^2)^{1/2}\|_p \le C_p \|S(f)\|_p$$

where v_0 is the distinguished minimal element of $\mathbf{T}_0$. The left hand side of the third inequality can be verified in the same way by using Corollary 2.23. ∎

This theorem together with Proposition 4.4 yield also that, in the regular case, $\|f^*\|_p \sim \|S(f)\|_p \sim \|f\|_{H_p^S}$ for $1 < p < \infty$ where H_p^S denotes the one-parameter martingale space. There is a counterexample (see Schipp [161]) which says that, in general, $\|f^*\|_1$ is not equivalent to $\|f\|_{H_1^*}$ and $\|S(f)\|_1$ is not equivalent to $\|f\|_{H_1^S}$ even in the regular case. However, it is still unknown whether $\|f^*\|_1$ is equivalent to $\|S(f)\|_1$.

For general martingales we are able to prove Burkholder-Gundy's inequality for $2 < p < \infty$, only.

Theorem 4.13. (**Burkholder-Gundy's inequality**) *Let* $f = (f_t, t \in \mathbf{T})$ *be a tree martingale and* $2 < p < \infty$. *Then*

$$c_p \|f\|_p \le \|S(f)\|_p \le C_p \|f\|_p \tag{4.26}$$

for all $f \in L_p$.

This theorem is to be proved as Theorem 4.7 and is due to Fridli and Schipp [78]. An additional lemma is needed while Lemmas 4.8 and 4.9 have to be modified. Set

$$f_t^* := \sup_{u \le t} |f_u| \qquad (t \in \mathbf{T}).$$

Lemma 4.14. *Suppose that* $f = (f_v, v \in \mathbf{T})$ *is a tree martingale and*

$$\epsilon_v := \chi(x < f_{v+}^* \le 2x)$$

for any real number $x > 0$. *Then*

$$S_\epsilon(f) \le \sup_{v \in \mathbf{T}} \alpha_v S_{\epsilon;v}(f) + 4x\chi(f^* > x)$$

where

$$\alpha_v := \chi\big(|f_v| > x, |f_u| \le x \ (\forall u < v)\big) \qquad (v \in \mathbf{T}).$$

Though, opposed to Lemma 4.8, ϵ_v is here $\mathcal{F}_{v+}$ measurable, the proof remains the same because this fact was not utilized at all.

The following one-parameter result is due to Burkholder [19] for $p = 1$ and to Fridli and Schipp [78] for $1 < p < \infty$.

Lemma 4.15. *Let* $(\mathcal{F}_n, n \in \mathbf{N})$ *be a non-decreasing sequence of* σ*-algebras and* $f \in L_{2p}$ $(1 \le p < \infty)$. *For any* $y \ge 0$ *we define a stopping time by*

$$\nu := \inf\{n \in \mathbf{N} : f_n^* > y\}.$$

Then

$$\|S_{\nu-1}(f)\|_{2p} \le C_p y^{1/2} \|f\|_p^{1/2}$$

where, in this lemma, $S(f)$ *denotes the one-parameter quadratic variation of* f.

Proof. It is easy to see that

$$S_{n-1}^2(f) + f_{n-1}^2 = 2f_n f_{n-1} - 2g_n \qquad (n \ge 1)$$

where $g_n := \sum_{k=1}^n f_{k-1}(f_k - f_{k-1})$. Thus

$$S_{\nu-1}^2(f) \le 2f_\nu f_{\nu-1} - 2g_\nu.$$

Observe that $|f_{\nu-1}| \le y$ and

$$E(g_\nu) = E\Big(\sum_{k=1}^\infty \chi(\nu \ge k) f_{k-1} E_{k-1}(f_k - f_{k-1})\Big) = 0.$$

The desired inequality is proved for $p = 1$.

Assume that $1 < p < \infty$. We have

$$\|S_{\nu-1}(f)\|_{2p} \leq (2\|f_\nu f_{\nu-1}\|_p + 2\|g_\nu\|_p)^{1/2}.$$

The martingale f_ν is the martingale transform of f by the sequence $(\chi(\nu \geq k))$ and g_ν is the one of f by $(\chi(\nu \geq k)f_{k-1})$. Using Theorem 2.55 and the fact that $|f_{k-1}| \leq y$ whenever $\nu \geq k$ we can finish the proof of the lemma. ∎

Note that

$$\begin{aligned} S_{\nu-1}(f) &= (\sum_{k=0}^{\nu-1} |f_k - f_{k-1}|^2)^{1/2} \\ &= (\sum_{k=0}^{\infty} \chi(\nu > k)|f_k - f_{k-1}|^2)^{1/2} \\ &= (\sum_{k=0}^{\infty} \chi(f_k^* \leq y)|f_k - f_{k-1}|^2)^{1/2}. \end{aligned}$$

An analogous result to Lemma 4.9 is going to be shown. Set again

$$Y := \sup_{v \in \mathbf{T}} \alpha_v S_{\epsilon;v}(f).$$

Lemma 4.16. *If the conditions of Lemma 4.14 are satisfied then for $2, p \leq q < \infty$, $1 \leq p$ and $z > 0$*

$$P(Y > zx) \leq C_{p,q} z^{-q} x^{-p-q/2} \|f\|_{\mathbf{M}^{pq}}^p$$

whenever $\|f\|_\infty \leq 1$.

Proof. Applying Lemma 4.15 we obtain for $p_0 \geq 2$ that

$$\begin{aligned} E_v[S_{\epsilon;v}(f)^{p_0}] &= E_v\Big[\Big(\sum_{u \geq v} \epsilon_u |d_u f|^2\Big)^{p_0/2}\Big] \\ &\leq C_{p_0} x^{p_0/2}. \end{aligned}$$

As in the proof of Lemma 4.9 we conclude that

$$\begin{aligned} P(Y > zx) &\leq (zx)^{-q} E\Big[\Big(\sum_{v \in \mathbf{T}} \alpha_v S_{\epsilon;v}(f)^{p_0}\Big)^{q/p_0}\Big] \\ &\leq C_q (zx)^{-q} E\Big[\Big(\sum_{v \in \mathbf{T}} \alpha_v E_v[S_{\epsilon;v}(f)^{p_0}]\Big)^{q/p_0}\Big] \\ &\leq C_q z^{-q} x^{-q/2} E\Big[\Big(\sum_{v \in \mathbf{T}} \alpha_v\Big)^{q/p_0}\Big] \end{aligned}$$

for $p_0 \geq q \geq 2$. Setting $p_0 := q^2/p \geq q \geq 2$ and using (4.16) we can complete the proof of Lemma 4.16. ∎

The proof of Theorem 4.13 is similar to the one of Theorem 4.7, however, it needs to be worked out a little bit more carefully.

Proof of Theorem 4.13. The left hand side of (4.26) is obvious (see the proof of Corollary 4.12).

For the other side we shall show that for any $y > 0$ and $2 < p_1 < \infty$

$$y^{p_1} P(S(\chi(H)) > 9y) \le C_{p_1} \|\chi(H)\|_{p_1}^{p_1} \tag{4.27}$$

where H is an arbitrary $\mathcal{A}$ measurable set. During the proof let us denote the characteristic function of H by f. Set

$$\epsilon_t^k := \chi(2^k < f_{t+}^* \le 2^{k+1}) \qquad (t \in \mathbf{T}; k \in \mathbf{Z})$$

and observe that

$$S(f) \le \sum_{k \in \mathbf{Z}} S_{\epsilon^k}(f).$$

Lemma 4.14 implies for ϵ^k and $x = 2^k$ that

$$S_{\epsilon^k}(f) \le \sup_{v \in \mathbf{T}} \alpha_v^k S_{\epsilon^k;v}(f) + 2^{k+2}\chi(f^* > 2^k)$$

where

$$\alpha_v^k := \chi(|f_v| > 2^k, |f_u| \le 2^k \ (\forall u < v)) \qquad (v \in \mathbf{T}).$$

Setting

$$Y_k := \sup_{v \in \mathbf{T}} \alpha_v^k S_{\epsilon^k;v}(f) \qquad (k \in \mathbf{Z}),$$

by Lemma 4.16, we obtain for any $k \in \mathbf{Z}$ and $z_k > 0$ that

$$P(Y_k > z_k 2^k) \le C_{p,q} z_k^{-q} 2^{-k(p+q/2)} \|f\|_{\mathbf{M}^{pq}}^p$$

if $1 \le p$ and $2, p \le q < \infty$. Choosing again $j \in \mathbf{Z}$ such that $2^j < y \le 2^{j+1}$, as in the proof of Theorem 4.7 we get that

$$\chi(f^* \le y) S(f) \le \sum_{k \le j} Y_k + 8y.$$

Thus

$$\begin{aligned} P(S(f) > 9y) &\le P(f^* > y) + P(S(f) > 9y, f^* \le y) \\ &\le y^{-p} \|f^*\|_p^p + P(\sum_{k \le j} Y_k > 2^j). \end{aligned} \tag{4.28}$$

Define z_k like in (4.23). If we choose $\beta = (q - 2p)/(4q)$, we get

$$\begin{aligned} z_k^{-q} 2^{-k(p+q/2)} &\le C_{p,q} 2^{-j(p+q/2)} 2^{(p+q/2)(j-k)+q(\beta-1)(j-k)} \\ &\le C_{p,q} y^{-p-q/2} 2^{(q-2p)(k-j)/4}. \end{aligned}$$

If $q > 2p$ then

$$\begin{aligned} P(\sum_{k \le j} Y_k > 2^j) &\le C_{p,q} y^{-p-q/2} \|f\|_{\mathbf{M}^{pq}}^p \sum_{k \le j} 2^{(q-2p)(k-j)/4} \\ &\le C_{p,q} y^{-p-q/2} \|f\|_{\mathbf{M}^{pq}}^p. \end{aligned} \tag{4.29}$$

Suppose that $y \geq 1$, $1 < p < \infty$ and $(1/p, 1/q) \in \Delta$. The fact that $y^{-p-q/2} \leq y^{-p}$ as well as (4.29) and Theorem 4.5 imply

$$P(\sum_{k \leq j} Y_k > 2^j) \leq C_{p,q} y^{-p} \|f\|_p^p.$$

Applying this together with (4.28) and Proposition 4.4 we get the inequality (4.27) for $p_1 = p > 1$ in case $y \geq 1$.

Of course,

$$y^{-p-q/2} \|f\|_p^p = y^{-p-q/2} \|f\|_{p+q/2}^{p+q/2}$$

because f is a characteristic function. However, in consequence of the assumptions $q > 2p$ and $(1/p, 1/q) \in \Delta$ one can show the inequality (4.27) in this way for $p_1 > 3$, only.

Assume that $y < 1 \leq q_0 \leq p_1$ and choose $i \in \mathbb{Z}$ such that $2^i < y^{q_0} \leq 2^{i+1}$. On the one hand,

$$P(f^* > y^{q_0}) \leq y^{-q_0} \|f^*\|_{q_0}^{q_0} \leq y^{-p_1} \|f^*\|_{p_1}^{p_1}. \tag{4.30}$$

On the other hand,

$$\begin{aligned} \chi(f^* \leq y^{q_0}) S(f) &\leq \sum_{k \leq i} Y_k + 8 y^{q_0} \\ &\leq \sum_{k \leq i} Y_k + 8y \end{aligned}$$

and so

$$P(S(f) > 9y, f^* \leq y^{q_0}) \leq P(\sum_{k \leq i} Y_k > 2^j).$$

Set

$$2^{j-i} 2^i c_\beta 2^{\beta(k-j)} = c_\beta 2^{j-i} 2^{(\beta-1)(k-i)} \cdot 2^k =: z_k \cdot 2^k.$$

Choosing $\beta = (q-2p)/(4q)$ we get

$$\begin{aligned} z_k^{-q} 2^{-k(p+q/2)} &\leq C_{p,q} 2^{-i(p+q/2)} 2^{-(j-i)q} 2^{(p+q/2)(i-k)+q(\beta-1)(i-k)} \\ &\leq C_{p,q} y^{(-p-q/2)q_0} y^{-q+qq_0} 2^{(q-2p)(k-j)/4}. \end{aligned}$$

Assume again that $q > 2p > 2$ with $(1/p, 1/q) \in \Delta$ and set $q_0 = 2$. Therefore

$$P(\sum_{k \leq i} Y_k > 2^j) \leq C_{p,q} y^{-2p} \|f\|_p^p = C_{p,q} y^{-2p} \|f\|_{2p}^{2p}.$$

Writing $p_1 = 2p$ and taking into account (4.30) we get

$$\begin{aligned} P(S(f) > 9y) &\leq P(f^* > y^2) + P(S(f) > 9y, f^* \leq y^2) \\ &\leq C_{p,q} y^{-p_1} \|f\|_{p_1}^{p_1} \end{aligned}$$

which completes the proof of (4.27).

The inequality (4.27) yields that the sublinear operator S is of *restricted weak type* (L_{p_1}, L_{p_1}) $(2 < p_1 < \infty)$. Based on Stein-Weiss's theorem (see e.g. Bennett,

Sharpley [6] p. 233), which says that in this case the operator S is of type (L_p, L_p) $(2 < p < \infty)$, the proof of Theorem 4.13 has been finished. ∎

4.2. CONVERGENCE THEOREMS

In this section convergence theorems for Vilenkin systems are demonstrated. Special trees generated by Vilenkin systems are considered. Let us introduce the index set

$$\mathbf{T} := \mathcal{I} := \{[kP_n, (k+1)P_n) \cap \mathbf{N} : k, n \in \mathbf{N}\}.$$

The ordering in $\mathcal{I}$ is defined by set inclusion. Obviously, $\mathcal{I}_0 = \mathbf{N}$. For $I = [kP_n, (k+1)P_n) \cap \mathbf{N} \in \mathcal{I}$, the σ-algebra $\mathcal{F}_I$ is defined by $\mathcal{F}_n$ generated by the Vilenkin system belonging to (P_n) (see (1.5)). Therefore, $(\mathcal{F}_I, I \in \mathcal{I})$ can linearly be ordered. The projections

$$P_I f := \sum_{j \in I} \hat{f}(j) w_j \qquad (I \in \mathcal{I})$$

are to be investigated. For a function $f \in L_1$ we suppose that $\hat{f}(0) = 0$. By (1.7), it is easy to see that

$$\begin{aligned} P_I f &= \sum_{j \in [kP_n, (k+1)P_n)} E(f \overline{w}_j) w_j \\ &= \sum_{i=0}^{P_n - 1} E[(f \overline{w}_{kP_n}) \overline{w}_i] w_{kP_n} w_i = w_{kP_n} E_n(f \overline{w}_{kP_n}) \end{aligned}$$

whenever $I = [kP_n, (k+1)P_n) \cap \mathbf{N} \in \mathcal{I}$. It can be proved in the same way that, for an arbitrary $m \in I$,

$$P_I f = w_m E_n(f \overline{w}_m).$$

This implies that $(\mathcal{F}_I, P_I; I \in \mathcal{I})$ is a tree basis, indeed.

The partial sums

$$R_m f = \sum_{k=0}^{m-1} \hat{f}(k) w_k$$

of the Vilenkin-Fourier series of $f \in L_1$ can be expressed as a martingale transform of the tree martingale $(E_I f, I \in \mathcal{I})$. For this set

$$m(n) := \sum_{k=n}^{\infty} m_k P_k, \qquad I_n(m) := [m(n), m(n) + P_n) \qquad (n \in \mathbf{N})$$

for $m \in \mathbf{N}$ with the expansion

$$m = \sum_{k=0}^{\infty} m_k P_k \qquad (0 \le m_k < p_k). \tag{4.31}$$

Notice that m is contained by $I_n(m)$. For $I = I_n(m)$, set

$$T^I := T^{I_n(m)} := \sum_{\substack{[m(n+1),m(n))\supset J\in\mathcal{I} \\ |J|=P_n}} P_J. \tag{4.32}$$

Since $I_n(m) = I_n(\tilde{m})$ implies $m(n+1) = \tilde{m}(n+1)$, the sequence of operators $T = (T^I, I \in \mathcal{I})$ is well defined. Note that in (4.32) there are m_n summands. In case the Vilenkin system is bounded, it is easy to show that these operators T^I $(I \in \mathcal{I})$ satisfy the conditions in Definition 4.2. It is easy to see that, for the Walsh system,

$$T^{I_n(m)}(d_{I_n(m)}f) = m_n d_{I_n(m)}f.$$

Observe that

$$[0,m) = \bigcup_{n=0}^{\infty} [m(n+1), m(n))$$

which implies

$$\begin{aligned} R_m f &= \sum_{n=0}^{\infty} \sum_{k\in[m(n+1),m(n))} \hat{f}(k) w_k \\ &= \sum_{n=0}^{\infty} T^{I_n(m)} f \\ &= \sum_{\{m\}\le I} T^I(d_I f) \end{aligned} \tag{4.33}$$

where $(d_I f, I \in \mathcal{I})$ is the martingale difference sequence of the tree martingale $f = (E_I f, I \in \mathcal{I})$. Thus the maximal function of the partial sums of the Vilenkin-Fourier series of $f \in L_1$ can be estimated by the maximal function of the martingale transform of the tree martingale f, namely,

$$R^* f := \sup_{m\in\mathbb{N}} |R_m f| \le T^* f.$$

In the proof of the a.e. convergence of the Vilenkin-Fourier series the following lemma will be used (see also Schipp, Wade, Simon, Pál [167]).

Lemma 4.17. *Suppose that X_p is a closed subspace of L_p and X_0 is dense in X_p. Let U and U_n $(n \in \mathbb{N})$ be linear operators from X_p to L_p for some $1 \le p < \infty$ such that U is bounded and, for each $f \in X_0$, $U_n f \to Uf$ a.e. on $[0,1)$ as $n \to \infty$. Set*

$$U^* f := \sup_{n\in\mathbb{N}} |U_n f| \qquad (f \in X_p).$$

If there is a constant $C > 0$ that is independent of f and n such that

$$\sup_{y>0} y^p P(U^* f > y) \le C\|f\|_p^p \qquad (f \in X_p)$$

then

$$Uf = \lim_{n\to\infty} U_n f$$

a.e. on $[0,1)$ *for every* $f \in X_p$.

Proof. Fix $f \in X_p$ and set

$$\gamma := \limsup_{n\to\infty} |U_n f - Uf|.$$

It is sufficient to show that $\gamma = 0$ a.e.

Choose $f_m \in X_0$ $(m \in \mathbb{N})$ such that $\|f - f_m\|_p \to 0$ as $m \to \infty$. Observe that

$$\begin{aligned}\gamma &\le \limsup_{n\to\infty} |U_n(f - f_m)| + \limsup_{n\to\infty} |U_n(f_m) - Uf_m| + |U(f_m - f)| \\ &\le U^*(f_m - f) + |U(f_m - f)|\end{aligned}$$

for all $m \in \mathbb{N}$. Henceforth, for all $y > 0$ and $m \in \mathbb{N}$ we have

$$\begin{aligned}P(\gamma > 2y) &\le P(U^*(f_m - f) > y) + P(|U(f_m - f)| > y) \\ &\le Cy^{-p}\|f_m - f\|_p^p + y^{-p}\|U\|^p\|f_m - f\|_p^p.\end{aligned}$$

Since $f_m \to f$ in X_p as $m \to \infty$, it follows that

$$P(\gamma > 2y) = 0$$

for all $y > 0$. So we can conclude that $\gamma = 0$ a.e. ∎

If the Vilenkin system is bounded then the stochastic basis $\mathcal{F}$ is regular (see Section 1.2). On the other hand, it can be shown that each martingale is previsible, too. Indeed, observe that

$$P_I f = \sum_{\substack{I \supset J \in \mathcal{I} \\ |J| = P_{n-1}}} P_J f \tag{4.34}$$

where $|I| = P_n$ and that the number of the summands is bounded by p_{n-1}. Hence

$$|P_I f| \le \sum_{\substack{I \supset J \in \mathcal{I} \\ |J| = P_{n-1}}} |P_J f| \tag{4.35}$$

which yields previsibility. Applying Lemma 4.17 to $U_m := R_m$, $U := \mathrm{id}$, $X_p := L_p$ and to the set of the Vilenkin polynomials X_0 and considering Corollary 4.12 we get the following result which is the main one of this chapter.

Corollary 4.18. *Suppose that the Vilenkin system is bounded,* $1 < p < \infty$ *and* $f \in L_p$. *Then*

$$\sup_{y>0} yP(R^* f > y)^{1/p} \le \|R^* f\|_p \le C_p \|f\|_p.$$

Consequently, for every $f \in L_p$ *with* $p > 1$ *we have*

$$R_m f \to f \quad \text{a.e. as} \quad m \to \infty.$$

Corollary 4.18 does not hold for $p = 1$, moreover, there exists a counterexample such that f is even in H_1^* (see Schipp, Wade, Simon, Pál [167]). For special functions it will be extended to the two-parameter case in Section 6.3.

A new convergence result for an unbounded Vilenkin system is going to be proved. A function $f \in L_1$ is said to be *lacunary* if there exists an $N \in \mathbf{N}$ such that $\hat{f}(m) = 0$ in case in the expansion (4.31) there is at least one m_k that is greater than N. It is easy to see that (4.34) holds also for an unbounded Vilenkin system and, morover, that the number of the summands in (4.35) is bounded by $N + 1$. Thus (4.25) is valid which yields that the tree martingale generated by a lacunary function is previsible. Obviously, the operators T^I $(I \in \mathcal{I})$ satisfy the conditions in Definition 4.2 in this case, too. So we have the following corollary.

Corollary 4.19. *Suppose that the Vilenkin system is arbitrary,* $1 < p < \infty$ *and the function* $f \in L_p$ *is lacunary. Then*

$$\sup_{y>0} yP(R^*f > y)^{1/p} \leq \|R^*f\|_p \leq C_p\|f\|_p.$$

Consequently, for every lacunary function $f \in L_p$ *with* $p > 1$ *we have*

$$R_m f \to f \quad a.e. \quad as \quad m \to \infty.$$

Note that this corollary is still unknown for general functions.

Of course, it follows from Corollary 4.18 and from Banach-Steinhaus's theorem that $R_m f$ converges to f also in L_p norm $(1 < p < \infty)$ as $m \to \infty$ whenever the Vilenkin system is bounded. We shall prove this result for arbitrary Vilenkin systems. Note that for an unbounded Vilenkin system and for a general f the operators T^I $(I \in \mathcal{I})$ defined in (4.32) do not satisfy the condition (iv) of Definition 4.2. However, the weaker inequality

$$E_n|T^I f|^2 \leq CE_n|f|^2 \qquad (f \in L_1)$$

follows from Bessel's inequality. This result can be extended to each $1 < p < \infty$. Since

$$m - (m_n - l)P_n \in \left[m(n+1) + lP_n, m(n+1) + (l+1)P_n\right) \qquad (0 \leq l < m_n),$$

the operator T^I can be written in the following form:

$$\begin{aligned}
T^I f &= \sum_{l=0}^{m_n-1} \sum_{k=m(n+1)+lP_n}^{m(n+1)+(l+1)P_n-1} \hat{f}(k) w_k \\
&= \sum_{l=0}^{m_n-1} w_{m-(m_n-l)P_n} E_n(f \overline{w}_{m-(m_n-l)P_n}) \\
&= w_m \sum_{l=0}^{m_n-1} \overline{w}_{(m_n-l)P_n} E_n[(f\overline{w}_m) w_{(m_n-l)P_n}] \\
&= w_m \sum_{l=1}^{m_n} \overline{r}_n^{m_n-l} E_n[(f\overline{w}_m) r_n^{m_n-l}].
\end{aligned}$$

The inequality

$$\|R_m f\|_p \le C_p \|f\|_p \qquad (1 < p < \infty)$$

for Fourier series (see e.g. Zygmund [205] p. 266) and an inequality relative to the discrete Fourier series presented in [206] (p. 28) imply that

$$E_n |T^I f|^p \le C_p E_n |f|^p \qquad (f \in L_p) \tag{4.36}$$

where C_p is independent of n and f. Using this we can prove the following norm convergence result due to Schipp [163] and Young [203] (see also Simon [173]).

Theorem 4.20. *Suppose that the Vilenkin system is arbitrary,* $1 < p < \infty$ *and* $f \in L_p$. *Then*

$$\|R_m f\|_p \le C_p \|f\|_p$$

where C_p *is independent of* m *and* f. *Consequently, for every* $f \in L_p$ *with* $1 < p < \infty$ *we have*

$$R_m f \to f \quad \text{in } L_p \text{ norm as} \quad m \to \infty.$$

Proof. First suppose that $2 \le p < \infty$. It follows from (4.33) and from Corollary 2.36 that

$$\begin{aligned}
\|R_m f\|_p &= \|\sum_{\{m\}\le I} T^I(d_I f)\|_p \\
&\le C_p \|(\sum_{\{m\}\le I} E_I |T^I(d_I f)|^2)^{1/2}\|_p + C_p \|\sup_{\{m\}\le I} |T^I(d_I f)|\|_p \\
&\le C_p \|(\sum_{\{m\}\le I} E_I |T^I(d_I f)|^2)^{1/2}\|_p + C_p (E[\sum_{\{m\}\le I} |T^I(d_I f)|^p])^{1/p} \\
&\le C_p \|(\sum_{\{m\}\le I} E_I |T^I(d_I f)|^2)^{1/2}\|_p + C_p (E[\sum_{\{m\}\le I} E_I |T^I(d_I f)|^p])^{1/p}.
\end{aligned}$$

The inequality (4.36) implies

$$\|R_m f\|_p \le C_p \|(\sum_{\{m\}\le I} E_I |d_I f|^2)^{1/2}\|_p + C_p \|(\sum_{\{m\}\le I} |d_I f|^p)^{1/p}\|_p$$
$$\le C_p \|(\sum_{\{m\}\le I} E_I |d_I f|^2)^{1/2}\|_p + C_p \|(\sum_{\{m\}\le I} |d_I f|^2)^{1/2}\|_p.$$

Using Burkholder-Gundy's inequality (Theorem 2.12) and Theorem 2.11. (ii) we conclude

$$\|R_m f\|_p \le C_p \|f \overline{w}_m\|_p = C_p \|f\|_p.$$

To prove the theorem for $1 < p \le 2$ we need a duality argument. It can easily be shown that

$$E[(R_m f)\overline{g}] = E[f \overline{R_m g}].$$

Henceforth

$$\|R_m f\|_p = \sup_{\|g\|_q \le 1} |E[(R_m f)\overline{g}]|$$
$$= \sup_{\|g\|_q \le 1} |E[f \overline{R_m g}]|$$
$$\le \|f\|_p \|R_m g\|_q \le C_q \|f\|_p$$

where $1/p + 1/q = 1$. The inequality in Theorem 4.20 is proved. The convergence result follows easily from Banach-Steinhaus theorem. ∎

This result fails to hold for $p = 1$ and $p = \infty$ (see Simon [173]). Note that the convergence results in this section can be extended to the so-called product systems (cf. Schipp [163]). Theorem 4.20 can simply be generalized for two parameters (cf. Corollary 6.18).

CHAPTER 5

REAL INTERPOLATION

In this chapter the interpolation spaces between the martingale Hardy-Lorentz spaces are identified. For this we shall need some additional definitions.

The distribution function $m(y,f)$ of a measurable function f is defined by

$$m(y,f) := P(\{x : |f(x)| > y\}) \qquad (y \geq 0).$$

Clearly, $m(y,f)$ is non-increasing and continuous on the right. It is well-known that

$$\|f\|_p = (p \int_0^\infty y^{p-1} m(y,f)\, dy)^{1/p} \qquad (0 < p < \infty) \tag{5.1}$$

and

$$\|f\|_\infty = \inf\{y : m(y,f) = 0\}. \tag{5.2}$$

Using the distribution function $m(y,f)$ we introduce the *weak* L_p spaces denoted by L_p^*. The space L_p^* $(0 < p < \infty)$ consists of all measurable functions f for which

$$\|f\|_{L_p^*} := \sup_{y>0} y\, m(y,f)^{1/p} < \infty$$

while we set $L_\infty^* = L_\infty$. Note that L_p^* is a quasi-normed space. It is easy to see that $L_p \subset L_p^*$ for each $0 < p \leq \infty$.

The spaces L_p^* are special cases of the more general Lorentz spaces $L_{p,q}$. In their definition another concept is used. For a measurable function f the *non-increasing rearrangement* is defined by

$$\tilde{f}(t) := \inf\{y : m(y,f) \leq t\}. \tag{5.3}$$

It is easy to see that $\tilde{f}$ is non-increasing, continuous on the right and it is equimeasurable with f, namely,

$$m(x,\tilde{f}) = m(x,f) \qquad (x \geq 0). \tag{5.4}$$

Indeed, by the definition of $\tilde{f}$ we have $\tilde{f}(m(x,f)) \leq x$ and thus $m(x,\tilde{f}) \leq m(x,f)$. On the other hand, since $\tilde{f}$ is continuous on the right, $\tilde{f}(m(x,\tilde{f})) \leq x$ and so $m(x,f) \leq m(x,\tilde{f})$.

Note that if $\tilde{f}$ is continuous at a point t then $y = \tilde{f}(t)$ is equivalent to $t = m(y,f)$.

Lorentz space $L_{p,q}$ is defined as follows: for $0 < p < \infty, 0 < q < \infty$

$$\|f\|_{p,q} := \left(\int_0^\infty \tilde{f}(t)^q t^{q/p} \frac{dt}{t}\right)^{1/q}$$

while for $0 < p \leq \infty$

$$\|f\|_{p,\infty} := \sup_{t>0} t^{1/p} \tilde{f}(t).$$

Let

$$L_{p,q} := L_{p,q}(\Omega, \mathcal{A}, P) := \{f : \|f\|_{p,q} < \infty\}.$$

One can show the following equalities

$$L_{p,p} = L_p, \quad L_{p,\infty} = L_p^* \qquad (0 < p \leq \infty).$$

The first statement is a simple consequence of (5.1), (5.2) and (5.4). To prove the second one for $0 < p < \infty$ we can establish that $\tilde{f}(t) = y$ implies $m(y, f) \leq t$. Thus $ym(y,f)^{1/p} \leq t^{1/p}\tilde{f}(t)$ and so $\|f\|_{L_p^*} \leq \|f\|_{p,\infty}$. On the other hand, for a given $\epsilon > 0$ we can choose t such that $\tilde{f}$ is continuous in t and $\|f\|_{p,\infty} \leq t^{1/p}\tilde{f}(t) + \epsilon$. Set $y = \tilde{f}(t)$. Therefore $m(y, f) = t$ and

$$\|f\|_{p,\infty} \leq t^{1/p}\tilde{f}(t) + \epsilon = ym(y,f)^{1/p} + \epsilon \leq \|f\|_{L_p^*} + \epsilon$$

which proves the second equality.

We shall verify in (5.15) that Lorentz spaces $L_{p,q}$ increase as the second exponent q increases, namely, for $0 < p < \infty$ and $0 < q_1 \leq q_2 \leq \infty$ one has $L_{p,q_1} \subset L_{p,q_2}$. Furthermore, inclusion relations amongst the $L_{p,q}$ spaces, varying with p, are like those amongst the L_p spaces. The second exponent q is not involved. More exactly, one can show that $L_{r,s} \subset L_{p,q}$ for $0 < p < r \leq \infty$ and $0 < q, s \leq \infty$.

For $0 < p, q \leq \infty$ the *martingale Hardy-Lorentz spaces* $H_{p,q}^s$, $H_{p,q}^S$, $H_{p,q}^*$ $\mathcal{P}_{p,q}$ and $\mathcal{Q}_{p,q}$ are defined by the norms

$$\|f\|_{H_{p,q}^s} := \|s(f)\|_{p,q}, \quad \|f\|_{H_{p,q}^S} := \|S(f)\|_{p,q},$$

$$\|f\|_{H_{p,q}^*} := \|f^*\|_{p,q}$$

and

$$\|f\|_{\mathcal{P}_{p,q}} := \|\lim_{n\to\infty} \lambda_n\|_{p,q}, \quad \|f\|_{\mathcal{Q}_{p,q}} := \|\lim_{n\to\infty} \rho_n\|_{p,q},$$

respectively, where λ_n and ρ_n are the predictable, non-decreasing least majorant of (f_n) and of $(S_n(f))$, respectively. Note that in case $p = q$ the usual definitions of Hardy spaces $H_{p,p}^s = H_p^s$, $H_{p,p}^S = H_p^S$, $H_{p,p}^* = H_p^*$, $\mathcal{P}_{p,p} = \mathcal{P}_p$ and $\mathcal{Q}_{p,p} = \mathcal{Q}_p$ are obtained.

The basic definitions and theorems of interpolation theory in the real method are given without proofs as follows. (For the details see Bennett, Sharpley [6] and Bergh, Löfström [8].) Suppose that A_0 and A_1 are quasi-normed spaces embedded continuously in a topological vector space A. In the real method of interpolation, the ***interpolation spaces*** between A_0 and A_1 are defined by the means of an interpolating function $K(t, f, A_0, A_1)$. If $f \in A_0 + A_1$, define

$$K(t, f, A_0, A_1) := \inf_{f=f_0+f_1} \{\|f_0\|_{A_0} + t\|f_1\|_{A_1}\}$$

where the infimum is taken over all choices of f_0 and f_1 such that $f_0 \in A_0$, $f_1 \in A_1$ and $f = f_0 + f_1$. The interpolation space $(A_0, A_1)_{\theta,q}$ is defined as the space of all functions f in $A_0 + A_1$ such that

$$\|f\|_{(A_0,A_1)_{\theta,q}} := \left(\int_0^\infty [t^{-\theta} K(t, f, A_0, A_1)]^q \frac{dt}{t}\right)^{1/q} < \infty$$

where $0 < \theta < 1$ and $0 < q \leq \infty$. We use the conventions $(A_0, A_1)_{0,q} = A_0$ and $(A_0, A_1)_{1,q} = A_1$ for each $0 < q \leq \infty$.

Suppose that B_0 and B_1 are also quasi-normed spaces embedded continuously in a topological vector space B. A map

$$T : A_0 + A_1 \longrightarrow B_0 + B_1$$

is said to be *quasilinear* from (A_0, A_1) to (B_0, B_1) if for given $a \in A_0 + A_1$ and $a_i \in A_i$ with $a_0 + a_1 = a$ there exist $b_i \in B_i$ satisfying

$$Ta = b_0 + b_1$$

and

$$\|b_i\|_{B_i} \leq K_i \|a_i\|_{A_i} \qquad (K_i > 0, i = 0, 1).$$

The following theorem shows that the boundedness of a quasilinear operator is hereditary for the interpolation spaces.

Theorem 5.1. *If $0 < q \leq \infty$, $0 \leq \theta \leq 1$ and T is a quasilinear map from (A_0, A_1) to (B_0, B_1) then*

$$T : (A_0, A_1)_{\theta,q} \longrightarrow (B_0, B_1)_{\theta,q}$$

and

$$\|Ta\|_{(B_0,B_1)_{\theta,q}} \leq K_0^{1-\theta} K_1^\theta \|a\|_{(A_0,A_1)_{\theta,q}}.$$

The reiteration theorem below is one of the most important general results in interpolation theory. It says that the interpolation space of two interpolation spaces is also an interpolation space of the original spaces.

Theorem 5.2. **(Reiteration theorem)** *Suppose that $0 \leq \theta_0 < \theta_1 \leq 1$, $0 < q_0, q_1 \leq \infty$ and $X_i = (A_0, A_1)_{\theta_i,q_i}$ $(i = 0, 1)$. If $0 < \eta < 1$ and $0 < q \leq \infty$ then*

$$(X_0, X_1)_{\eta,q} = (A_0, A_1)_{\theta,q} \tag{5.5}$$

where

$$\theta = (1 - \eta)\theta_0 + \eta\theta_1.$$

If, in addition, A_0 and A_1 are complete and $0 < \theta_0 = \theta_1 = \rho < 1$ then

$$((A_0, A_1)_{\rho,q_0}, (A_0, A_1)_{\rho,q_1})_{\eta,q} = (A_0, A_1)_{\rho,q} \tag{5.6}$$

where

$$\frac{1}{q} = \frac{1-\eta}{q_0} + \frac{\eta}{q_1}.$$

Theorem 5.3. **(Duality theorem)** *Suppose that A_i are Banach spaces and $A_0 \cap A_1$ is dense in A_i $(i = 0, 1)$. If $0 < \theta < 1$ and $1 \le q < \infty$ then*

$$(A_0, A_1)'_{\theta,q} = (A'_0, A'_1)_{\theta,q'}$$

where $1/q + 1/q' = 1$ and X' denotes the dual space of a Banach space X.

Theorem 5.4. **(Wolff)** *Let A_1, A_2, A_3 and A_4 be quasi-Banach spaces satisfying $A_1 \cap A_4 \subset A_2 \cap A_3$. Suppose that*

$$A_2 = (A_1, A_3)_{\phi,q}, \qquad A_3 = (A_2, A_4)_{\psi,r}$$

for any $0 < \phi, \psi < 1$ and $0 < q, r \le \infty$. Then

$$A_2 = (A_1, A_4)_{\rho,q}, \qquad A_3 = (A_1, A_4)_{\theta,r}$$

where

$$\rho = \frac{\phi\psi}{1 - \phi + \phi\psi}, \qquad \theta = \frac{\psi}{1 - \phi + \phi\psi}.$$

The proofs of these theorems can be found e.g. in Bennett, Sharpley [6] and in Bergh, Löfström [8].

In the sequel the following two inequalities are needed many times. This theorem can also be found in Stein, Weiss [179].

Theorem 5.5. **(Hardy's inequality)** *If $1 \le q \le \infty$, $r > 0$ and g is a non-negative function defined on $(0, \infty)$ then*

$$\text{(i)} \qquad \Big(\int_0^\infty \Big(\int_0^t f(u)\,du\Big)^q t^{-r} \frac{dt}{t}\Big)^{1/q} \le \frac{q}{r} \Big(\int_0^\infty [tf(t)]^q t^{-r} \frac{dt}{t}\Big)^{1/q}$$

and

$$\text{(ii)} \qquad \Big(\int_0^\infty \Big(\int_t^\infty f(u)\,du\Big)^q t^{r} \frac{dt}{t}\Big)^{1/q} \le \frac{q}{r} \Big(\int_0^\infty [tf(t)]^q t^{r} \frac{dt}{t}\Big)^{1/q}.$$

Proof. Observe that the measure

$$d\mu := \frac{r}{q} t^{-r/q} u^{r/q-1}\,du$$

is a probability measure on $[0, t]$ for a fixed t. Applying Jensen's inequality with $\phi(x) = |x|^q$ we obtain

$$\begin{aligned}
\Big(\int_0^t f(u)\,du\Big)^q &= \Big(\frac{q}{r}\Big)^q t^r \Big(\int_0^t f(u) u^{1-r/q}\,d\mu\Big)^q \\
&\le \Big(\frac{q}{r}\Big)^q t^r \int_0^t f(u)^q u^{q-r}\,d\mu \\
&= \Big(\frac{q}{r}\Big)^{q-1} t^{r-r/q} \int_0^t f(u)^q u^{q-r+r/q-1}\,du.
\end{aligned}$$

Henceforth

$$\begin{aligned}\int_0^\infty \Big(\int_0^t f(u)\,du\Big)^q t^{-r-1}\,dt \\ &\le \Big(\frac{q}{r}\Big)^{q-1}\int_0^\infty t^{-1-r/q}\Big(\int_0^t f(u)^q u^{q-r+r/q-1}\,du\Big)\,dt \\ &= \Big(\frac{q}{r}\Big)^{q-1}\int_0^\infty [uf(u)]^q u^{-r+r/q-1}\Big(\int_u^\infty t^{-1-r/q}\,dt\Big)\,du \\ &= \Big(\frac{q}{r}\Big)^{q}\int_0^\infty [uf(u)]^q u^{-r-1}\,du\end{aligned}$$

which proves (i).

We show that (i) implies (ii). Applying (i) to the function $g(u) := u^{-2}f(u^{-1})$ we can see that

$$\begin{aligned}\int_0^\infty \Big(\int_0^t g(u)\,du\Big)^q t^{-r-1}\,dt &= \int_0^\infty \Big(\int_{1/t}^\infty f(v)\,dv\Big)^q t^{-r-1}\,dt \\ &= \int_0^\infty \Big(\int_s^\infty f(v)\,dv\Big)^q s^{r-1}\,ds\end{aligned}$$

is less than or equal to

$$\Big(\frac{q}{r}\Big)^q \int_0^\infty [tg(t)]^q t^{-r-1}\,dt = \Big(\frac{q}{r}\Big)^q \int_0^\infty [uf(u)]^q u^{r-1}\,du.$$

This completes the proof of (ii). ∎

It is known that the interpolation spaces of the L_p spaces are Lorentz spaces and that the interpolation spaces of Lorentz spaces are Lorentz spaces, too (see e.g. Bergh, Löfström [8]):

Theorem 5.6. *If* $0 < r < \infty$, $0 < \theta < 1$ *and* $r \le q \le \infty$ *then*

$$(L_r, L_\infty)_{\theta,q} = L_{p,q}, \qquad \frac{1}{p} = \frac{1-\theta}{r}.$$

Proof. First we prove the equivalence

$$K(t, f, L_r, L_\infty) \sim \Big(\int_0^{t^r} \tilde f(s)^r\,ds\Big)^{1/r}. \tag{5.7}$$

For a fixed t take

$$f_0(x) := \begin{cases} f(x) - \tilde f(t^r) f(x)/|f(x)| & \text{if } |f(x)| > \tilde f(t^r) \\ 0 & \text{else}\end{cases}$$

and $f_1 := f - f_0$. Set $E := \{|f| > \tilde{f}(t^r)\}$. It is easy to see that $P(E) \leq t^r$ and $\tilde{f}$ is constant on $[P(E), t^r]$. Henceforth

$$\begin{aligned} K(t, f, L_r, L_\infty) &\leq \|f_0\|_r + t\|f_1\|_\infty \\ &= \Big(\int_E [|f| - \tilde{f}(t^r)]^r \, dP\Big)^{1/r} + t\tilde{f}(t^r) \\ &= \Big(\int_0^{P(E)} [\tilde{f}(s) - \tilde{f}(t^r)]^r \, ds\Big)^{1/r} + \Big(\int_0^{t^r} [\tilde{f}(t^r)]^r \, ds\Big)^{1/r} \\ &= \Big(\int_0^{t^r} [\tilde{f}(s) - \tilde{f}(t^r)]^r \, ds\Big)^{1/r} + \Big(\int_0^{t^r} [\tilde{f}(t^r)]^r \, ds\Big)^{1/r} \\ &\leq C\Big(\int_0^{t^r} \tilde{f}(s)^r \, ds\Big)^{1/r} \end{aligned}$$

which shows the first part of (5.7).

For the converse inequality assume that $f = f_0 + f_1$ with $f_0 \in L_r$ and $f_1 \in L_\infty$. Using the inequality $m(y_0 + y_1, f) \leq m(y_0, f_0) + m(y_1, f_1)$ we obtain

$$\tilde{f}(s) \leq \tilde{f}_0((1 - \epsilon)s) + \tilde{f}_1(\epsilon s) \qquad (0 < \epsilon < 1).$$

As $\tilde{f}_1$ is non-increasing, we can conclude that

$$\begin{aligned} \Big(\int_0^{t^r} \tilde{f}(s)^r \, ds\Big)^{1/r} &\leq C_r\Big(\int_0^{t^r} [\tilde{f}_0((1 - \epsilon)s)]^r \, ds\Big)^{1/r} + C_r\Big(\int_0^{t^r} [\tilde{f}_1(\epsilon s)]^r \, ds\Big)^{1/r} \\ &\leq C_r\Big(\int_0^{\infty} [\tilde{f}_0((1 - \epsilon)s)]^r \, ds\Big)^{1/r} + C_r t\tilde{f}_1(0) \\ &= C_r[(1 - \epsilon)^{-1/r}\|f_0\|_r + t\|f_1\|_\infty]. \end{aligned}$$

Tending with ϵ to zero we have finished the proof of (5.7).

By (5.7) we have

$$\begin{aligned} \|f\|_{(L_r, L_\infty)_{\theta,q}} &= \Big(\int_0^\infty [t^{-\theta} K(t, f, L_r, L_\infty)]^q \frac{dt}{t}\Big)^{1/q} \\ &\leq C\Big(\int_0^\infty t^{-\theta q}\Big[\int_0^{t^r} \tilde{f}(s)^r \, ds\Big]^{q/r} \frac{dt}{t}\Big)^{1/q} \\ &= C\Big(\int_0^\infty t^{-\theta q/r}\Big[\int_0^{t} \tilde{f}(s)^r \, ds\Big]^{q/r} \frac{dt}{t}\Big)^{1/q}. \end{aligned}$$

Since $q \leq r$ we can apply Theorem 5.5 (i):

$$\|f\|_{(L_r, L_\infty)_{\theta,q}} \leq C\Big(\int_0^\infty t^{q/r - \theta q/r} \tilde{f}(t)^q \frac{dt}{t}\Big)^{1/q} = C\|f\|_{p,q}.$$

Conversely, using (5.7) and the fact that $\tilde{f}$ is non-increasing we obtain

$$\|f\|_{(L_r, L_\infty)_{\theta,q}} \geq C_r\Big(\int_0^\infty t^{-\theta q} t^q [\tilde{f}(t^r)]^q \frac{dt}{t}\Big)^{1/q} \geq C_r\|f\|_{p,q}.$$

The proof of the theorem is complete. ∎

Applying the reiteration theorem we get the following general result.

Corollary 5.7. *Suppose that* $0 < \eta < 1$ *and* $0 < p_0, p_1, q_0, q_1, q \leq \infty$. *If* $p_0 \neq p_1$ *then*

$$(L_{p_0,q_0}, L_{p_1,q_1})_{\eta,q} = L_{p,q}, \qquad \frac{1}{p} = \frac{1-\eta}{p_0} + \frac{\eta}{p_1}.$$

In a special case

$$(L_{p_0}, L_{p_1})_{\eta,p} = L_p, \qquad \frac{1}{p} = \frac{1-\eta}{p_0} + \frac{\eta}{p_1}.$$

Furthermore, for $0 < p < \infty$,

$$(L_{p,q_0}, L_{p,q_1})_{\eta,q} = L_{p,q}, \qquad \frac{1}{q} = \frac{1-\eta}{q_0} + \frac{\eta}{q_1}.$$

Proof. Let $0 < r \leq p_0, p_1, q_0, q_1, q$ and $1/p_i = (1-\theta_i)/r$ $(i = 0, 1)$, $\theta = (1-\theta_0)\theta_0 + \eta\theta_1$. Notice that $1/p = (1-\theta)/r$. If $p_0 \neq p_1$, by (5.5), we have

$$(L_{p_0,q_0}, L_{p_1,q_1})_{\eta,q} = ((L_r, L_\infty)_{\theta_0,q_0}, (L_r, L_\infty)_{\theta_1,q_1})_{\eta,q} = (L_r, L_\infty)_{\theta,q} = L_{p,q}.$$

In case $p_0 = p_1 = p$ we use (5.6). ■

The next theorem is an extension of Hardy's inequality to all $0 < q \leq \infty$ and was proved by Riviere and Sagher ([149] Theorem 8).

Theorem 5.8. *Let* $f \geq 0$ *be a non-increasing function on* $(0, \infty)$ *and* $0 < q \leq \infty$, $0 < s < q$. *Then*

$$\Big(\int_0^\infty (1/t \int_0^t f(u)\,du)^q t^s \,\frac{dt}{t}\Big)^{1/q} \leq C_{q,s}\Big(\int_0^\infty f(t)^q t^s \,\frac{dt}{t}\Big)^{1/q}.$$

Proof. Consider the sublinear operator

$$Tf(t) := \frac{1}{t}\int_0^t f(u)\,du.$$

Hölder's inequality implies

$$Tf(t) \leq t^{-1/p}\|f\|_p.$$

Hence

$$\widetilde{Tf}(t) \leq t^{-1/p}\|f\|_p$$

and

$$T : L_p \longrightarrow L_{p,\infty} \qquad (1 \leq p \leq \infty)$$

is bounded. The operator T is clearly quasilinear. Applying Theorem 5.1 and Corollary 5.7 we get that

$$T : L_{p,q} \longrightarrow L_{p,q} \tag{5.8}$$

is also bounded if $1 < p < \infty$ and $0 < q \leq \infty$. As f is non-increasing, so is $1/v \int_0^v f(u)\, du$. Thus $\tilde{f}(t) = f(t)$ and the non-increasing rearrangement of the function $1/v \int_0^v f(u)\, du$ at a point t is $1/t \int_0^t f(u)\, du$. Therefore, (5.8) yields that

$$\Big(\int_0^\infty (1/t \int_0^t f(u)\, du)^q t^{q/p} \frac{dt}{t}\Big)^{1/q} \leq C_{q,p}\Big(\int_0^\infty f(t)^q t^{q/p} \frac{dt}{t}\Big)^{1/q}.$$

Substituting q/p for s we get the desired inequality. ∎

The interpolation spaces between the classical Hardy spaces were identified by Fefferman, Riviere and Sagher in [68] and [149]; it was shown that

$$(\mathcal{H}_{p_0,q_0}, \mathcal{H}_{p_1,q_1})_{\theta,q} = \mathcal{H}_{p,q}, \qquad \frac{1}{p} = \frac{1-\theta}{p_0} + \frac{\theta}{p_1},$$

$$0 < \theta < 1,\ 0 < p_0 < p_1 < \infty,\ 0 < q_i, q \leq \infty$$

where $\mathcal{H}_{p,q} = L_{p,q}$ if $p > 1$. It is interpolated between the classical $\mathcal{H}_{p_0,q_0}$ and $\mathcal{BMO}$ spaces in Hanks [91] and Bennett, Sharpley [6]:

$$(\mathcal{H}_{p_0,q_0}, \mathcal{BMO})_{\theta,q} = \mathcal{H}_{p,q}, \qquad \frac{1}{p} = \frac{1-\theta}{p_0},$$

$$0 < \theta < 1,\ 0 < p_0 < \infty,\ 0 < q_0, q \leq \infty.$$

In the following two sections we shall show similar results for the martingale Hardy-Lorentz spaces.

5.1. INTERPOLATION BETWEEN ONE-PARAMETER MARTINGALE HARDY SPACES

In this section the interpolation spaces between the one-parameter martingale Hardy spaces, between Hardy and BMO and, moreover, between L_p and BMO spaces are identified. Theorem 2.26 is generalized, more precisely, it is proved that the dual of $H^s_{p,q}$ is $H^s_{p',q'}$ where $1 < p < \infty$, $1 \leq q < \infty$, $1/p + 1/p' = 1$ and $1/q + 1/q' = 1$.

First a new decomposition theorem for martingales is given. The proofs of the following two theorems are based on the atomic decomposition.

Theorem 5.9. *Let $f \in H^s_p$, $y > 0$ and fix $0 < p \leq 1$. Then f can be decomposed into the sum of two martingales g and h such that*

$$\|g\|_{H^s_\infty} \leq 4y$$

and

$$\|h\|_{H^s_p} \leq C_p\Big(\int_{\{s(f)>y\}} s(f)^p\, dP\Big)^{1/p}$$

where the positive constant C_p depends only on p.

Similar theorems can be formulated for the spaces $\mathcal{P}_p$ and $\mathcal{Q}_p$.

Theorem 5.10. *An analogous result to Theorem 5.9 holds if we replace H_p^s resp. $s(f)$ by $\mathcal{P}_p$ resp. λ_∞ or by $\mathcal{Q}_p$ resp. ρ_∞ where (λ_n) resp. (ρ_n) is the predictable, non-decreasing least majorant of (f_n) resp. of $(S_n(f))$ and, moreover, $\lambda_\infty = \lim_{n\to\infty} \lambda_n$ and $\rho_\infty = \lim_{n\to\infty} \rho_n$.*

Proof of Theorems 5.9 and 5.10. Choose $N \in \mathbb{Z}$ such that $2^{N-1} < y \leq 2^N$. Take the same stopping times ν_k, atoms a^k and real numbers μ_k $(k \in \mathbb{Z})$ as in Theorem 2.2. Set

$$g_n := \sum_{k=-\infty}^{N} \mu_k a_n^k$$

and

$$h_n := \sum_{k=N+1}^{\infty} \mu_k a_n^k.$$

It was proved in Theorem 2.2 that $f_n = g_n + h_n$ for all $n \in \mathbb{N}$ and $g = f^{\nu_{N+1}}$. By the definition of ν_{N+1} we got in Theorem 2.2 that

$$s(g) = s(f^{\nu_{N+1}}) \leq 2^{N+1} \leq 4y$$

which proves the first inequality of the theorem.

On the other hand, the inequality

$$\|h\|_{H_p^s}^p \leq \sum_{k=N+1}^{\infty} |\mu_k|^p = C_p \sum_{k=N+1}^{\infty} (2^k)^p P\left(s(f) > 2^k\right)$$

follows from Theorem 2.2. Similarly to (2.5), by Abel rearrangement, we obtain

$$\|h\|_{H_p^s}^p \leq C_p \int_{\{s(f)>2^N\}} s(f)^p \, dP \leq C_p \int_{\{s(f)>y\}} s(f)^p \, dP.$$

By the help of Theorem 2.3, Theorem 5.10 can be proved in the same way. ∎

The interpolation spaces between the martingale Hardy spaces having atomic decomposition can be identified.

Theorem 5.11. *If $0 < \theta < 1$, $0 < p_0 \leq 1$ and $0 < q \leq \infty$ then*

$$(X_{p_0}, X_\infty)_{\theta,q} = X_{p,q}, \qquad \frac{1}{p} = \frac{(1-\theta)}{p_0}$$

where X denotes one of the spaces H^s, $\mathcal{P}$ and $\mathcal{Q}$. (Note that $\mathcal{P}_\infty = L_\infty$ and $\mathcal{Q}_\infty = H_\infty$.)

Proof. We are going to show the theorem for the spaces H^s. The proof for the other two spaces is very similar. Let $f \in H_{p,q}^s$ and $\tilde{s}$ be the non-increasing rearrangement of $s = s(f)$. Set $1/\alpha = 1/p_0$ and choose y in Theorem 5.9 such that, for a fixed $t \in [0,1]$, $y = \tilde{s}(t^\alpha)$. For this y let us denote the two martingales in Theorem 5.9 by g_t and h_t. By the definition of the functional K,

$$K(t, f, H_{p_0}^s, H_\infty^s) \leq \|h_t\|_{H_{p_0}^s} + t\|g_t\|_{H_\infty^s}. \tag{5.9}$$

By Theorem 5.9 we get that

$$\|h_t\|_{H^s_{p_0}} \le C\left(\int_{\{s>\tilde{s}(t^\alpha)\}} s^{p_0}\,dP\right)^{1/p_0} = C\left(\int_0^{t^\alpha} \tilde{s}(x)^{p_0}\,dx\right)^{1/p_0}. \tag{5.10}$$

Consequently,

$$\int_0^1 \left(t^{-\theta}\|h_t\|_{H^s_{p_0}}\right)^q \frac{dt}{t} \le C\int_0^1 t^{-\theta q}\left(\int_0^{t^\alpha} \tilde{s}(x)^{p_0}\,dx\right)^{q/p_0} \frac{dt}{t}$$
$$\le C\int_0^1 t^{(1-\theta)q/p_0}\left(\frac{1}{t}\int_0^t \tilde{s}(x)^{p_0}\,dx\right)^{q/p_0} \frac{dt}{t}.$$

Using Theorem 5.8 we obtain

$$\int_0^1 \left(t^{-\theta}\|h_t\|_{H^s_{p_0}}\right)^q \frac{dt}{t} \le C\int_0^1 t^{(1-\theta)q/p_0}\tilde{s}(t)^q \frac{dt}{t} = C\|s(f)\|^q_{p,q}.$$

Furthermore,

$$\int_0^1 \left(t^{1-\theta}\|g_t\|_{H^s_\infty}\right)^q \frac{dt}{t} \le C\int_0^1 t^{(1-\theta)q}\tilde{s}(t^\alpha)^q \frac{dt}{t}.$$

Substituting $u = t^\alpha$ we can see that

$$\int_0^1 \left(t^{1-\theta}\|g_t\|_{H^s_\infty}\right)^q \frac{dt}{t} \le C\int_0^1 u^{(1-\theta)q/p_0}\tilde{s}(u)^q \frac{du}{u} = C\|s(f)\|^q_{p,q}.$$

Henceforth

$$\|f\|_{(H^s_{p_0},H^s_\infty)_{\theta,q}} = \left(\int_0^1 [t^{-\theta}K(t,f,H^s_{p_0},H^s_\infty)]^q \frac{dt}{t}\right)^{1/q} \le C\|f\|_{H^s_{p,q}}.$$

To prove the converse consider the sublinear operator $T : f \mapsto s(f)$. By the definition $T : H^s_\infty \longrightarrow L_\infty$ and $T : H^s_{p_0} \longrightarrow L_{p_0}$ are bounded. Therefore, by Theorems 5.1 and 5.6

$$T : (H^s_{p_0}, H^s_\infty)_{\theta,q} \longrightarrow (L_{p_0}, L_\infty)_{\theta,q} = L_{p,q}$$

is bounded, too, that is to say $f \in (H^s_{p_0}, H^s_\infty)_{\theta,q}$ implies

$$\|f\|_{H^s_{p,q}} = \|s(f)\|_{p,q} \le C\|f\|_{(H^s_{p_0},H^s_\infty)_{\theta,q}}.$$

The proof of Theorem 5.11 is complete if $0 < q < \infty$. With a fine modification of the previous proof the theorem can be shown in case $q = \infty$, too. ■

Applying the reiteration theorem we get the following result (see Weisz [195]).

Corollary 5.12. *Suppose that* $0 < \eta < 1$ *and* $0 < p_0, p_1, q_0, q_1, q \le \infty$. *If* $p_0 \ne p_1$ *then*

$$(X_{p_0,q_0}, X_{p_1,q_1})_{\eta,q} = X_{p,q}, \qquad \frac{1}{p} = \frac{1-\eta}{p_0} + \frac{\eta}{p_1}.$$

In a special case

$$(X_{p_0}, X_{p_1})_{\eta,p} = X_p, \qquad \frac{1}{p} = \frac{1-\eta}{p_0} + \frac{\eta}{p_1}.$$

Furthermore, for $0 < p < \infty$,

$$(X_{p,q_0}, X_{p,q_1})_{\eta,q} = X_{p,q}, \qquad \frac{1}{q} = \frac{1-\eta}{q_0} + \frac{\eta}{q_1}$$

where X *denotes again one of the spaces* H^s, $\mathcal{P}$ *and* $\mathcal{Q}$.

In Section 2.2 it is proved that if $\mathcal{F}$ is regular then $H_p^s \sim \mathcal{P}_p \sim \mathcal{Q}_p$ for each $0 < p < \infty$ and $H_p^s \sim L_p$ for all $1 < p < \infty$. So, applying Theorems 5.6 and 5.11 we obtain

Corollary 5.13. *If* $\mathcal{F}$ *is regular and* $0 < q \le \infty$ *then*

$$H_{p,q}^s \sim \mathcal{P}_{p,q} \sim \mathcal{Q}_{p,q}$$

for all $0 < p < \infty$ *and*

$$H_{p,q}^s \sim \mathcal{P}_{p,q} \sim \mathcal{Q}_{p,q} \sim L_{p,q}$$

for all $1 < p < \infty$.

We shall verify in the next section (Corollary 5.22) that this result holds also for the space H_p^S, namely, $H_{p,q}^S \sim H_{p,q}^s$ for every $0 < p < \infty$ and $0 < q \le \infty$.

The interpolation spaces between H_p^s and BMO_2 are to be identified (see Weisz [194]).

Theorem 5.14. *If* $0 < \theta < 1$, $0 < q \le \infty$ *and* $0 < r < \infty$ *then*

$$(H_r^s, BMO_2)_{\theta,q} = H_{p,q}^s, \qquad \frac{1}{p} = \frac{1-\theta}{r}.$$

Proof. By the equivalence amongst BMO_2 and BMO_p^s $(0 < p < \infty)$ it is obvious that

$$\|f\|_{BMO_2} \le C\|f\|_{BMO_1^s} \le C\|\sup_{n\in\mathbb{N}} E_n s(f)\|_\infty$$
$$\le C\|s(f)\|_\infty = C\|f\|_{H_\infty^s}.$$

Thus

$$\|f\|_{(H_r^s, BMO_2^s)_{\theta,q}} \le C\|f\|_{(H_r^s, H_\infty^s)_{\theta,q}} = C\|f\|_{H_{p,q}^s}$$

where $1/p = (1-\theta)/r$.

To see the converse consider the operator $T_u^s : f \mapsto f_u^s$ for a fixed $0 < u < r$. Remember that the sharp functions are introduced in Definition 2.52. It is easy to show that T_u^s is sublinear and in Theorem 2.53 it was proved that $T_u^s : H_r^s \longrightarrow L_r$ is bounded. Since the BMO_r^s norms are all equivalent, it is easy to see that the operator $T_u^s : BMO_2^s \longrightarrow L_\infty$ is bounded, too. Using Theorem 5.1, by the sublinearity of T_u^s, we get that

$$T_u^s : (H_r^s, BMO_2^s)_{\theta,q} \longrightarrow (L_r, L_\infty)_{\theta,q} = L_{p,q}$$

is bounded as well where $1/p = (1-\theta)/r$. Henceforth, by Theorem 2.53, one can see that $f \in (H_r^s, BMO_2^s)_{\theta,p}$ implies

$$\|f\|_{H_p^s} \le C_p \|f_u^s\|_p \le C_p \|f\|_{(H_r^s, BMO_2^s)_{\theta,p}}$$

which proves the theorem for $p = q$, namely,

$$(H_r^s, BMO_2)_{\theta,p} = H_p^s, \qquad \frac{1}{p} = \frac{1-\theta}{r}. \tag{5.11}$$

Applying the reiteration theorem we get that

$$(H_r^s, BMO_2)_{\theta,q} = \left(H_r^s, (H_r^s, BMO_2)_{\theta_1, r/(1-\theta_1)}\right)_{\eta,q}$$

where $0 < \theta_1, \eta < 1$ and $\theta_1 \eta = \theta$. Therefore, by (5.11) and Corollary 5.12 we obtain

$$(H_r^s, BMO_2)_{\theta,q} = \left(H_r^s, H_{r/(1-\theta_1)}^s\right)_{\eta,q} = H_{p',q}^s$$

where

$$\frac{1}{p'} = \frac{1-\eta}{r} + \frac{1-\theta_1}{r}\eta = \frac{1-\theta_1\eta}{r} = \frac{1-\theta}{r}.$$

So $p = p'$ and the proof is complete. ∎

In the regular case $H_p^s \sim H_p^*$ $(0 < p < \infty)$ and $BMO_2 \sim BMO_2^-$, so the following result is valid.

Corollary 5.15. *If $\mathcal{F}$ is regular, $0 < \theta < 1$, and $0 < r < \infty$ then*

$$(H_r^*, BMO_2^-)_{\theta,p} = H_p^*, \qquad \frac{1}{p} = \frac{1-\theta}{r}.$$

This Corollary was proved in a special case (for d-martingales) by Janson and Jones in [108] with the complex method. Moreover, they have shown, that Corollary 5.15 is not valid for a general $\mathcal{F}$.

As a further application of the reiteration theorem we get the following

Corollary 5.16. *If $0 < \theta < 1$, $0 < p_0 < \infty$ and $0 < q_0, q \le \infty$ then*

$$\left(H_{p_0,q_0}^s, BMO_2\right)_{\theta,q} = H_{p,q}^s, \qquad \frac{1}{p} = \frac{1-\theta}{p_0}.$$

With the duality theorem a slightly more general result than Theorem 2.26 can be proved.

Theorem 5.17. *The dual of $H_{p,q}^s$ is $H_{p',q'}^s$ where $1 < p < \infty$, $1 \le q < \infty$, $1/p + 1/p' = 1$ and $1/q + 1/q' = 1$.*

Proof. By the duality theorem we get that

$$(H_1^s, L_2)'_{\theta,q} = (BMO_2, L_2)_{\theta,q'}.$$

However, by Corollaries 5.12 and 5.16, this yields

$$(H^s_{p,q})' = (L_2, BMO_2)_{1-\theta,q'} = H^s_{p',q'}$$

where $1/p = (1-\theta) + \theta/2$ and $1/p' = \theta/2$, thus $1/p + 1/p' = 1$. Writing $p = q$ we have

$$(H^s_p)' = H^s_{p'} \qquad (1 < p \le 2). \tag{5.12}$$

The space H^s_p is uniformly convex for $2 \le p < \infty$ (see Theorem 2.25), consequently, it is reflexive. From this it can easily be shown that (5.12) is true for $1 < p < \infty$, too. Applying again the duality theorem we get for $p < r < \infty$ that

$$\begin{aligned}(H^s_{p,q})' &= (H^s_1, H^s_r)'_{\theta,q} = (BMO_2, H^s_{r'})_{\theta,q'} \\ &= (H^s_{r'}, BMO_2)_{1-\theta,q'} = H^s_{p',q'}\end{aligned}$$

where $1/p = (1-\theta) + \theta/r$, $1/r + 1/r' = 1$, $1/q + 1/q' = 1$ and $1/p' = \theta/r'$, namely, $1/p + 1/p' = 1$. The proof of Theorem 5.17 is complete. ■

Note that one can similarly show the duality between $L^s_{p,q}$ and $L^s_{p',q'}$ where $1 < p < \infty$, $1 \le q < \infty$, $1/p + 1/p' = 1$ and $1/q + 1/q' = 1$, too.

The interpolation spaces of L_p and BMO_1^- are going to be calculated.

Theorem 5.18. *If* $0 < \theta < 1$, $0 < p_0 < \infty$ *and* $0 < q, q_0 \le \infty$ *then*

$$(L_{p_0,q_0}, BMO_1^-)_{\theta,q} = L_{p,q}, \qquad \frac{1}{p} = \frac{1-\theta}{p_0}. \tag{5.13}$$

Proof. Considering the operator $T^\sharp : f \mapsto f^\sharp$ we can establish that, by Doob's inequality, $T^\sharp : L_1 \longrightarrow L_{1,\infty}$ and $T^\sharp : BMO_1^- \longrightarrow L_\infty$ are bounded. Since $\|f\|_{BMO_1^-} \le 2\|f\|_\infty$, the equivalence

$$(L_1, BMO_1^-)_{\theta,q} = L_{p,q}, \qquad \frac{1}{p} = 1-\theta$$

can be proved as Theorem 5.14. By reiteration,

$$(L_{p_0,q_0}, BMO_1^-)_{\theta,q} = L_{p,q}, \qquad \frac{1}{p} = \frac{1-\theta}{p_0}$$

where $1 < p_0 < \infty$ and $0 < q_0 \le \infty$. We are going to apply Wolff's theorem. Set $A_1 = L_{p_0,q_0}$ for any $0 < p_0 \le 1$, $A_2 = L_{p_1,q_1}$ for any $1 < p_1 < \infty$, $A_3 = L_{p,q}$ for any $p_1 < p < \infty$ and $A_4 = BMO_1^-$. By Corollary 5.7 we can apply Theorem 5.4 and so we get (5.13) for $1 < p < \infty$. Let us apply again Wolff's theorem. Now set $A_1 = L_{p_0,q_0}$ for any $0 < p_0 < 1$, $A_2 = L_{p,q}$ for any $p_0 < p \le 1$, $A_3 = L_{p_1,q_1}$ for any $1 < p_1 < \infty$ and $A_4 = BMO_1^-$. Applying (5.13) to $1 < p < \infty$ together with Corollary 5.7 and Theorem 5.4 we obtain (5.13) for all $0 < p < \infty$. The proof is complete. ■

This theorem can be found in Schipp, Wade, Simon, Pál [167] for L_2 and BMO_1 with the complex method.

5.2. INTERPOLATION BETWEEN TWO-PARAMETER MARTINGALE HARDY SPACES

The results of the previous section are generalized for two parameters.

Let us begin with a decomposition theorem for the martingales of the two-parameter H_p^s space. Since we work with $(p,2)$ atoms in the two-parameter case, instead of the H_∞^s norm the L_2 norm of the martingale g in the decomposition can be estimated. The proofs of the next two theorems are more complex than the ones in the one-parameter case and are due to the author [195].

Theorem 5.19. *Let $f \in H_p^s$, $y > 0$ and fix $0 < p \le 1$. Then f can be decomposed into the sum of two martingales g and h such that*

$$\|g\|_2 \le C_2 \left[\left(\int_{\{s(f)\le y\}} s(f)^2 \, dP \right)^{1/2} + yP(s(f) > y)^{1/2} \right]$$

and

$$\|h\|_{H_p^s} \le C_p \left(\int_{\{s(f)>y\}} s(f)^p \, dP \right)^{1/p}$$

where the positive constant C_p depends only on p.

Proof. Let $N \in \mathbf{Z}$ and $z \in (1/2, 1]$ such that $y = z2^N$. Similarly to the proof of Theorem 3.2 we introduce the following sets and stopping times for all $k \in \mathbf{Z}$:

$$F_k := \{s(f) > z2^k\},$$

$$\nu_k := \inf\{n \in \mathbf{N}^2 : E_n\chi(F_k) > 1/2\}.$$

Furthermore, set again

$$\mu_k := \sqrt{2} z 2^{k+3} P(\nu \ne \infty)^{1/p}$$

and

$$a_n^k := \frac{1}{\mu_k}(f_n^{\nu_{k+1}} - f_n^{\nu_k}).$$

We recall that a^k is a $(p,2)$ atom with respect to the stopping time ν_k ($k \in \mathbf{Z}$) and that $f_n = \sum_{k=-\infty}^{\infty} \mu_k a_n^k$ for all $n \in \mathbf{N}^2$. Set

$$g_n := \sum_{k=-\infty}^{N} \mu_k a_n^k$$

and

$$h_n := \sum_{k=N+1}^{\infty} \mu_k a_n^k.$$

Since each a^k is a $(p,2)$ atom, therefore, by Theorem 3.2, we get the inequality

$$\|h\|_{H_p^s}^p \le \sum_{k=N+1}^{\infty} |\mu_k|^p = \sum_{k=N+1}^{\infty} C_p (z2^k)^p P(\nu_k \ne \infty).$$

As in the proof of Theorem 3.2 we obtain

$$\sum_{k=N+1}^{\infty} |\mu_k|^p \le C_p \sum_{k=N+1}^{\infty} (z2^k)^p P\left(\sup_{n\in\mathbf{N}^2} E_n\chi(F_k) > 1/2\right)$$
$$\le C_p \sum_{k=N+1}^{\infty} (z2^k)^p 4E\left[\sup_{n\in\mathbf{N}^2} (E_n\chi(F_k))^2\right]$$
$$\le C_p \sum_{k=N+1}^{\infty} (z2^k)^p P\left(s(f) > z2^k\right).$$

By Abel rearrangement,

$$\|h\|_{H_p^s}^p \le C_p \int_{\{s(f)>z2^N\}} s(f)^p \, dP \le C_p \int_{\{s(f)>y\}} s(f)^p \, dP.$$

Further on let us consider the martingale g. Analogously to the first inequality of Theorem 5.9 it is proved in Theorem 3.2 that

$$\|g\|_2 = \|f^{\nu_{N+1}}\|_2 \le \sqrt{2}\cdot 2^{N+1}.$$

However, this estimation is unsufficient to prove the next theorem, a finer one is needed. One can derive from (3.13) that

$$\|g\|_2^2 = \int_\Omega |\sum_{k=-\infty}^{N} \mu_k a^k|^2 \, dP$$
$$= \sum_{k=-\infty}^{N} \int_\Omega |\mu_k a^k|^2 \, dP$$
$$\le \sum_{k=-\infty}^{N} C_2 (z2^k)^2 P(\nu \ne \infty)^{2/p} P(\nu \ne \infty)^{1-2/p}.$$

Similarly to the previous part the following inequality can be shown:

$$\|g\|_2^2 \le C_2 \left[\int_{\{s(f)\le z2^N\}} s(f)^2 \, dP + z^2 4^{N+1} P\left(s(f) > z2^N\right)\right]$$
$$\le C_2 \left[\int_{\{s(f)\le y\}} s(f)^2 \, dP + y^2 P\left(s(f) > y\right)\right].$$

This completes the proof of Theorem 5.19. ■

After these a result that is analogous to Theorem 5.11 can be proved for two parameters.

Theorem 5.20. *If* $0 < p_0 \le 1$, $0 < \theta < 1$ *and* $0 < q \le \infty$ *then*

$$\left(H_{p_0}^s, L_2\right)_{\theta,q} = H_{p,q}^s, \qquad \frac{1}{p} = \frac{1-\theta}{p_0} + \frac{\theta}{2}.$$

Proof. Let $f \in H^{\theta}_{p,q}$. Denote by $\tilde{s}$ again the non-increasing rearrangement of $s = s(f)$. Let $1/\alpha = 1/p_0 - 1/2$ and, for a fixed t, $y = \tilde{s}(t^\alpha)$. The corresponding martingales in Theorem 5.19 are h_t and g_t. We get from (5.10) that

$$\int_0^1 \left(t^{-\theta}\|h_t\|_{H^{s}_{p_0}}\right)^q \frac{dt}{t} \le C\int_0^1 t^{-\theta q}\left(\int_0^{t^\alpha} \tilde{s}(x)^{p_0}\,dx\right)^{q/p_0} \frac{dt}{t}$$

$$\le C\int_0^1 t^{(1-\theta)q/p_0+\theta q/2}\left(\frac{1}{t}\int_0^t \tilde{s}(x)^{p_0}\,dx\right)^{q/p_0} \frac{dt}{t}.$$

Using Theorem 5.8 we obtain

$$\int_0^1 \left(t^{-\theta}\|h_t\|_{H^{s}_{p_0}}\right)^q \frac{dt}{t} \le C\int_0^1 t^{(1-\theta)q/p_0+\theta q/2}\tilde{s}(t)^q \frac{dt}{t} = C\|s(f)\|^q_{p,q}.$$

On the other hand, by Theorem 5.19,

$$\|g_t\|_2 \le C\left(\int_{\{s\le \tilde{s}(t^\alpha)\}} s^2\,dP\right)^{1/2} + C\tilde{s}(t^\alpha)P\left(s > \tilde{s}(t^\alpha)\right)^{1/2}.$$

Since the distributions of s and $\tilde{s}$ are identical and $\tilde{s}$ is non-increasing,

$$P(s > \tilde{s}(t^\alpha)) = P(\tilde{s} > \tilde{s}(t^\alpha)) \le t^\alpha.$$

It follows easily that

$$\|g_t\|_2 \le C\left(\int_{t^\alpha}^1 \tilde{s}(x)^2\,dx\right)^{1/2} + C\tilde{s}(t^\alpha)t^{\alpha/2}$$

and

$$\int_0^1 \left(t^{1-\theta}\|g_t\|_2\right)^q \frac{dt}{t} \le C\int_0^1 t^{(1-\theta)q}\left(\int_{t^\alpha}^1 \tilde{s}(x)^2\,dx\right)^{q/2} \frac{dt}{t}$$

$$+C\int_0^1 t^{(1-\theta)q}\tilde{s}(t^\alpha)^q t^{\alpha q/2}\frac{dt}{t}$$

$$=: (A) + (B).$$

First let us estimate (B) by replacing $u = t^\alpha$:

$$(B) \le C\int_0^1 u^{(1-\theta)q/p_0-(1-\theta)q/2}\tilde{s}(u)^q u^{q/2}\frac{du}{u} = C\|s(f)\|^q_{p,q}.$$

Before estimating (A) we prove that for $0 < r \le 2$

$$\left(\int_u^1 \left(x^{1/2}\tilde{s}(x)\right)^2 \frac{dx}{x}\right)^{1/2} \le C\left(\int_{u/4}^1 \left(x^{1/2}\tilde{s}(x)\right)^r \frac{dx}{x}\right)^{1/r}. \tag{5.14}$$

Indeed, choose $N \in \mathbf{Z}$ such that $2^N \le u < 2^{N+1}$; then

$$\left(\int_u^1 \left(x^{1/2}\tilde{s}(x)\right)^2 \frac{dx}{x}\right)^{r/2} \le \left[\sum_{k=N+1}^{0} \tilde{s}(2^{k-1})^2 \int_{2^{k-1}}^{2^k} dx\right]^{r/2}$$
$$\le \sum_{k=N+1}^{0} \tilde{s}(2^{k-1})^r 2^{(k-1)r/2}$$
$$\le C \sum_{k=N+1}^{0} \int_{2^{k-2}}^{2^{k-1}} x^{r/2-1}\tilde{s}(x)^r\, dx$$
$$\le C \int_{u/4}^1 \left(x^{1/2}\tilde{s}(x)\right)^r \frac{dx}{x}.$$

This completes the proof of (5.14).

Note that the inequality

$$\|\cdot\|_{p,q_2} \le C\|\cdot\|_{p,q_1} \qquad (q_1 \le q_2) \tag{5.15}$$

can be proved in the same way.

Let us begin to estimate (A). Replace again $u = t^\alpha$ and use (5.14):

$$(A) \le C \int_0^1 u^{(1-\theta)q/p_0-(1-\theta)q/2}\left(\int_u^1 \tilde{s}(x)^2\, dx\right)^{q/2} \frac{du}{u}$$
$$\le C \int_0^1 u^{q/p-q/2}\left(\int_{u/4}^1 \left(x^{1/2}\tilde{s}(x)\right)^r \frac{dx}{x}\right)^{q/r} \frac{du}{u}$$
$$\le C \int_0^1 u^{q/p-q/2}\left(\int_u^1 x^{r/2-1}\tilde{s}(x)^r\, dx\right)^{q/r} \frac{du}{u}$$

where $r \le \min(2, q)$. Applying Hardy's inequality (Theorem 5.5) we can conclude that

$$(A) \le C \int_0^1 \left(u u^{r/2-1}\tilde{s}(u)^r\right)^{q/r} u^{q/p-q/2}\frac{du}{u}$$
$$= C \int_0^1 \tilde{s}(u)^q u^{q/p} \frac{du}{u}$$
$$= C\|s(f)\|_{p,q}^q.$$

Consequently, by (5.9),

$$\|f\|_{(H_{p_0}^s, L_2)_{\theta,q}} = \left(\int_0^1 [t^{-\theta}K(t,f,H_{p_0}^s,L_2)]^q \frac{dt}{t}\right)^{1/q} \le C\|f\|_{H_{p,q}^s}.$$

For the converse observe that $T : L_2 \longrightarrow L_2$ and $T : H_{p_0}^s \longrightarrow L_{p_0}$ are bounded where $T : f \mapsto s(f)$. Therefore

$$T : (H_{p_0}^s, L_2)_{\theta,q} \longrightarrow (L_{p_0}, L_2)_{\theta,q} = L_{p,q}$$

is bounded as well. In other words,

$$\|f\|_{H^s_{p,q}} = \|s(f)\|_{p,q} \le C\|f\|_{(H^s_{p_0},L_2)_{\theta,q}}$$

which proves the desired result for $0 < q < \infty$. In case $q = \infty$ the theorem can be shown similarly. ∎

Applying the reiteration theorem we get the following result:

Corollary 5.21. *Suppose that* $0 < \eta < 1$ *and* $0 < q_0, q_1, q \le \infty$. *If* $0 < p_0 < p_1 < 2$ *or if* $p_1 = 2$ *and* $q_1 = 2$ *then*

$$(H^s_{p_0,q_0}, H^s_{p_1,q_1})_{\eta,q} = H^s_{p,q}, \qquad \frac{1}{p} = \frac{1-\eta}{p_0} + \frac{\eta}{p_1}.$$

Furthermore, for $0 < p < 2$,

$$(H^s_{p,q_0}, H^s_{p,q_1})_{\eta,q} = H^s_{p,q}, \qquad \frac{1}{q} = \frac{1-\eta}{q_0} + \frac{\eta}{q_1}.$$

In a special case

$$(H^s_{p_0}, H^s_{p_1})_{\eta,p} = H^s_p, \qquad \frac{1}{p} = \frac{1-\eta}{p_0} + \frac{\eta}{p_1}$$

where $0 < p_0 < p_1 \le 2$.

From now on suppose that $\mathcal{F}$ is regular. Since, in this case, H^S_p has also an atomic decomposition (see Theorem 3.3), all the theorems of this section can be proved for H^S_p, too. Using the equivalences $H^S_p \sim L_p$ $(1 < p < \infty)$ and $H^S_p \sim H^s_p$ $(0 < p < \infty)$ we obtain the following corollary.

Corollary 5.22. *If* $\mathcal{F}$ *is regular and* $0 < q \le \infty$ *then*

$$H^s_{p,q} \sim H^S_{p,q} \qquad (0 < p < \infty)$$

and

$$H^s_{p,q} \sim H^S_{p,q} \sim L_{p,q} \qquad (1 < p < \infty).$$

The next result is an extension of Corollary 5.21 to every $0 < p < \infty$.

Corollary 5.23. *Suppose that* $\mathcal{F}$ *is regular,* $0 < \eta < 1$ *and* $0 < p_0, p_1, q_0, q_1, q \le \infty$. *If* $p_0 \ne p_1$ *then*

$$(H^S_{p_0,q_0}, H^S_{p_1,q_1})_{\eta,q} = H^S_{p,q}, \qquad \frac{1}{p} = \frac{1-\eta}{p_0} + \frac{\eta}{p_1}. \tag{5.16}$$

In special cases

$$(H^S_{p_0}, H^S_{p_1})_{\eta,p} = H^S_p, \qquad \frac{1}{p} = \frac{1-\eta}{p_0} + \frac{\eta}{p_1}$$

and, for $1 < p_1 \le \infty$,

$$(H^S_1, L_{p_1})_{\eta,p} = L_p, \qquad \frac{1}{p} = 1 - \eta + \frac{\eta}{p_1}.$$

Only in this theorem we set $H_\infty^S = L_\infty$. Furthermore, if $0 < p < \infty$ then

$$(H_{p,q_0}^S, H_{p,q_1}^S)_{\eta,q} = H_{p,q}^S, \qquad \frac{1}{q} = \frac{1-\eta}{q_0} + \frac{\eta}{q_1}.$$

Proof. (5.16) follows from Corollary 5.21 for $0 < p_1 < 2$. Since we can interpolate between the $H_p^S \sim L_p$ $(1 < p \leq \infty)$ spaces (see Corollary 5.7), we get (5.16) for all $0 < p_1 \leq \infty$ by applying Wolff's theorem twice. ∎

A similar result for the two-parameter classical Hardy spaces can be found in Lin [119].

In the two-parameter case we are able to identify the interpolation spaces between H_p^s and BMO_2 and between L_p and BMO_2 for a regular stochastic basis, only.

Corollary 5.24. *Suppose that $\mathcal{F}$ is regular. If $0 < \theta < 1$, $0 < p_1 < \infty$ and $0 < q, q_1 \leq \infty$ then*

$$(BMO_2, L_{p_1,q_1})_{\theta,q} = L_{p,q}, \qquad \frac{1}{p} = \frac{\theta}{p_1}$$

and

$$(BMO_2, H_{p_1,q_1}^S)_{\theta,q} = H_{p,q}^S, \qquad \frac{1}{p} = \frac{\theta}{p_1}.$$

Proof. From (5.16) and from the duality theorem one can conclude that

$$(BMO_2, L_2)_{\theta,q} = L_{p,q}, \qquad \frac{1}{p} = \frac{\theta}{2}.$$

Applying now Corollaries 5.23, 5.7 and Wolff's theorem we can complete the proof. ∎

CHAPTER 6

INEQUALITIES FOR VILENKIN-FOURIER COEFFICIENTS

In this chapter the notations of Section 1.2 will be used. The one- and two-parameter Vilenkin systems will be considered on the unit interval and on the unit square, respectively, with the σ-algebras generated by them and defined in (1.5) and (1.6), respectively. For the sake of shortness the theorems will be formulated and proved for two parameters, only.

In Section 6.1 a Hardy type inequality is shown for the Vilenkin-Fourier coefficients (see Weisz [193]). This inequality was proved by Hardy and Littlewood [93] and by Coifman and Weiss [52] for the one-parameter trigonometric Fourier coefficients. An inequality relative to the l_2 norm of the so-called "defective" Vilenkin-Fourier coefficients is also verified. In Section 6.2 the dual inequalities to the ones of Section 6.1 are presented. In Section 6.3 the Hardy type inequality that was given in Section 6.1 is extended to all parameters p in case the Vilenkin system is bounded and the sequence of the Vilenkin-Fourier coefficients is monotone. The converse inequality is also verified. For the one-parameter Walsh system the last two results are due to Móricz [136]. Under the conditions mentioned above Carleson's theorem holds also for the two-parameter Vilenkin-Fourier series, namely, for $f \in H_1^*$ the Vilenkin-Fourier series $R_{n,m}f$ tends to f a.e. and in H_1^* norm as $(n,m) \to \infty$. Some of the theorems of this chapter can be found in Weisz [193].

6.1. HARDY TYPE INEQUALITIES

Hausdorff-Young's theorem which is a partial generalization of Parseval's formula and Riesz-Fischer's theorem can also be proved for two parameters. For this let us introduce the notation $\hat{f} := (\hat{f}(n), n \in \mathbb{N}^j)$ $(j = 1,2)$.

The next theorem can be found in Edwards [66] for the one-parameter trigonometric Fourier system and can be shown in the same way for the two-parameter Vilenkin-Fourier coefficients.

Theorem 6.1. (Hausdorff-Young) *Suppose that* $1 \le p \le 2$ *and* $1/p + 1/p' = 1$ $(j = 1,2)$.

(i) *If* $f \in L_p$ *then*

$$\|\hat{f}\|_{l_{p'}} \le \|f\|_p.$$

(ii) *If* $b = (b_n, n \in \mathbb{N}^j) \in l_p$ *then the sequence*

$$R_n = \sum_{m \ll n} b_m w_m$$

converges in $L_{p'}$ norm as $n \to \infty$ to a function f for which

$$\|f\|_{p'} \le \|b\|_{l_p}.$$

Proof. (i) This result can easily be derived from the interpolation theorems. Define the operator T by $f \mapsto \hat{f}$. Obviously,

$$T : L_1 \longrightarrow l_\infty, \qquad T : L_2 \longrightarrow l_2$$

is bounded in both cases with its norms less than or equal to 1. One can see from Theorem 5.1 that

$$T : (L_1, L_2)_{\theta,q} \longrightarrow (l_\infty, l_2)_{\theta,q}$$

is bounded as well with its norm not greater than 1 where $0 < \theta < 1$ and $0 < q \le \infty$. However, by Corollary 5.7, $(L_1, L_2)_{\theta,q} = L_{p,q}$ and $(l_\infty, l_2)_{\theta,q} = l_{r,q}$ with $1/p = 1 - \theta + \theta/2$ and $1/r = \theta/2$. Observe that $p < r$. Setting $q = r$ and considering (5.15) we obtain

$$\|\hat{f}\|_{l_r} \le \|f\|_{p,r} \le \|f\|_{p,p}$$

which proves (i) because $r = p'$.

To prove (ii) we define

$$Tb := \sum_{n \in \mathbb{N}^j} b_n w_n$$

for any sequence b having finite support. It is easy to see that the extension of the operator T is of types (l_1, L_∞) and (l_2, L_2) with its norms less than or equal to 1. As in the first part one can verify that the operator T is bounded form l_p to $L_{p'}$ with its norm not greater than 1. The continuity of T shows at once that $R_n \to f$ in $L_{p'}$ norm as $n \to \infty$. ∎

There are several results in Fourier analysis that hold for L_p $(1 < p < \infty)$, fail to be true for L_1 and remain true for H_1^*. It has been proved that H_1^* contains all L_p $(1 < p < \infty)$ spaces. So H_1^* can be regarded as a good substitute for L_1 that is endowed with many of the properties enjoyed by the L_p spaces $(1 < p < \infty)$. Let us see three examples. The spaces L_p $(1 < p < \infty)$ are reflexive while L_1 is usually not a dual space. We have verified that, in a special case, H_1^* is "almost" reflexive, more exactly, the dual of VMO_2^- is H_1^*. The second example is the Hardy type inequality

$$\left(\sum_{k=1}^{\infty} k^{p-2} |\hat{f}(k)|^p\right)^{1/p} \le C_p \|f\|_{H_p^*} \qquad (0 < p \le 2). \tag{6.1}$$

This can easily be proved for the L_p spaces $(1 < p \le 2)$ (see the last part of the proof of Theorem 6.3) while it does not hold for L_1 (see Zygmund [205] p. 184 in the trigonometric case). Moreover, in case $p = 1$, (6.1) can be shown for the classical

Hardy space and for H_1^* generated by a bounded Vilenkin system. The third example is an inequality due to Paley: for the one-parameter Walsh coefficients

$$\sum_{k=1}^{\infty} |\hat{f}(2^k)|^2 < \infty$$

if $f \in L_p$ $(1 < p < \infty)$. This result either does not hold for $p = 1$. However, it can be verified for H_1^*. This inequality is generalized in Theorem 6.7.

From the following theorems it can be seen that the space L_p $(1 < p < \infty)$ resp. H_1^* is needed to be replaced by $H_p^\bullet$ $(1 < p < \infty)$ resp. $H_1^\bullet$ in case the Vilenkin system is unbounded.

The inequality (6.1) can be found in the paper written by Coifman and Weiss [52] for the one-parameter trigonometric Fourier coefficients, for the classical Hardy space and for $p = 1$ and, moreover, in Hardy, Littlewood [93] for $1 < p \leq 2$. For a one-parameter bounded Vilenkin system and for $p = 1$ the inequality (6.1) was first proved by Chao [37]. For another investigation of this case see Fridli, Simon [79], Ladhawala [112] and Schipp, Wade, Simon, Pál [167]. In Fridli, Simon [79] a similar inequality is proved for $p = 1$ and for a one-parameter unbounded Vilenkin system, however, the Hardy space used there is different from the ones above and is not considered in this book. Moreover, it is proved there that for a one-parameter unbounded Vilenkin system (6.1) does not hold.

Proposition 6.2. *If* $\sup_{k\in\mathbb{N}} p_k = \infty$ *or* $\sup_{k\in\mathbb{N}} q_k = \infty$ *then there exists a function* $f \in H_1^*$ *such that*

$$\sum_{k=1}^{\infty}\sum_{l=1}^{\infty} \frac{|\hat{f}(k,l)|}{kl} = \infty. \tag{6.2}$$

Proof. First we prove this theorem in the one-parameter case. Assume that $\sup_{k\in\mathbb{N}} p_k = \infty$. Without loss of generality one may suppose that $p_k \geq 3$ $(k \in \mathbb{N})$. Let $r_k := [p_k/2] + 1$ $(k \in \mathbb{N})$ and define the functions f_k $(k \in \mathbb{N})$ as follows:

$$f_k(x) := \begin{cases} P_{k+1} & \text{if } x \in [P_{k+1}^{-1}, 2P_{k+1}^{-1}) \\ -P_{k+1} & \text{if } x \in [r_k P_{k+1}^{-1}, (r_k+1)P_{k+1}^{-1}) \\ 0 & \text{otherwise} \end{cases}$$

where $x \in [0,1)$. It is easy to show that the supports of the functions f_k resp. f_k^* are disjoint and $\|f_k\|_1 = \|f_k^*\|_1 = 2$ $(k \in \mathbb{N})$. Let $\lambda_k > 0$ $(k \in \mathbb{N})$ and $\sum_{k=0}^{\infty} \lambda_k < \infty$. Consider the function $f := \sum_{k=0}^{\infty} \lambda_k f_k$. We can establish that

$$\|f\|_1 = \|f^*\|_1 = 2\sum_{k=0}^{\infty} \lambda_k < \infty.$$

Clearly,

$$\sum_{k=1}^{\infty} \frac{|\hat{f}(k)|}{k} = \sum_{k=0}^{\infty} \sum_{j=1}^{p_k-1} \sum_{s=0}^{P_k-1} \frac{|\hat{f}(jP_k+s)|}{jP_k+s}.$$

For fixed j, k and s we have

$$|\hat{f}(jP_k+s)| = |\sum_{t=0}^{\infty} \lambda_t \hat{f}_t(jP_k+s)| = \lambda_k |\hat{f}_k(jP_k+s)| = \lambda_k \Big| \int_0^{P_k^{-1}} f_k \bar{r}_k^j \, dP \Big|$$

since the absolute value of a Vilenkin function w_n is 1. Moreover, it is easy to see that

$$\begin{aligned}\lambda_k \Big| \int_0^{P_k^{-1}} f_k \bar{r}_k^j \, dP \Big| &= \lambda_k \Big| \exp \frac{2\pi \imath j}{p_k} - \exp \frac{2\pi \imath j r_k}{p_k} \Big| \\ &= \lambda_k \Big| 1 - \exp \frac{2\pi \imath j [p_k/2]}{p_k} \Big| \\ &= 2\lambda_k \Big| \sin \frac{\pi j [p_k/2]}{p_k} \Big| \geq C\lambda_k\end{aligned}$$

if $r_k \geq j \equiv 1 \pmod 2$. Henceforth

$$\begin{aligned}\sum_{k=1}^{\infty} \frac{|\hat{f}(k)|}{k} &\geq \sum_{k=0}^{\infty} \sum_{\substack{j=1 \\ j \equiv 1(2)}}^{r_k} \sum_{s=0}^{P_k-1} \frac{C\lambda_k}{jP_k+s} \\ &\geq C \sum_{k=0}^{\infty} \lambda_k \sum_{\substack{j=1 \\ j \equiv 1(2)}}^{r_k} \frac{1}{j} \\ &\geq C \sum_{k=0}^{\infty} \lambda_k \log p_k.\end{aligned}$$

Since $\sup_{k \in \mathbf{N}} p_k = \infty$, we can choose a sequence $(n_k, k \in \mathbf{N})$ of indices such that

$$\sum_{k=0}^{\infty} \frac{1}{\log p_{n_k}} < \infty.$$

Setting $\lambda_{n_k} := \frac{1}{\log p_{n_k}}$ and $\lambda_j = 0$ for $j \neq n_k$ $(k \in \mathbf{N})$ we have $\sum_{k=1}^{\infty} \frac{|\hat{f}(k)|}{k} = \infty$.

In the two-parameter case take an arbitrary function $g \in H_1^*(q_k)$ where $H_1^*(q_k)$ is generated by a Vilenkin system having the basic sequence (q_k). It can easily be verified that the function $f(x)g(y)$ is in H_1^* and, moreover, that

$$\sum_{k=1}^{\infty} \sum_{l=1}^{\infty} \frac{|\widehat{fg}(k,l)|}{kl} = \sum_{k=1}^{\infty} \frac{|\hat{f}(k)|}{k} \sum_{l=1}^{\infty} \frac{|\hat{g}(l)|}{l} = \infty$$

which completes the proof. ∎

The inequality (6.1) will be proved for $0 < p \leq 2$ both in the one- and in the two-parameter case, however, for the space H_p^s instead of H_p^* (see Weisz [193]).

Theorem 6.3. *For an arbitrary martingale $f \in H_p^s$*

$$\Big(\sum_{n=1}^{\infty} \sum_{m=1}^{\infty} \frac{|\hat{f}(n,m)|^p}{(nm)^{2-p}} \Big)^{1/p} \leq C_p \|f\|_{H_p^s} \qquad (0 < p \leq 2). \tag{6.3}$$

Proof. (i) First suppose that $0 < p \leq 1$. Consider the stopping times ν_k, the real numbers μ_k and the $(p,2)$ atoms a^k $(k \in \mathbf{Z})$ as in Theorem 3.2. Then

$$(\sum_{k\in\mathbf{Z}} |\mu_k|^p)^{1/p} \leq C_p \|f\|_{H_p^\bullet}.$$

Since the sets $\{\nu_k \ll n \not\gg \nu_{k+1}\}$ are disjoint for a fixed n, the series (3.13) can be integrated term by term. Hence

$$|\hat{f}(n,m)| \leq \sum_{k\in\mathbf{Z}} |\mu_k||\hat{a}^k(n,m)|.$$

Since $0 < p \leq 1$,

$$\sum_{n=1}^{\infty}\sum_{m=1}^{\infty} \frac{|\hat{f}(n,m)|^p}{(nm)^{2-p}} \leq \sum_{k\in\mathbf{Z}} |\mu_k|^p \sum_{n=1}^{\infty}\sum_{m=1}^{\infty} \frac{|\hat{a}^k(n,m)|^p}{(nm)^{2-p}}.$$

Because of this the only thing we need to prove is that for an arbitrary (p,2) atom a one has

$$\sum_{n=1}^{\infty}\sum_{m=1}^{\infty} \frac{|\hat{a}(n,m)|^p}{(nm)^{2-p}} \leq C_p. \tag{6.4}$$

If ν is the stopping time belonging to a fixed atom a then the support of a^* is obviously $F := \{\nu \neq \infty\}$. We recall that the (set) atoms of the σ-algebra $\mathcal{F}_{n,m}$ are identical with the rectangles in (1.6). For the time being let $m \in \mathbf{N}$ be fixed. To this m, let us choose n such that there exists an atom $A \in \mathcal{F}_{n,m}$ for which $A \subset F$ while for an arbitrary atom $B \in \mathcal{F}_{n-1,m}$ one has $B \cap F^c \neq \emptyset$ (F^c denotes the complement of F); denote this number by $N(m)$. If there is no such number n then say $N(m) = \infty$. The sequence $(N(m))_{m\in\mathbf{N}}$ is obviously non-increasing. Moreover, let

$$m_1 := \min\{m : N(m) < \infty\}, \qquad n_1 := N(m_1).$$

We define the sequence (n_k, m_k) recursively (if it does exist):

$$m_k := \min\{m : N(m) < n_{k-1}\}, \qquad n_k := N(m_k).$$

The sequence (n_k) is decreasing while the sequence (m_k) is increasing, consequently, we have finite pairs (n_k, m_k), only. Let us denote the number of these pairs by K. Set

$$G := \{(n_k, m_k) : 1 \leq k \leq K\}, \qquad H := \{(P_{n_k}, Q_{m_k}) : 1 \leq k \leq K\}.$$

If $G \not\leq (n,m)$ then the construction implies that there is not any atom $A \in \mathcal{F}_{n,m}$ such that $A \subset F$, consequently, for all $\omega \in \Omega$ we have $(n,m) \notin \nu(\omega)$. Thus for all ω

$$G \leq \nu(\omega).$$

If $G \not< (n,m)$ then for all ω one has $\nu(\omega) \not< (n,m)$. Therefore, using the definition of the atom we can see that $a_{n,m}(\omega) = 0$ $(\omega \in \Omega)$ if $G \not< (n,m)$. Next, it is easy to show that $\hat{a}(n,m) = 0$ if $H \not\leq (n,m)$. So, by Hölder's inequality,

$$(6.5)\quad \sum_{n=1}^{\infty}\sum_{m=1}^{\infty} \frac{|\hat{a}(n,m)|^p}{(nm)^{2-p}} = \sum_{H\leq(n,m)} \frac{|\hat{a}(n,m)|^p}{(nm)^{2-p}}$$

$$\leq \Big(\sum_{H\leq(n,m)} |\hat{a}(n,m)|^2\Big)^{p/2}\Big(\sum_{H\leq(n,m)} \frac{1}{(nm)^2}\Big)^{(2-p)/2}.$$

It is obvious by the definition of the atom and from Parseval's inequality that

$$(6.6)\qquad \Big(\sum_{H\leq(n,m)} |\hat{a}(n,m)|^2\Big)^{p/2} = \|a\|_2^p \leq P(F)^{p/2-1}.$$

We shall show that

$$(6.7)\qquad \sum_{H\leq(n,m)} \frac{1}{(nm)^2} \leq 2P(F).$$

Combining (6.5), (6.6) and (6.7) we obtain (6.4) from which the theorem follows.

Notice that

$$(6.8)\qquad \sum_{k=n}^{m-1} \frac{1}{k^2} \leq \int_{n-1}^{m-1} \frac{1}{x^2}\,dx = \frac{1}{n-1} - \frac{1}{m-1} \leq \frac{2}{n} - \frac{2}{m} \qquad (n \geq 3).$$

It is easy to show that the inequality $\sum_{k=n}^{m-1} \frac{1}{k^2} \leq \frac{2}{n} - \frac{2}{m}$ holds also for $n = 1, 2$. Using (6.8) we get immediately that

$$(6.9)\qquad \frac{1}{2}\sum_{H\leq(n,m)} \frac{1}{(nm)^2} \leq \sum_{k=1}^{K} Q_{m_k}^{-1}(P_{n_k}^{-1} - P_{n_{k-1}}^{-1}) =: |H|$$

where $P_{n_0}^{-1} := 0$. By the construction of the set H, for $1 \leq k \leq K$, there exists an atom $A_k \in \mathcal{F}_{n_k,m_k}$ such that $A_k \subset F$. Let $A := \cup_{k=1}^{K} A_k$, then $A \subset F$. It will be shown that

$$(6.10)\qquad |H| \leq P(A).$$

Let us choose an atom $B_k \in \mathcal{F}_{n_k,m_k}$ for all $1 \leq k \leq K$ such that the intersection of two arbitrary atoms B_{k_1} and B_{k_2} is nonempty. Then it can easily be verified that the Lebesgue measure of the set $B := \cup_{k=1}^{K} B_k$ is equal to $|H|$. Let $C := \cup_{k=1}^{K} C_k$ where $C_k \in \mathcal{F}_{n_k,m_k}$ is an atom. By induction on K we shall see that C has minimal area if and only if the intersection of two arbitrary atoms C_{k_1} and C_{k_2} is nonempty. For $k = 1$ or $k = 2$ it is trivial. Let $1 \leq l < K$ be the minimal index for which

$C_{l+1} \cap C_l = \emptyset$. If there is no such index then the intersection of two arbitrary sets C_{k_1} and C_{k_2} is nonempty. It can be seen that

$$P\left(C_l - \bigcup_{\substack{k=1\\k\neq l}}^{K} C_k\right) > P\left(B_l - \bigcup_{\substack{k=1\\k\neq l}}^{K} B_k\right).$$

Nevertheless, by the induction hypothesis we have

$$P\left(\bigcup_{\substack{k=1\\k\neq l}}^{K} C_k\right) \geq P\left(\bigcup_{\substack{k=1\\k\neq l}}^{K} B_k\right),$$

consequently, $P(C) > P(B)$. Thus we have proved (6.10) and (6.7), too, so the proof of the theorem is complete for $0 < p \leq 1$.

The proof is essentially simpler in the one-parameter case. Since in this case instead of H we have simply P_i for any i, we do not need the construction of the sets G and H. The inequality (6.7) follows immediately from (6.8) and so the theorem is proved (cf. Theorem 6.5).

(ii) Secondly, let $1 < p \leq 2$. Introduce on $\mathbf{P}^2$ the measure $\eta(n,m) := 1/(n^2m^2)$. If

$$Tf(n,m) = nm\hat{f}(n,m)$$

then it follows from Parseval's formula and from the previous theorem (for $p = 1$) that both operators

$$T : L_2 \longrightarrow L_2(\mathbf{P}^2, \eta) \tag{6.11}$$

and

$$T : H_1^s \longrightarrow L_1(\mathbf{P}^2, \eta)$$

are bounded. By Theorem 5.1 the operator

$$T : (H_1^s, L_2)_{\theta,p} \longrightarrow \big(L_1(\mathbf{P}^2,\eta), L_2(\mathbf{P}^2,\eta)\big)_{\theta,p} \qquad (0 < \theta < 1)$$

is bounded as well. However, on the one hand,

$$\big(L_1(\mathbf{P}^2,\eta), L_2(\mathbf{P}^2,\eta)\big)_{\theta,p} = L_p(\mathbf{P}^2,\eta)$$

(see Corollary 5.7) and, on the other hand, it was proved in Theorem 5.20 that

$$(H_1^s, L_2)_{\theta,p} = H_p^s$$

where, in both cases, $0 < \theta < 1$ and $1/p = (1-\theta) + \theta/2$. This completes the proof of Theorem 6.3.

In the one-parameter case (6.3) can be proved also for the L_p spaces $(1 < p \leq 2)$ even if the Vilenkin system is unbounded. In this case it can be shown that T is of type $(L_1, L_{1,\infty}(\mathbf{P},\eta))$ (this is probably not true for two parameters), namely, for all $y \in \mathbf{R}$

$$y\eta(|Tf| > y) \leq C\|f\|_1. \tag{6.12}$$

Indeed,

$$\eta(|Tf| > y) = \sum \frac{1}{n^2}$$

where the sum is taken over all n for which $|n\hat{f}(n)| > y$. From this it comes that $n > y/\|f\|_1$. Thus, by (6.8),

$$\eta(|Tf| > y) \le \sum_{n > y/\|f\|_1} \frac{1}{n^2} \le \frac{2\|f\|_1}{y}$$

and this verifies (6.12). From (6.12), (6.11) and from Corollary 5.7 the boundedness of the operator

$$T : L_p \longrightarrow L_p(\mathbf{P}^2, \eta) \qquad (1 < p \le 2)$$

follows which implies the theorem. It is still an open question whether (6.3) holds for any L_p $(1 < p < 2)$ space in the two-parameter unbounded case. ∎

Of course, a similar theorem can be proved even if we do not suppose for a martingale f that $f_{n,0} = f_{0,n} = 0$.

Next we extend Theorem 6.3 to a rearrangement of the Vilenkin-Fourier coefficients. Let α resp. β be permutations on $[P_n, P_{n+1}) \cap \mathbf{N}$ resp. on $[Q_n, Q_{n+1}) \cap \mathbf{N}$ for every $n \in \mathbf{N}$. It can easily be seen that the condition $H \le (n,m)$ is equivalent to $H \le (\alpha(n), \beta(m))$ $(n, m \in \mathbf{N})$ where the set H is constructed in the proof of Theorem 6.3. So the following corollary can be proved in the same way as Theorem 6.3.

Corollary 6.4. *Let α resp. β be permutations on $[P_n, P_{n+1}) \cap \mathbf{N}$ resp. on $[Q_n, Q_{n+1}) \cap \mathbf{N}$ for every $n \in \mathbf{N}$. For an arbitrary martingale $f \in H_p^s$*

$$\left(\sum_{n=1}^{\infty}\sum_{m=1}^{\infty} \frac{|\hat{f}[\alpha(n), \beta(m)]|^p}{(nm)^{2-p}}\right)^{1/p} \le C_p \|f\|_{H_p^s} \qquad (0 < p \le 2).$$

Note that this result can not directly be derived from the inequality (6.3) when the Vilenkin system is unbounded. It is easy to show that, in the one-parameter case, Corollary 6.4 is true even if we replace H_p^s by L_p $(1 < p \le 2)$.

Theorem 6.3 is going to be proved for the H_q^{sat} spaces.

Theorem 6.5. *For an arbitrary martingale $f \in H_q^{\mathrm{sat}}$*

$$\sum_{n=1}^{\infty}\sum_{m=1}^{\infty} \frac{|\hat{f}(n,m)|}{nm} \le \|f\|_{H_q^{\mathrm{sat}}} \qquad (1 < q \le \infty).$$

Proof. Since

$$\|f\|_{H_q^{\mathrm{sat}}} \le \|f\|_{H_r^{\mathrm{sat}}} \qquad (q < r),$$

the theorem is enough to be shown for $1 < q \le 2$. By the previous theorem we have to verify Theorem 6.5 for $(1, q)$ atoms, only. Let F be the support of the atom a and suppose that $F \in \mathcal{F}_{k,l}$. Then, similarly to (6.5),

$$\sum_{n=1}^{\infty}\sum_{m=1}^{\infty}\frac{|\hat{a}(n,m)|}{nm} = \sum_{(P_k,Q_l)\le(n,m)}\frac{|\hat{a}(n,m)|}{nm}$$

$$\le \Big(\sum_{(P_k,Q_l)\le(n,m)}|\hat{a}(n,m)|^{q'}\Big)^{1/q'}\Big(\sum_{(P_k,Q_l)\le(n,m)}\frac{1}{(nm)^q}\Big)^{1/q}$$

where $1/q + 1/q' = 1$. By the definition of the atom a and by Theorem 6.1 (i) it follows that

$$\Big(\sum_{(P_k,Q_l)\le(n,m)}|\hat{a}(n,m)|^{q'}\Big)^{1/q'} \le \|a\|_q \le P(F)^{1/q-1}.$$

Applying again the inequality

$$\sum_{n\ge P_k}\frac{1}{n^q} \le \int_{P_k-1}^{\infty}\frac{1}{x^q}\,dx \le \frac{(P_k-1)^{-q+1}}{q-1} \le C_q P_k^{-q+1}$$

we can conclude that

$$\Big(\sum_{(P_k,Q_l)\le(n,m)}\frac{1}{(nm)^q}\Big)^{1/q} \le C_q P(F)^{1/q'}.$$

The proof of the theorem is complete. ∎

Note that the inequality (6.3) can not be proved with similar methods for the spaces H_q^{at}.

Analogously to Corollary 6.4 the following result holds.

Corollary 6.6. *Let α resp. β be permutations on $[P_n, P_{n+1}) \cap \mathbb{N}$ resp. on $[Q_n, Q_{n+1}) \cap \mathbb{N}$ for every $n \in \mathbb{N}$. For an arbitrary martingale $f \in H_q^{\mathrm{sat}}$*

$$\sum_{n=1}^{\infty}\sum_{m=1}^{\infty}\frac{|\hat{f}[\alpha(n),\beta(m)]|}{nm} \le \|f\|_{H_q^{\mathrm{sat}}} \qquad (1 < q \le \infty).$$

Finally, we show that the l_2 norm of the "defective" Vilenkin coefficients can be estimated by the H_1^s norm of the function. This theorem was first proved by Gundy and Varopoulos [88] for the one-parameter trigonometric Fourier coefficients.

Theorem 6.7. *If $f \in L_1$ then*

$$\Big(\sum_{n=0}^{\infty}\sum_{m=0}^{\infty}\sum_{i=1}^{p_n-1}\sum_{j=1}^{q_m-1}|\hat{f}(iP_n, jQ_m)|^2\Big)^{1/2} \le C\|f\|_{H_1^s}. \tag{6.13}$$

Proof. First of all we show that if

$$g := \sum_{n=0}^{\infty}\sum_{m=0}^{\infty}\sum_{i=1}^{p_n-1}\sum_{j=1}^{q_m-1} \hat{g}(iP_n, jQ_m) w_{iP_n,jQ_m} \in L_2 \tag{6.14}$$

then

$$\|g\|_{BMO_2} \le \|g\|_2 = \Big(\sum_{n=0}^{\infty}\sum_{m=0}^{\infty}\sum_{i=1}^{p_n-1}\sum_{j=1}^{q_m-1} |\hat{g}(iP_n, jQ_m)|^2\Big)^{1/2}. \tag{6.15}$$

Using the definition of the stopped martingale we have

$$\begin{aligned}
&\|g\|_{BMO_2}\\
&= \sup_{\nu\in T} P(\nu \ne \infty)^{-1/2} \|g - g^{\nu}\|_2\\
&= \sup_{\nu\in T} P(\nu \ne \infty)^{-1/2} \Big(\sum_{n=0}^{\infty}\sum_{m=0}^{\infty} E\Big[\chi(\nu \ll (n+1, m+1)) |d_{n+1,m+1}g|^2\Big]\Big)^{1/2}\\
&= \sup_{\nu\in T} P(\nu \ne \infty)^{-1/2} \Big(\sum_{n=0}^{\infty}\sum_{m=0}^{\infty} E\Big[\chi(\nu \ll (n+1, m+1))\\
&\qquad\qquad E_{n,m}|d_{n+1,m+1}g|^2\Big]\Big)^{1/2}.
\end{aligned}$$

Since

$$d_{n+1,m+1}g = \sum_{i=1}^{p_n-1}\sum_{j=1}^{q_m-1} \hat{g}(iP_n, jQ_m) w_{iP_n,jQ_m},$$

we get by (1.10) that

$$E_{n,m}|d_{n+1,m+1}g|^2 = \sum_{i=1}^{p_n-1}\sum_{j=1}^{q_m-1} |\hat{g}(iP_n, jQ_m)|^2.$$

Hence

$$\begin{aligned}
&\|g\|_{BMO_2}\\
&\le \sup_{\nu\in T} P(\nu \ne \infty)^{-1/2} \Big(\sum_{n=0}^{\infty}\sum_{m=0}^{\infty} E\Big[\chi(\nu \ne \infty) \sum_{i=1}^{p_n-1}\sum_{j=1}^{q_m-1} |\hat{g}(iP_n, jQ_m)|^2\Big]\Big)^{1/2}\\
&= \Big(\sum_{n=0}^{\infty}\sum_{m=0}^{\infty}\sum_{i=1}^{p_n-1}\sum_{j=1}^{q_m-1} |\hat{g}(iP_n, jQ_m)|^2\Big)^{1/2}
\end{aligned}$$

which proves (6.15).

If $f \in L_2$ then by Riesz's representation theorem

$$\Big(\sum_{n=0}^{\infty}\sum_{m=0}^{\infty}\sum_{i=1}^{p_n-1}\sum_{j=1}^{q_m-1} |\hat{f}(iP_n, jQ_m)|^2\Big)^{1/2} = \sup_{g} E(fg)$$

where the supremum is taken over all g of the form (6.14) with $\|g\|_2 = 1$. Using (6.15) and Theorem 3.20 we obtain

$$|E(fg)| \le C\|f\|_{H_1^s}\|g\|_{BMO_2} \le C\|f\|_{H_1^s}$$

which, on the one hand verifies (6.13) for $f \in L_2$. On the other hand, Theorem 6.7 follows easily from the fact that L_2 is dense in H_1^s which was proved in Theorem 3.2. ■

The same result holds for a rearrangement of the Vilenkin-Fourier coefficients:

Corollary 6.8. *Let α resp. β be permutations on $[P_n, P_{n+1}) \cap \mathbf{N}$ resp. on $[Q_n, Q_{n+1}) \cap \mathbf{N}$ for every $n \in \mathbf{N}$. If $f \in L_1$ then*

$$\Big(\sum_{n=0}^{\infty}\sum_{m=0}^{\infty}\sum_{i=1}^{p_n-1}\sum_{j=1}^{q_m-1}|\hat f[\alpha(iP_n),\beta(jQ_m)]|^2\Big)^{1/2} \le C\|f\|_{H_1^s}.$$

It is still unknown whether (6.13) holds if we write on the right hand side of the inequality $C_p\|f\|_p$ instead of $C\|f\|_{H_1^s}$ whenever $f \in L_p$ $(1 < p < 2)$ and (p_n) or (q_n) is unbounded.

6.2. DUAL INEQUALITIES

In this section direct proofs for the dual inequalities of the theorems of the previous section are given. Let us denote the set of the sequences $\{(P_{n_k}, Q_{m_k}) : 1 \le k \le K\}$ by $\mathcal{H}$ where the sequence (n_k) is decreasing, the sequence (m_k) is increasing and $K \in \mathbf{N}$. Let us begin with the dual inequality of Theorem 6.3 for $0 < p \le 1$.

Theorem 6.9. *If $(b_{n,m}; n, m \in \mathbf{P})$ is a sequence of complex numbers such that*

$$M := \sup_{H\in\mathcal{H}} |H|^{-1/2-\alpha}\Big(\sum_{H\le(n,m)} |b_{n,m}|^2\Big)^{1/2} < \infty$$

then there exists $\phi \in \Lambda_2(\alpha)$ for which $\hat\phi(n,m) = b_{n,m}$ $(n, m \in \mathbf{P})$ and $\|\phi\|_{\Lambda_2(\alpha)} \le M$. (The definition of $|H|$ is given in (6.9).)

Proof. It follows from Riesz-Fischer's theorem that there exists a function $\phi \in L_2$ such that $\hat\phi(n,m) = b_{n,m}$ $(n, m \in \mathbf{P})$. For an arbitrary stopping time ν one has

$$\|\phi - \phi^\nu\|_2^2 = \sum_{i=1}^{\infty}\sum_{j=1}^{\infty} E[\chi(\nu \ll (i,j))|d_{i,j}\phi|^2].$$

Let us use again the sets G and H constructed for the stopping time ν in the proof of Theorem 6.3. It is clear that

$$\|\phi - \phi^\nu\|_2^2 = \sum_{G\ll(i,j)} E[\chi(\nu \ll (i,j))|d_{i,j}\phi|^2] \le \sum_{G\ll(i,j)} E(|d_{i,j}\phi|^2).$$

Expressing $d_{i,j}\phi$ as a linear combination of the functions $w_{n,m}$ one obtains

$$\|\phi - \phi^\nu\|_2^2 \leq \sum_{H \leq (n,m)} |b_{n,m}|^2.$$

As we could see in the proof of Theorem 6.3, $|H| \leq P(\nu \neq \infty)$, consequently,

$$P(\nu \neq \infty)^{-1/2-\alpha} \|\phi - \phi^\nu\|_2 \leq |H|^{-1/2-\alpha} \Big(\sum_{H \leq (n,m)} |b_{n,m}|^2 \Big)^{1/2},$$

that is to say $\|\phi\|_{\Lambda_2(\alpha)} \leq M$, which shows Theorem 6.9. ∎

Let us give the dual theorem of Theorem 6.3 for $1 < p \leq 2$, too. It was proved in Theorem 3.23 that the dual of H_p^s is H_q^s where $1/p + 1/q = 1$ and $1 < p < \infty$. The following theorem can be shown as the one-parameter corresponding result for the trigonometric system (see Edwards [66] p. 193).

Theorem 6.10. *If* $2 \leq q < \infty$ *and* $(b_{n,m}; n, m \in \mathbf{P})$ *is a sequence of complex numbers such that*

$$\sum_{n=1}^\infty \sum_{m=1}^\infty \frac{|b_{n,m}|^q}{(nm)^{2-q}} < \infty,$$

then the Vilenkin polynomials

$$f_{N,M} := \sum_{n=1}^{N-1} \sum_{m=1}^{M-1} b_{n,m} w_{n,m}$$

converge in H_q^s *norm as* $(N, M) \to \infty$ *to a function* $f \in H_q^s$ *satisfying* $\hat{f}(n,m) = b_{n,m}$ $(n, m \in \mathbf{P})$ *and*

$$\|f\|_{H_q^s} \leq C_q \Big(\sum_{n=1}^\infty \sum_{m=1}^\infty \frac{|b_{n,m}|^q}{(nm)^{2-q}} \Big)^{1/q}.$$

Proof. Assume that $n \in \mathbf{N}$ has the canonical decomposition

$$n = \sum_{k=0}^N n_k P_k \qquad (n_N \neq 0)$$

where $n_k \in \mathbf{N}$ and $n_k < p_k$. It is easy to see that setting

$$m := \sum_{k=0}^N (p_k - n_k) P_k$$

we have $w_m = \overline{w}_n$. Say $\alpha(n) = m$. Since $p_N - n_N \neq 0$, the hypothesis $P_N \leq n < P_{N+1}$ implies that $P_N \leq \alpha(n) < P_{N+1}$. Therefore the permutation α satisfies the condition in Corollary 6.4. Another permutation β can be defined in the same way with replacing P_k by Q_k $(k \in \mathbf{N})$.

Let $1/p+1/q=1$, so $1<p\leq 2$. Suppose that $g\in H_p^\bullet$ and $(N,M)\leq(N',M')$. Applying Hölder's inequality and Corollary 6.4 we can see that

$$\begin{aligned}
|E(f_{N,M}g)| &= \Big|\sum_{n=1}^{N-1}\sum_{m=1}^{M-1} b_{n,m}\hat{g}[\alpha(n),\beta(m)]\Big| \\
&= \Big|\sum_{n=1}^{N-1}\sum_{m=1}^{M-1}(nm)^{1-2/q}b_{n,m}(nm)^{1-2/p}\hat{g}[\alpha(n),\beta(m)]\Big| \\
&\leq \Big(\sum_{n=1}^{N-1}\sum_{m=1}^{M-1}\frac{|b_{n,m}|^q}{(nm)^{2-q}}\Big)^{1/q}\Big(\sum_{n=1}^{N-1}\sum_{m=1}^{M-1}\frac{|\hat{g}[\alpha(n),\beta(m)]|^p}{(nm)^{2-p}}\Big)^{1/p} \\
&\leq \Big(\sum_{n=1}^{N-1}\sum_{m=1}^{M-1}\frac{|b_{n,m}|^q}{(nm)^{2-q}}\Big)^{1/q} C_p\|g\|_{H_p^\bullet}.
\end{aligned}$$

By Theorem 3.23,

$$\|f_{N,M}\|_{H_q^\bullet}\leq C_p\Big(\sum_{n=1}^{N-1}\sum_{m=1}^{M-1}\frac{|b_{n,m}|^q}{(nm)^{2-q}}\Big)^{1/q}.$$

Similarly,

$$\|f_{N',M'}-f_{N,M}\|_{H_q^\bullet}\leq C_p\left(\sum_{(N,M)\not\gg(n,m)\ll(N',M')}\frac{|b_{n,m}|^q}{(nm)^{2-q}}\right)^{1/q}.$$

Henceforth the sequence $(f_{N,M})$ is a Cauchy one in H_q^s. Denoting its limit by f we obtain the desired inequality. The obvious inequality $\|f\|_1\leq\|f\|_{H_q^\bullet}$ shows that

$$\hat{f}(n,m)=\lim_{(N,M)\to\infty}\hat{f}_{N,M}(n,m)=b_{n,m}$$

which completes the proof. ■

Note that, in the one-parameter case, the same theorem holds for the L_q spaces $(2\leq q<\infty)$.

A result analogous to Theorem 6.9 will be proved for the $BMO_q^\square$ spaces.

Theorem 6.11. *If* $2\leq q<\infty$ *and* $(b_{n,m};n,m\in\mathbf{P})$ *is a sequence of complex numbers such that*

$$M:=\sup_{n,m\in\mathbf{N}}(P_nQ_m)^{1/q}\left(\sum_{(P_n,Q_m)\leq(k,l)}|b_{k,l}|^{q'}\right)^{1/q'}<\infty$$

then there exists $\phi\in BMO_q^\square$ *for which* $\hat{\phi}(k,l)=b_{k,l}$ $(k,l\in\mathbf{P})$ *and* $\|\phi\|_{BMO_q^\square}\leq M$.

Proof. By Theorem 6.1 (ii) there exists $\phi \in L_q$ such that $\hat{\phi}(k,l) = b_{k,l}$ $(k,l \in \mathbf{P})$. Let $n,m,l,j \in \mathbf{N}$ and introduce the Vilenkin polynomials

$$P_{l,j}^{(n,m)} := \sum_{k=0}^{P_n-1} \sum_{i=0}^{Q_m-1} \hat{\phi}(lP_n + k, jQ_m + i) w_{k,i}.$$

Then

$$\phi - E_{n,\infty}\phi - E_{\infty,m}\phi + E_{n,m}\phi = \sum_{l=1}^{\infty}\sum_{j=1}^{\infty} P_{l,j}^{(n,m)} w_{lP_n, jQ_m}.$$

As $P_{l,j}^{(n,m)}$ is $\mathcal{F}_{n,m}$ measurable, applying again Theorem 6.1 (ii), we get the following inequalities $(1/q + 1/q' = 1)$:

$$\begin{aligned}
(E_{n,m}|\phi - E_{n,\infty}\phi - E_{\infty,m}\phi + E_{n,m}\phi|^q)^{1/q} \\
&= \left(E_{n,m}\Big|\sum_{l=1}^{\infty}\sum_{j=1}^{\infty} P_{l,j}^{(n,m)} w_{lP_n,jQ_m}\Big|^q\right)^{1/q} \\
&\le \left(\sum_{l=1}^{\infty}\sum_{j=1}^{\infty} |P_{l,j}^{(n,m)}|^{q'}\right)^{1/q'} \\
&\le \left((P_nQ_m)^{q'-1} \sum_{(P_n,Q_m)\le(k,l)} |\hat{\phi}(k,l)|^{q'}\right)^{1/q'},
\end{aligned}$$

namely, $\|\phi\|_{BMO_q^{\square}} \le M$. This completes the proof of Theorem 6.11. ∎

This theorem can be found for the one-parameter dyadic martingales in Ladhawala [112] and Schipp, Wade, Simon, Pál [167] while another version for nonlinear martingales is written by Weisz [196].

The dual of Theorem 6.7 is formulated in (6.13).

6.3. CONVERGENCE OF VILENKIN-FOURIER SERIES

In this section, under certain conditions, Theorem 6.3 is extended to the case $p > 2$. Moreover, the converse inequality is proved, too. In the sequel it is supposed that the Vilenkin system is bounded, namely,

$$p_n = O(1), \qquad q_n = O(1). \tag{6.16}$$

If $f = (f_{n,m}; n,m \in \mathbf{N})$ is a martingale and $b_{k,l} := \hat{f}(k,l)$ then it is obvious that

$$f_{n,m} = \sum_{k=1}^{P_n-1}\sum_{i=1}^{Q_m-1} b_{k,l} w_{k,l}$$

and, conversely, an arbitrary sequence $(b_{k,l}; k,l \in \mathbf{P})$ defines a martingale. From this time on we consider only those martingales for which

$$b_{k,l} \to 0 \quad \text{as} \quad \max(k,l) \to \infty, \tag{6.17}$$

(6.18) $$\Re(b_{k,l}-b_{k+1,l}-b_{k,l+1}+b_{k+1,l+1})\geq 0$$

and

(6.18) $$\Im(b_{k,l}-b_{k+1,l}-b_{k,l+1}+b_{k+1,l+1})\geq 0 \qquad (k,l\in \mathbf{P})$$

where $\Re b$ and $\Im b$ denote the real and the imaginary part of a complex number b, respectively. It follows immediately from (6.17) and (6.18) that the sequences $(\Re b_{k,l})$, $(\Im b_{k,l})$ and $(|b_{k,l}|)$ are decreasing.

Theorem 6.12. *Under the conditions (6.16) and (6.18) let us suppose that $f\in H_p^\bullet$. Then*

(6.19) $$\Big(\sum_{n=1}^{\infty}\sum_{m=1}^{\infty}\frac{|\hat{f}(n,m)|^p}{(nm)^{2-p}}\Big)^{1/p}\leq C_p\|f\|_{H_p^\bullet} \qquad (0<p<\infty).$$

First we note that, in this case, by Theorems 3.9 and 3.19 the equivalence $H_p^s\sim H_p^\bullet$ $(0<p<\infty)$ is valid, so this theorem is really an extension of Theorem 6.3. Of course, (6.17) follows from the condition $f\in H_p^\bullet$ $(p\geq 1)$ (see e.g. Schipp, Wade, Simon, Pál [167]). Theorem 6.12 was shown by Móricz [136] for the one-parameter Walsh system and for $p\geq 1$. Our proof follows basically his verification.

Proof. The statement would be enough to be shown for $p>2$, however, the below version gives a proof for $p>1$. Let

$$F(u,v):=\int_0^u\int_0^v f(x,y)\,dx\,dy, \qquad F_1(u,v):=\int_0^u\int_0^v |f(x,y)|\,dx\,dy.$$

By a well known property of Dirichlet kernel functions mentioned in Section 1.2 one has

$$\begin{aligned}F(P_k^{-1},Q_l^{-1})&=P_k^{-1}Q_l^{-1}\int_0^1\int_0^1 f(x,y)\overline{D_{P_k,Q_l}(x,y)}\,dx\,dy\\&=P_k^{-1}Q_l^{-1}\sum_{i=1}^{P_k-1}\sum_{j=1}^{Q_l-1}b_{i,j}\end{aligned}$$

where $b_{i,j}:=\hat{f}(i,j)$. Since $(\Re b_{i,j})$ and $(\Im b_{i,j})$ are decreasing,

$$\Re F(P_k^{-1},Q_l^{-1})\geq \Re b_{P_k-1,Q_l-1}$$

and

$$\Im F(P_k^{-1},Q_l^{-1})\geq \Im b_{P_k-1,Q_l-1}.$$

Consequently,

$$|F(P_k^{-1},Q_l^{-1})|\geq |b_{P_k-1,Q_l-1}|.$$

By a simple calculation we get that

(6.20) $$\sum_{k=1}^{\infty}\sum_{l=1}^{\infty}|b_{P_k-1,Q_l-1}|^p(P_kQ_l)^{p-1}$$

$$\leq \sum_{k=1}^{\infty}\sum_{l=1}^{\infty} |F(P_k^{-1}, Q_l^{-1})|^p (P_k Q_l)^{p-1}$$

$$\leq C_p \sum_{k=1}^{\infty}\sum_{l=1}^{\infty} \int_{P_k^{-1}}^{P_{k-1}^{-1}} \int_{Q_l^{-1}}^{Q_{l-1}^{-1}} \left(\frac{F_1(x,y)}{xy}\right)^p dx\, dy$$

$$= C_p \int_0^1 \int_0^1 \left(\frac{F_1(x,y)}{xy}\right)^p dx\, dy.$$

Next, Hardy's inequality is used (see Theorem 5.5 (i)): if f and F_1 are functions of an only variable then

$$\int_0^1 \left(\frac{F_1(x)}{x}\right)^p dx \leq C_p \int_0^1 |f(x)|^p\, dx.$$

Here the condition $p > 1$ is utilized. Applying the previous inequality twice we get the two-variable version of this inequality:

$$\int_0^1 \int_0^1 \left(\frac{F_1(x,y)}{xy}\right)^p dx\, dy \leq C_p \int_0^1 \int_0^1 |f(x,y)|^p\, dx\, dy. \tag{6.21}$$

Using the monotony of the sequence $(|b_{n,m}|)$ we obtain the inequality

$$\sum_{n=1}^{\infty}\sum_{m=1}^{\infty} |b_{n,m}|^p (nm)^{p-2} \leq \sum_{k=1}^{\infty}\sum_{l=1}^{\infty} |b_{P_k-1,Q_l-1}|^p \sum_{n=P_k-1}^{P_{k+1}-2} \sum_{Q_l-1}^{Q_{l+1}-2} (nm)^{p-2}. \tag{6.22}$$

By a simple calculation it can be seen that

$$\sum_{n=P_k-1}^{P_{k+1}-2} n^{p-2} \leq C_p P_k^{p-1}.$$

Now

$$\sum_{n=1}^{\infty}\sum_{m=1}^{\infty} |b_{n,m}|^p (nm)^{p-2} \leq C_p \sum_{k=1}^{\infty}\sum_{l=1}^{\infty} |b_{P_k-1,Q_l-1}|^p (P_k Q_l)^{p-1} \leq C_p \|f\|_p^p$$

follows from (6.20), (6.21) and (6.22) and this proves the theorem. ∎

A sharper assertion than the converse of (6.19) will be shown. If

$$R_{n,m} := \sum_{k=1}^{n-1}\sum_{l=1}^{m-1} b_{k,l} w_{k,l}$$

then the following inequality holds, too.

Theorem 6.13. *Under the conditions (6.16), (6.17) and (6.18),*

$$\| \sup_{n,m\in\mathbb{N}} |R_{n,m}| \|_p \leq C_p \Big(\sum_{n=1}^{\infty}\sum_{m=1}^{\infty} \frac{|b_{n,m}|^p}{(nm)^{2-p}}\Big)^{1/p} \qquad (0 < p < \infty).$$

Theorems 6.12 and 6.13 were proved also by Móricz [134] for the two-parameter Walsh system and for $p \geq 1$. Roughly at the same time the author verified them for two-parameter bounded Vilenkin systems as well. For the proof a slightly modified version of Hardy's inequality is needed.

Lemma 6.14. *If $r > 1$, $0 < p < \infty$ and $(d_n, n \in \mathbf{P})$ is a non-negative, non-increasing sequence then*

$$\sum_{n=1}^{\infty} n^{-r}\left(\sum_{k=1}^{n+1} d_k\right)^p \leq C_p \sum_{n=1}^{\infty} d_n^p n^{p-r}.$$

Proof. This lemma follows easily from Theorem 5.8 with taking the function $f(x) := d_n$ for $x \in [n, n+1)$ $(n \in \mathbf{P})$. ∎

Proof of Theorem 6.13. Introduce the notations

$$\triangle_{10} b_{k,l} := b_{k,l} - b_{k+1,l}, \qquad \triangle_{01} b_{k,l} := b_{k,l} - b_{k,l+1}$$

and

$$\triangle_{11} b_{k,l} := b_{k,l} - b_{k+1,l} - b_{k,l+1} + b_{k+1,l+1}.$$

It is known (see e.g. Fridli, Simon [79]) that, for a bounded Vilenkin system,

$$|D_n(x)| \leq \frac{2}{x} \qquad (x \in [0,1), n \in \mathbf{N}).$$

Let $(x,y) \in [0,1)^2$ be given. Choose the numbers i and $j \in \mathbf{P}$ such that

$$\frac{1}{i+1} \leq x < \frac{1}{i}, \qquad \frac{1}{j+1} \leq y < \frac{1}{j}. \tag{6.23}$$

Then, for $n \geq i+2$ and $m \geq j+2$,

$$\begin{aligned} (6.24) \quad & |\Re R_{n,m}(x,y)| \\ & \leq |\sum_{k=1}^{i}\sum_{l=1}^{j} \Re(b_{k,l} w_{k,l}(x,y))| + \sum_{l=1}^{j} |\sum_{k=i+1}^{n-1} \Re(b_{k,l} w_{k,l}(x,y))| \\ & + \sum_{k=1}^{i} |\sum_{l=j+1}^{m-1} \Re(b_{k,l} w_{k,l}(x,y))| + |\sum_{k=i+1}^{n-1}\sum_{l=j+1}^{m-1} \Re(b_{k,l} w_{k,l}(x,y))|. \end{aligned}$$

The first term can simply be estimated by

$$\sum_{k=1}^{i}\sum_{l=1}^{j} |b_{k,l}|.$$

For the second term we have

$$|\sum_{k=i+1}^{n-1} \Re(b_{k,l} w_{k,l}(x,y))| \leq |\sum_{k=i+1}^{n-1} \Re(b_{k,l} w_k(x))| + |\sum_{k=i+1}^{n-1} \Im(b_{k,l} w_k(x))|.$$

By Abel rearrangement we get that

$$\sum_{k=i+1}^{n-1} \Re b_{k,l} \Re w_k(x)$$

$$= -\Re b_{i+1,l} \Re D_{i+1}(x) + \sum_{k=i+1}^{n-1} \Re D_{k+1} \triangle_{10} \Re b_{k,l} + \Re b_{n,l} \Re D_n(x),$$

hence

$$\begin{aligned} |\sum_{k=i+1}^{n-1} \Re b_{k,l} \Re w_k(x)| &\le \frac{2}{x}(|\Re b_{i+1,l}| + \sum_{k=i+1}^{n-1} |\triangle_{10} \Re b_{k,l}| + |\Re b_{n,l}|) \\ &= \frac{4}{x} \Re b_{i+1,l} \le \frac{4}{x} |b_{i+1,l}|. \end{aligned}$$

The same estimation can be done for the expression $|\sum_{k=i+1}^{n-1} \Im b_{k,l} \Im w_k(x))|$. Thus

$$|\sum_{k=i+1}^{n-1} \Re(b_{k,l} w_k(x))| \le \frac{8}{x} |b_{i+1,l}|.$$

Similarly,

$$|\sum_{k=i+1}^{n-1} \Im(b_{k,l} w_k(x))| \le \frac{8}{x} |b_{i+1,l}|.$$

Using this and (6.23) we can conclude that

$$\begin{aligned} \text{(6.25)} \qquad \sum_{l=1}^{j} |\sum_{k=i+1}^{n-1} \Re(b_{k,l} w_{k,l}(x,y))| &\le \frac{16}{x} \sum_{l=1}^{j} |b_{i+1,l}| \\ &\le 16(i+1) \sum_{l=1}^{j} |b_{i+1,l}| \\ &\le 16 \sum_{k=1}^{i+1} \sum_{l=1}^{j} |b_{k,l}|. \end{aligned}$$

The inequality

$$\text{(6.26)} \qquad \sum_{k=1}^{i} |\sum_{l=j+1}^{m-1} \Re(b_{k,l} w_{k,l}(x,y))| \le 16 \sum_{k=1}^{i} \sum_{l=1}^{j+1} |b_{k,l}|$$

can be shown analogously. In order to estimate the fourth term two expressions have to be considered again:

$$|\sum_{k=i+1}^{n-1} \sum_{l=j+1}^{m-1} \Re b_{k,l} \Re(w_{k,l}(x,y))|, \qquad |\sum_{k=i+1}^{n-1} \sum_{l=j+1}^{m-1} \Im b_{k,l} \Im(w_{k,l}(x,y))|.$$

Let the first one be majorized. With the two-parameter Abel rearrangement we obtain

$$\sum_{k=i+1}^{n-1}\sum_{l=j+1}^{m-1} \Re b_{k,l}\Re(w_{k,l}(x,y))$$
$$= \sum_{k=i+1}^{n-1}\sum_{l=j+1}^{m-1} \Re D_{k+1,l+1}(x,y)\Delta_{11}\Re b_{k,l}$$
$$+ \sum_{k=i+1}^{n-1} \Re D_{k+1,m}(x,y)\Delta_{10}\Re b_{k,m} - \sum_{k=i+1}^{n-1} \Re D_{k+1,j+1}(x,y)\Delta_{10}\Re b_{k,j+1}$$
$$+ \sum_{l=j+1}^{m-1} \Re D_{n,l+1}(x,y)\Delta_{01}\Re b_{n,l} - \sum_{l=j+1}^{m-1} \Re D_{i+1,l+1}(x,y)\Delta_{01}\Re b_{i+1,l}$$
$$+ \Re b_{n,m}\Re D_{n,m}(x,y) - \Re b_{n,j+1}\Re D_{n,j+1}(x,y)$$
$$- \Re b_{i+1,m}\Re D_{i+1,m}(x,y) + \Re b_{i+1,j+1}\Re D_{i+1,j+1}(x,y).$$

Consequently,

$$|\sum_{k=i+1}^{n-1}\sum_{l=j+1}^{m-1} \Re b_{k,l}\Re(w_{k,l}(x,y))|$$
$$\leq \frac{4}{xy}\Bigg(\sum_{k=i+1}^{n-1}\sum_{l=j+1}^{m-1} \Delta_{11}\Re b_{k,l} + \sum_{k=i+1}^{n-1} \Delta_{10}\Re b_{k,m} + \sum_{k=i+1}^{n-1} \Delta_{10}\Re b_{k,j+1}$$
$$+ \sum_{l=j+1}^{m-1} \Delta_{01}\Re b_{n,l} + \sum_{l=j+1}^{m-1} \Delta_{01}\Re b_{i+1,l}$$
$$+ \Re b_{n,m} + \Re b_{n,j+1} + \Re b_{i+1,m} + \Re b_{i+1,j+1}\Bigg)$$
$$= \frac{16}{xy}|b_{i+1,j+1}|.$$

By (6.23),

$$|\sum_{k=i+1}^{n-1}\sum_{l=j+1}^{m-1} \Re(b_{k,l}w_{k,l}(x,y))| \leq 32(i+1)(j+1)|b_{i+1,j+1}| \tag{6.27}$$
$$\leq 32\sum_{k=1}^{i+1}\sum_{l=1}^{j+1}|b_{k,l}|.$$

Combining the estimates (6.24)-(6.27) we get that, for $(n,m) \geq (i+2, j+2)$,

$$|R_{n,m}(x,y)| \leq 130\sum_{k=1}^{i+1}\sum_{l=1}^{j+1}|b_{k,l}|.$$

This inequality can be extended in the same way to $(n,m) \not\geq (i+2, j+2)$. Using this we obtain the inequality

$$\| \sup_{n,m\in\mathbb{N}} |R_{n,m}| \|_p^p = \sum_{i=1}^{\infty}\sum_{j=1}^{\infty} \int_{1/(i+1)}^{1/i} \int_{1/(j+1)}^{1/j} \Big(\sup_{n,m\in\mathbb{N}} |R_{n,m}(x,y)| \Big)^p \, dx\, dy$$
$$\leq C_p \sum_{i=1}^{\infty}\sum_{j=1}^{\infty} \frac{1}{i^2 j^2} \Big(\sum_{k=1}^{i+1}\sum_{l=1}^{j+1} |b_{k,l}| \Big)^p .$$

Finally, applying Lemma 6.14 ($r = 2$) twice we can conclude that

$$\sum_{i=1}^{\infty}\sum_{j=1}^{\infty} \frac{1}{i^2 j^2} \Big(\sum_{k=1}^{i+1}\sum_{l=1}^{j+1} |b_{k,l}| \Big)^p \leq C_p \sum_{k=1}^{\infty}\sum_{l=1}^{\infty} |b_{k,l}|^p (kl)^{p-2}$$

which completes the proof of Theorem 6.13. ■

The next consequence follows easily from the last theorems.

Corollary 6.15. *If (6.16), (6.17) and (6.18) are satisfied and $0 < p < \infty$ is fixed then the following three conditions are equivalent:*

$$\sup_{n,m\in\mathbb{N}} |R_{n,m}| \in L^p, \quad f \in H_p^*, \quad \sum_{n=1}^{\infty}\sum_{m=1}^{\infty} \frac{|b_{n,m}|^p}{(nm)^{2-p}} < \infty.$$

From this it follows immediately that the following generalization of a theorem relative to the one-parameter Walsh system due to Balasov and Móricz (see [1], [136] Theorem A, Corollary 2) holds (we recall that $H_p^* \sim L_p$ for $1 < p < \infty$).

Corollary 6.16. *Suppose that (6.16), (6.17) and (6.18) are satisfied and $1 \leq p < \infty$. Then the following three statements are equivalent:*

(i) *For the partial sums $R_{n,m}$ of the series*

$$\sum_{k=1}^{\infty}\sum_{l=1}^{\infty} b_{k,l} w_{k,l}$$

we have $\sup_{n,m\in\mathbb{N}} |R_{n,m}| \in L^p$.

(ii) *The previous series is the Vilenkin-Fourier series of a function $f \in H_p^*$.*

(iii) *For the sequence $(b_{n,m})$ we have*

$$\sum_{n=1}^{\infty}\sum_{m=1}^{\infty} \frac{|b_{n,m}|^p}{(nm)^{2-p}} < \infty.$$

The inequality

$$\| \sup_{n,m\in\mathbb{N}} |R_{n,m}| \|_p \leq C_p \|f\|_{H_p^*} \qquad (0 < p < \infty) \tag{6.28}$$

can be obtained from Corollary 6.15. Since the Vilenkin polynomials are dense in H_p^*, by (6.28), by Banach-Steinhaus's theorem and by Lemma 4.17 the following generalization of Carleson's theorem is reached.

Corollary 6.17. *Let $p_n = O(1)$ and $q_n = O(1)$. If $f \in L^p$ $(p > 1)$ or $f \in H_1^*$ such that (6.18) is satisfied then $R_{n,m}f \to f$ a.e. and also in H_p^* norm $(p \geq 1)$ as $(n, m) \to \infty$.*

The papers written by D'jachenko [59] and Móricz [135] are dealing with similar problems for the two-parameter trigonometric system.

For one parameter Móricz [136] has proved that under the conditions (6.16), (6.17) and (6.18) there exists a function $f \in L_1$ the partial sums of which converge in L_1 norm though $f \notin H_1^*$. Of course, this result holds for two parameters, too. The L_1 convergence of the one- and the two-parameter Walsh series is also investigated in Móricz, Schipp [138] and [139].

Note that the a.e. convergence of the Walsh series $\sum_{k=1}^{\infty}\sum_{l=1}^{\infty} b_{k,l}w_{k,l}$ follows also from the conditions (6.17) and (6.18) (see Móricz, Schipp, Wade [141] or Móricz [133]). Under these conditions with $\sum_{n=1}^{\infty}\sum_{m=1}^{\infty} \frac{|b_{n,m}|}{nm} < \infty$ the same authors ([141]) have verified the L_1 convergence of the previous series, too.

Finally, we note that, by applying the one-parameter result twice, we get the L_p $(1 < p < \infty)$ convergence of the two-parameter Vilenkin-Fourier series both for bounded and unbounded Vilenkin systems independently of the monotony of the coefficients.

Corollary 6.18. *For an arbitrary Vilenkin system and for every $f \in L_p$ with $1 < p < \infty$ we have that $R_{n,m}f \to f$ in L_p norm as $(n, m) \to \infty$.*

REFERENCES

[1] Balasov, L.A.: Series with respect to the Walsh system with monotone coefficients. Sibirsk. Mat. Ž. 12, 25-39 (1971) (in Russian)

[2] Banuelos, R.: A note on martingale transforms and A_p-weights. Studia Math. 85, 125-135 (1987)

[3] Bassily, N.L., Mogyoródi, J.: On the BMO_Φ-spaces with general Young function. Ann. Univ. Sci. Budap. Rolando Eötvös, Sect. Math. 27, 215-227 (1984)

[4] Bassily, N.L., Mogyoródi, J.: On the $\mathcal{K}_\Phi$-spaces with general Young function Φ. Ann. Univ. Sci. Budap. Rolando Eötvös, Sect. Math. 27, 205-214 (1984)

[5] Benedeck, A., Panzone, R.: The spaces L^p with mixed norm. Duke Math. J. 28, 301-324 (1961)

[6] Bennett, C., Sharpley, R.: Interpolation of operators, Pure and Applied Mathematics, vol. 129. Academic Press, New York 1988

[7] Bennett, C., Sharpley, R.: Weak type inequalities for H^p and BMO, Harmonic Analysis in Euclidean Spaces. (Proceedings of Symposia in Pure Mathematics, vol. 35. Part 1, pp. 201-229) American Mathematical Society 1979

[8] Bergh, J., Löfström, J.: Interpolation spaces, an introduction. Berlin, Heidelberg, New York: Springer 1976

[9] Bernard, A.: Espaces H_1 de martingales à deux indices. Dualité avec les martingales de type BMO. Bull. Sc. math. 103, 297-303 (1979)

[10] Bernard, A., Maisonneuve, B.: Décomposition atomique de martingales de la classe H^1, Séminaire de Probabilités XI. (Lect. Notes Math., vol. 581, pp. 303-323) Berlin, Heidelberg, New York: Springer 1977

[11] Bichteler, K.: Stochastic integration and L^p-theory of semimartingales. Annals of Prob. 9, 49-89 (1981)

[12] Billard, P.: Sur la convergence presque partout des séries de Fourier-Walsh des fonctions de l'espace $L^2[0,1]$. Studia Math. 28, 363-388 (1967)

[13] Blasco, O., Xu, Q.: Interpolation between vector-valued Hardy spaces. J. Func. Anal. 102, 331-359 (1991)

[14] Brossard, J.: Comparaison des "normes" L_p du processus croissant et de la variable maximale pour les martingales régulières à deux indices. Théorème local correspondant. Annals of Prob. 8, 1183-1188 (1980)

[15] Brossard, J.: Régularité des martingales à deux indices et inégalités de normes, Processus Aléatoires à Deux Indices. (Lect. Notes Math., vol. 863, pp. 91-121) Berlin, Heidelberg, New York: Springer 1981

[16] Brossard, J., Chevalier, L.: Calcul stochastique et inégalités de norme pour les martingales bi-Browniennes. Application aux fonctions bi-harmoniques. Ann. Inst. Fourier, Grenoble 30, 97-120 (1980)

[17] Burkholder, D.L.: Brownian motion and classical analysis. Proc. Symp. Pure Math. 31, 5-14 (1977)

[18] Burkholder, D.L.: Differential subordination of harmonic functions and martingales, Proceedings of the Seminar on Harmonic Analysis and Partial Differential Equations (El Escorial, Spain, 1987) (Lect. Notes Math., vol. 1384, pp. 1-23) Berlin, Heidelberg, New York: Springer 1989

[19] Burkholder, D.L.: Distribution function inequalities for martingales. Annals of Prob. 1, 19-42 (1973)

[20] Burkholder, D.L.: Exit times of Brownian motion, harmonic majorization and Hardy spaces. Adv. in Math. 26, 182-205 (1977)

[21] Burkholder, D.L.: Explorations in martingale theory and its applications, Saint-Flour lectures (1989) (Lect. Notes Math., vol. 1464, pp. 1-66) Berlin, Heidelberg, New York: Springer 1991

[22] Burkholder, D.L.: Martingale transforms. Ann. Math. Stat. 37, 1494-1504 (1966)

[23] Burkholder, D.L.: On the number of escapes of a martingale and its geometrical significance. Almost Everywhere Convergence, Proceedings of the Conference on Almost Everywhere Convergence in Probability and Ergodic Theory (Columbus, Ohio, 1988). Academic Press, 1988, pp.159-177

[24] Burkholder, D.L.: One-sided maximal functions and H^p. J. Func. Anal. 18, 429-454 (1975)

[25] Burkholder, D.L.: Sharp inequalities for martingales and stochastic integrals. Asterisque 157-158, 75-94 (1988)

[26] Burkholder, D.L., Davis, B.J., Gundy, R.F.: Integral inequalities for convex functions of operators on martingales. Proc. Sixth Berkeley Symp. Math. Stat. and Prob. 223-240 (1972)

[27] Burkholder, D.L., Gundy, R.F.: Extrapolation and interpolation of quasi-linear operators on martingales. Acta Math. 124, 249-304 (1970)

[28] Cairoli, R.: Sur la convergence des martingales indexees par $\mathrm{N} \times \mathrm{N}$. Séminaire de Probabilités XIII. (Lect. Notes Math., vol. 721, pp. 162-173) Berlin, Heidelberg, New York: Springer 1979

[29] Cairoli, R.: Une inégalité pour martingale à indices multiples et res applications. Séminaire de Probabilités IV. (Lect. Notes Math., vol. 124, pp. 1-27) Berlin, Heidelberg, New York: Springer 1970

[30] Cairoli, R., Walsh, J.B.: Régions d'arrêt, localisations et prolongements de martingales. Z. Wahrscheinlichkeitstheorie Verw. Geb. 44, 279-306 (1974)

[31] Cairoli, R., Walsh, J.B.: Stochastic integrals in the plane. Acta Math. 134, 111-183 (1975)

[32] Carleson, L.: A Counterexample for Measures Bounded on H^p spaces for the bidisk. Mittag-Leffler Report 7 (1974)

[33] Carleson, L.: On convergence and growth of partial sums of Fourier series. Acta Math. 116, 135-157 (1966)

[34] Chang, S-Y.A., Fefferman, R.: A continuous version of duality of H^1 with BMO on the bidisc. Ann. of Math. 112, 179-201 (1980)

[35] Chang, S-Y.A., Fefferman, R.: The Calderon-Zygmund decomposition on product domains. Amer. J. Math. 104, 455-468 (1982)

[36] Chao, J.A.: Conjugate characterizations of H^1 dyadic martingales. Math. Ann. 240, 63-67 (1979)

[37] Chao, J.A.: Hardy spaces on regular martingales, Martingale theory in harmonic analysis and Banach spaces. (Lect. Notes Math., vol. 939, pp. 18-28) Berlin, Heidelberg, New York: Springer 1982

[38] Chao, J.A.: H^p and BMO regular martingales, Harmonic Analysis. (Lect. Notes Math., vol. 908, pp. 274-284) Berlin, Heidelberg, New York: Springer 1982

[39] Chao, J.A.: H^p spaces of conjugate systems on local fields. Studia Math. 49, 267-287 (1974)

[40] Chao, J.A.: Lusin area functions on local fields. Pacific J. Math. 59, 383-390 (1975)

[41] Chao, J.A., Janson, J.: A note on H^1 q-martingales. Pacific J. Math. 97, 307-317 (1981)

[42] Chao, J.A., Long, R.L.: Martingale transforms and Hardy spaces. Probab. Th. Rel. Fields 91, 399-404 (1992)

[43] Chao, J.A., Long, R.L.: Martingale transforms with unbounded multipliers. Proc. Amer. Math. Soc. 114, 831-838 (1992)

[44] Chao, J.A., Taibleson, M.H.: A sub-regularity inequality for conjugate systems on local fields. Studia Math. 46, 249-257 (1973)

[45] Chao, J.A., Taibleson, M.H.: Generalized conjugate systems on local fields. Studia Math. 64, 213-225 (1979)

[46] Chevalier, L.: Démonstration atomique des inégalités de Burkholder- Davis-Gundy. Ann. Scient. Univ. Clermont 67, 19-24 (1979)

[47] Chevalier, L.: L^p-inequalities for two-parameter martingales, Stochastic Integrals. (Lect. Notes Math., vol. 851, pp. 470-475) Berlin, Heidelberg, New York: Springer 1981

[48] Chevalier, L.: Martingales continues à deux paramètres. Bull. Sc. math. 106, 19-82 (1982)

[49] Chevalier, L.: Un nouveau type d'inégalités pour les martingales discretes. Z. Wahrscheinlichkeitstheorie verw. Gebiete 49, 249-255 (1979)

[50] Chung, K.L.: A course in probability theory. Probability and Math. Stat. Academic Press, New York 1968

[51] Coifman, R.R.: A real variable characterization of H^p. Studia Math. 51, 269-274 (1974)

[52] Coifman, R.R., Weiss, G.: Extensions of Hardy spaces and their use in analysis. Bull. Amer. Math. Soc. 83, 569-645 (1977)

[53] Conference on Harmonic Analysis, in Honor of Antoni Zygmund, 1981, Chicago. Ed.: Beckner, W., Calderon, P., Fefferman, R., Jones, P.W., Wadsworth International Group, Belmont, California 1983

[54] Dam, B.K.: The dual of the martingale Hardy space $\mathcal{H}_\Phi$ with general Young function Φ. Anal. Math. 14, 287-294 (1988)

[55] Davis, B.J.: On the integrability of the martingale square function. Israel J. Math. 8, 187-190 (1970)

[56] Décamp, É.: Caractérisations des espaces BMO de martingales dyadiques à deux indices, et de fonctions biharmoniques sur $\mathbf{R}^2 \times \mathbf{R}^2$. These de doctorat de troisieme cycle de mathematiques pures, Universite Scientifique et Medicale de Grenoble 1979

[57] Dellacherie, C., Meyer, P.-A.: Probabilities and potential A. North-Holland Math. Studies 29, North-Holland 1978

[58] Dellacherie, C., Meyer, P.-A.: Probabilities and potential B. North-Holland Math. Studies 72, North-Holland 1982

[59] D'jachenko, M.I.: Multiple trigonometric series with lexicographically monotone coefficients. Anal. Math. 16, 173-190 (1990)

[60] Doleans, C.: Variation quadratique des martingales continues à droite. Annals of Math. Stat. 40, 284-289 (1969)

[61] Doob, J.L.: Semimartingales and subharmonic functions. Trans. Amer. Math. Soc. 77, 86-121 (1954)

[62] Dunford, N., Schwartz, J.T.: Linear operators, Part I. Interscience Publishers, Inc., New York 1958

[63] Duren, P.: Theory of H^p spaces. Academic Press, New York 1970

[64] Durrett, R.: Brownian motion and martingales in analysis. Wadsworth Advanced Books and Software, Belmont, California 1984

[65] Edwards, R.E.: Fourier series, A modern introduction. vol. 1, Berlin, Heidelberg, New York: Springer 1982

[66] Edwards, R.E.: Fourier series, A modern introduction. vol. 2, Berlin, Heidelberg, New York: Springer 1982

[67] Fefferman, C.: Characterization of bounded mean oscillation. Bull. Amer. Math. Soc. 77, 587-588 (1971)

[68] Fefferman, C., Riviere, N.M., Sagher, Y.: Interpolation between H^p spaces: the real method. Trans. Amer. Math. Soc. 191, 75-81 (1974)

[69] Fefferman, C., Stein, E.M.: H^p spaces of several variables. Acta Math. 129, 137-194 (1972)

[70] Fefferman, R.: Bounded mean oscillation on the polydisk. Ann. of Math. 110, 395-406 (1979)

[71] Fefferman, R.: Calderon-Zygmund theory for product domains: H^p spaces. Proc. Nat. Acad. Sci. USA 83, 840-843 (1986)

[72] Fefferman, R.: Harmonic analysis on product spaces. Ann. of Math. 126, 109-130 (1987)

[73] Fine, N.J.: On the Walsh functions. Trans. Amer. Math. Soc. 65, 372-414 (1949)

[74] Fine, N.J.: The generalized Walsh functions. Trans. Amer. Math. Soc. 69, 66-77 (1950)

[75] Frangos, N.E., Imkeller, P.: Quadratic variation for a class of $L\log^+ L$-bounded two-parameter martingales. Ann. of Prob. 15, 1097-1111 (1987)

[76] Frangos, N.E., Imkeller, P.: Some inequalities for strong martingales. Ann. Inst. Henri Poincare 24, 395-402 (1988)

[77] Frangos, N.E., Imkeller, P.: The continuity of the quadratic variation of two-parameter martingales. Stochastic Processes and their Applications 29, 267-279 (1988)

[78] Fridli, S., Schipp, F.: Tree-martingales. Proc. 5th Pannonian Symp. on Math. Stat., Visegrád, Hungary (1985), 53-63

[79] Fridli, S., Simon, P.: On the Dirichlet kernels and a Hardy space with respect to the Vilenkin system. Acta Math. Hung. 45, 223-234 (1985)

[80] Garcia-Cuerva, J., Rubio de Francia, J.L.: Weighted norm inequalities and related topics. North-Holland Math. Studies 116, North-Holland 1985

[81] Garnett, J.B.: Bounded analytic functions. Academic Press, New York 1981

[82] Garsia, A.M.: Martingale inequalities, Seminar Notes on Recent Progress. Math. Lecture notes series. New York: Benjamin Inc. 1973

[83] Getoor, R.K., Sharpe, M.J.: Conformal martingales. Invent. math. 16, 271-308 (1972)

[84] Gosselin, J.: Almost everywhere convergence of Vilenkin-Fourier series. Trans. Amer. Math. Soc. 185, 345-370 (1973)

[85] Gundy, R.F.: A decomposition for L^1-bounded martingales. Ann. Math. Stat. 39, 134-138 (1968)

[86] Gundy, R.F.: Inégalités pour martingales à un et deux indices: L'espace H_p, Ecole d'Eté de Probabilités de Saint-Flour VIII-1978. (Lect. Notes Math., vol. 774, pp. 251-331) Berlin, Heidelberg, New York: Springer 1980

[87] Gundy, R.F.: Local convergence of a class of martingales in multidimensional time. Ann. of Prob. 8, 607-614 (1980)

[88] Gundy, R.F., Varopoulos, N.Th.: A martingale that occurs in harmonic analysis. Ark. Math. 14, 179-187 (1976)

[89] Gundy, R.F., Stein, E.M.: H^p theory for the poly-disc. Proc. Nat. Acad. Sci. USA 76, 1026-1029 (1979)

[90] Halmos, P.R.: Measure Theory. (Graduate Texts in Math., vol. 18) Berlin, Heidelberg, New York: Springer 1974

[91] Hanks, R.: Interpolation by the real method between BMO, L^α $(0 < \alpha < \infty)$ and H^α $(0 < \alpha < \infty)$. Indiana Univ. Math. J. 26, 679-689 (1977)

[92] Hardy, G.H.: Notes on some points in the integral calculus (LXIV), Messenger for Mathematics. 57, 12-16 (1928)

[93] Hardy, G.H., Littlewood, J.E.: Some new properties of Fourier constants. J. London Math. Soc. 6, 3-9 (1931)

[94] Herz, C.: Bounded mean oscillation and regulated martingales. Trans. Amer. Math. Soc. 193, 199-215 (1974)

[95] Herz, C.: H_p-spaces of martingales, $0 < p \leq 1$. Z. Wahrscheinlichkeitstheorie Verw. Geb. 28, 189-205 (1974)

[96] Hunt, R.A.: On the convergence of Fourier series, Orthogonal expansions and their continuous analogues (Proc. Conf. Edwardsville, Ill., 1967) S. Illinois Univ. Press Carbondale, Ill. 235-255 (1968)

[97] Imkeller, P.: A class of two-parameter stochastic integrators. Stochastics and Stochastics Reports. 27, 167-188 (1989)

[98] Imkeller, P.: A note on the localization of two-parameter processes. Probab. Th. Rel. Fields. 73, 119-125 (1986)

[99] Imkeller, P.: A stochastic calculus for continuous N-parameter strong martingales. Stochastic Processes and their Applications 20, 1-40 (1985)

[100] Imkeller, P.: On changing time for two-parameter strong martingales: a counterexample. Ann. of Prob. 14, 1080-1084 (1986)

[101] Imkeller, P.: On inequalities for two-parameter martingales. Math. Ann. 284, 329-341 (1989)

[102] Imkeller, P.: Regularity and integrator properties of variation processes of two-parameter martingales with jumps, Séminaire de Probabilités XXIII. (Lect. Notes Math., vol. 1372,) Berlin, Heidelberg, New York: Springer

[103] Imkeller, P.: Stochastic integrals of point processes and the decomposition of two-parameter martingales. J. Mult. Anal. 30, 98-123 (1989)

[104] Imkeller, P.: The transformation theorem for two-parameter pure jump martingales. Probab. Th. Rel. Fields. 89, 261-283 (1991)

[105] Imkeller, P.: Two-parameter martingales and their quadratic variation. (Lect. Notes Math., vol. 1308) Berlin, Heidelberg, New York: Springer 1988

[106] Janson, S.: Characterizations of H^1 by singular integral transforms on martingales and $\mathbf{R}^n$. Math. Scand. 41, 140-152 (1977)

[107] Janson, S.: On functions with conditions on the mean oscillation. Ark. Math. 14, 189-196 (1976)

[108] Janson, S., Jones, P.: Interpolation between H^p spaces: the complex method. J. Func. Analysis 48, 58-80 (1982)

[109] John, F., Nirenberg, L.: On functions of bounded mean oscillation. Comm. Pure Appl. Math. 14, 415-426 (1961)

[110] Kasin, B.S., Saakjan, A.A.: Orthogonal series. Izdat. Nauka, Moscow, 1984

[111] Koosis, P.: Introduction to H_p spaces. London Math. Soc. Lecture Note Series 40, Cambridge Univ. Press 1980

[112] Ladhawala, N.R.: Absolute summability of Walsh-Fourier series. Pacific J. Math. 65, 103-108 (1976)

[113] Ladhawala, N.R., Pankratz, D.C.: Almost everywhere convergence of Walsh-Fourier series of H^1 functions. Studia Math. 59, 85-92 (1976)

[114] Ledoux, M.: Classe $L \log L$ et martingales fortes à paramètre bidimensionnel. Ann. Inst. H. Poincaré Sect. B 17, 275-280 (1981)

[115] Ledoux, M.: Inégalités de Burkholder pour martingales indexes par $N \times N$, Processus Aléatoires à Deux Indices. (Lect. Notes Math. 863, pp. 122-127) Berlin, Heidelberg, New York: Springer 1981

[116] Ledoux, M.: Transformées de Burkholder et sommabilité de martingales à deux paramètres. Math. Z. 181, 529-535 (1982)

[117] Lenglart, E., Lepingle, D., Pratelli, M.: Presentation unifiée de certaines inégalités de la theory des martingales, Séminaire de Probabilités XIV. (Lect. Notes Math., vol. 784, pp. 26-48) Berlin, Heidelberg, New York: Springer 1980

[118] Lepingle, D.: Une inégalite de martingales, Séminaire de Probabilités XII.(Lect. Notes Math., vol. 649, pp. 134-137) Berlin, Heidelberg, New York: Springer 1978

[119] Lin, K.-C.: Interpolation between Hardy spaces on the bidisc. Studia Math. 84, 89-96 (1986)

[120] Lindenstrauss, J., Tzafriri, L.: Classical Banach spaces I, II, Berlin, Heidelberg, New York: Springer 1973, 1979

[121] Long, J.-L.: Sur l'espace H_p de martingales régulières $(0 < p \leq 1)$. Ann. Inst. Henri Poincare 17, 123-142 (1981)

[122] Marcinkiewicz, J., Zygmund, A.: Quelques théorèmes sur les fonctions indépendantes. Studia Math. 7, 104-120 (1938)

[123] Merzbach, E.: Stopping for two-dimensional stochastic processes. Stoch. Proc, Appl. 10, 49-63 (1980)

[124] Metivier, M.: Semimartingales, a course on stochastic processes. de Gruyter Studies in Math. 2, Walter de Gruyter 1982

[125] Metraux, C.: Quelques inégalités pour martingales à parametre bidimensional, Séminaire de Probabilités XII. (Lect. Notes Math., vol. 649, pp. 170-179) Berlin, Heidelberg, New York: Springer 1978

[126] Meyer, P.-A.: Le dual de H^1 est BMO, Séminaire de Probabilités VII. (Lect. Notes Math., vol. 321, pp. 136-145) Berlin, Heidelberg, New York: Springer 1973

[127] Meyer, P.-A.: Martingales and stochastic integrals. (Lect. Notes Math., vol. 284) Berlin, Heidelberg, New York: Springer 1972

[128] Meyer, P.-A.: Un cours les integrales stochastiques, Séminaire de Probabilités X. (Lect. Notes Math., vol. 511, pp. 245-400) Berlin, Heidelberg, New York: Springer 1976

[129] Millet, A., Sucheston, L.: Convergence of classes of amarts indexed by directed sets. Can. J. Math. 32, 86-125 (1980)

[130] Millet, A., Sucheston, L.: On regularity of multiparameter amarts and martingales. Z. Wahrscheinlichkeitstheorie verw. Gebiete 56, 21-45 (1981)

[131] Mogyoródi, J.: Maximal inequalities, convexity inequality and their duality I, II. Anal. Math. 7, 131-140, 185-197 (1981)

[132] Mogyoródi, J.: Linear functionals on Hardy spaces. Ann. Univ. Sci. Budap. Rolando Eötvös, Sect. Math. 27, 161-174 (1983)

[133] Móricz, F.: Double Walsh series with coefficients of bounded variation. Anal. Math. (to appear)

[134] Móricz, F.: On double cosine, sine and Walsh series with monotone coefficients. Proc. Amer. Math. Soc. 109, 417-425 (1990)

[135] Móricz, F.: On the maximum of the rectangular partial sums of double trigonometric series with non-negative coefficients. Anal. Math. 15, 283-290 (1989)

[136] Móricz, F.: On Walsh series with coefficients tending monotonically to zero. Acta Math. Acad. Sci. Hung. 38, 183-189 (1981)

[137] Móricz, F.: The regular convergence of multiple series. ISNM 60, Functional Analysis and Approximation. 1981 Birkhauser Verlag, Basel 203-218

[138] Móricz, F., Schipp, F.: On the integrability and L^1-convergence of double Walsh series. Acta Math. Hung. 57, 371-380 (1991)

[139] Móricz, F., Schipp, F.: On the integrability and L^1-convergence of Walsh series with coefficients of bounded variation. J. Math. Anal. Appl. 146, 99-109 (1990)

[140] Móricz, F., Schipp, F., Wade, W.R.: Cesàro summability of multiple Walsh-Fourier series. Trans. Amer. Math. Soc. 329, 131-140 (1992)

[141] Móricz, F., Schipp, F., Wade, W.R.: On the integrability of double Walsh series with special coefficients. Mich. Math. J. 37, 191-201 (1990)

[142] Neveu, J.: Discrete-parameter martingales. Nort-Holland 1971

[143] Nualart, D.: On the quadratic variation of two-parameter continuous martingales. Ann. of Prob. 12, 445-457 (1984)

[144] Nualart, D.: Variations quadratiques et inégalités pour les martingales à deux indices. Stochastics 15, 51-63 (1985)

[145] Paley, R.E.A.C.: A remarkable system of orthogonal functions. Proc. Lond. Math. Soc. 34, 241-279 (1932)

[146] Petersen, K.E.: Brownian motion, Hardy spaces and bounded mean oscillation. London Math. Soc. Lecture Note Series 28, Cambridge Univ. Press 1977

[147] Pratelli, M.: Sur certains espaces de martingales localement de carré intégrable, Séminaire de Probabilités X. (Lect. Notes Math., vol. 511, pp. 401-413) Berlin, Heidelberg, New York: Springer 1976

[148] Processus Aléatoires à Deux Indices. (Lect. Notes Math., vol. 863) Berlin, Heidelberg, New York: Springer 1981

[149] Riviere, N.M., Sagher, Y.: Interpolation between L^∞ and H^1, the real method. J. Func. Analysis 14, 401-409 (1973)

[150] Rosenthal, H.P.: On the subspaces of L^p $(p > 2)$ spanned by sequences of independent random variables. Israel J. Math. 8, 273-303 (1970)

[151] Rudin, W.: Functional Analysis, McGraw Hill Book Company, New York, N.Y., 1973

[152] Sanz, M.: r-variations for two-parameter continuous martingales and Ito's formula. Stoch. Proc. Appl. 32, 69-92 (1989)

[153] Sarason, D.: Functions of vanishing mean oscillation. Trans. Amer. Math. Soc. 207, 391-405 (1975)

[154] Schipp, F.: Fourier series and martingale transforms. Linear Spaces and Approximation. Ed. P.L.Butzer and B.Sz.-Nagy. Birkhauser Verlag Basel , 571-581 (1978)

[155] Schipp, F.: Haar and Walsh series and martingales. Coll. Math. Soc. J.Bolyai 49. Haar Memorial Conference, Budapest, 775-785 (1985)

[156] Schipp, F.: Martingale Hardy spaces. Internat. Conference on Approximation Theory (Kiev, 1983), Nauka, 510-515 (1987)

[157] Schipp, F.: Martingales with directed index set. First Pannonian Symp. on Math. Stat., Lecture Notes in Statistics 8, 254-261 Berlin, Heidelberg, New York: Springer

[158] Schipp, F.: Maximal inequalities. Proc. Internat. Conference on Approximation and Function Spaces, Gdansk (1979), PWN-Warszava, North-Holland, Amsterdam, NewYork, Oxford, 629-644.

[159] Schipp, F.: Multiple Walsh analysis. Theory and Applications of Gibbs Derivatives, Kupari-Dubrovnik, Yugoslawia 73-90, Ed.: Butzer, P.L., Stankovic, R.S., Mathematical Institut, Beograd, (1989)

[160] Schipp, F.: On a generalization of the martingale maximal theorem. Approximation Theory, Banach Center Publications 4, Warsaw, 207-212 (1979)

[161] Schipp, F.: On a Paley-type inequality. Acta Sci. Math. 45, 357-364 (1983)

[162] Schipp, F.: On Carleson's method. Coll. Math. Soc. J. Bolyai 19. Fourier Anal. and Approximation Theory, Budapest 679-695 (1976)

[163] Schipp, F.: On L_p-norm convergence of series with respect to product systems. Anal. Math. 2, 49-64 (1976)

[164] Schipp, F.: Pointwise convergence of expansions with respect to certain product systems. Anal. Math. 2, 65-76 (1976)

[165] Schipp, F.: The dual space of martingale VMO space. Proc. Third Pannonian Symp. Math. Stat., Visegrád, Hungary, 305-315 (1982)

[166] Schipp, F.: Universal contractive projections and a.e. convergence. Probability Theory and Applications, Essays to the Memory of József Mogyoródi. Eds.: J. Galambos, I. Kátai. Kluwer Academic Publishers, Dordrecht, Boston, London, 1992, 47-75

[167] Schipp, F., Wade, W.R., Simon, P., Pál, J.: Walsh series: An introduction to dyadic harmonic analysis. Adam Hilger, Bristol-New York 1990

[168] Sekiguchi, T.: Remarks on Cairoli's condition. Tohoku Math. J. 24, 457-461 (1972)

[169] Sharpley, R.: On the atomic decomposition of H^1 and interpolation. Proc. Amer. Math. Soc. 97, 186-188 (1986)

[170] Simon, P.: Investigations with respect to the Vilenkin system. Annales Univ. Sci. Budapest Eötvös, Sect. Math. 28, 87-101 (1985)

[171] Simon, P.: On the concept of a conjugate function. Colloq. Math. Soc. J. Bolyai 1. Fourier Analysis and Approximation Theory, Budapest, 747-755 (1978)

[172] Simon, P.: Verallgemeinerte Walsh-Fourierreihen I. Annales Univ. Sci. Budapest Eötvös, Sect. Math. 16, 103-113 (1973)

[173] Simon, P.: Verallgemeinerte Walsh-Fourierreihen II. Acta Math. Acad. Sci. Hungar. 27, 329-341 (1976)

[174] Sjölin, P.: An inequality of Paley and convergence a.e. of Walsh-Fourier series. Arkiv för Math. 8, 551-570 (1969)

[175] de Souza, G.S.: Spaces formed by special atoms. State Univ. of New York at Albany. PH.D. 1980

[176] de Souza, G.S.: The dyadic special atom space, Harmonic Analysis. (Lect. Notes Math., vol. 908, pp. 297-299) Berlin, Heidelberg, New York: Springer 1982

[177] Stein, E.M.: Singular Integrals and Differentiability Properties of Functions. Princeton Univ. Press, Princeton, N.J., 1970.

[178] Stein, E.M.: Topics in harmonic analysis. Princeton 1970

[179] Stein, E.M., Weiss, G.: Introduction to Fourier analysis on Euclidean spaces. Princeton, New Jersey. Princeton University Press 1971

[180] Taibleson, M.H.: Fourier Analysis on Local Fields. Princeton Univ. Press, Princeton, N.J. 1975

[181] Taibleson, M.H., Weiss, G.: Spaces generated by blocks, Probability Theory and Harmonic Analysis, ed: Chao, J.A., Woyczynski, A., (Pure and Applied Math.) 209-226 (1986)

[182] Taibleson, M.H., Weiss, G.: The molecular characterization of certain Hardy spaces. Asterisque 77, 67-150 (1980)

[183] Torchinsky, A.: Real-variable methods in harmonic analysis. Academic Press, New York 1986

[184] Uchiyama, A.: A constructive proof of the Fefferman-Stein decomposition of BMO on simple martingales. Conference on Harmonic Analysis, in Honor of Antoni Zygmund, 1981, Chicago. Ed.: Beckner, W., Calderon, P., Fefferman, R., Jones, P.W., Wadsworth International Group, Belmont, California 1983, 495-505

[185] Uchiyama, A.: The singular integral characterization of H^p on simple martingales. Proc. Amer. Math. Soc. 88, 617-621 (1983)

[186] Vilenkin, N.Ja.: On a class of complete orthonormal systems. Izv. Akad. Nauk. SSSR, Ser. Math. 11, 363-400 (1947)

[187] Walsh, J.B.: Convergence and regularity of multiparameter strong martingales. Z. Wahrscheinlichkeitstheorie Verw. Geb. 46, 177-192 (1974)

[188] Walsh, J.B.: Martingales with a multidimensional parameter and stochastic integrals in the plane, Lectures in Probability and Statistics. (Lect. Notes Math., vol. 1215, pp. 329-491) Berlin, Heidelberg, New York: Springer 1986

[189] Weisz, F.: Atomic Hardy spaces. Anal. Math. 20, (to appear)

[190] Weisz, F.: Conjugate martingale transforms. Studia Math. 103, 207-220 (1992)
[191] Weisz, F.: Convergence of singular integrals. Annales Univ. Sci. Budapest, Sect. Math. 32, 243-256 (1989)
[192] Weisz, F.: Dyadic martingale Hardy and VMO spaces on the plane. Acta Math. Hung. 56, 143-154 (1990)
[193] Weisz, F.: Inequalities relative to two-parameter Vilenkin-Fourier coefficients. Studia Math. 99, 221-233 (1991)
[194] Weisz, F.: Interpolation between martingale Hardy and BMO spaces, the real method. Bull. Sc. math. 116, 145-158 (1992)
[195] Weisz, F.: Interpolation between two-parameter martingale Hardy spaces, the real method. Bull. Sc. math. 115, 253-264 (1991)
[196] Weisz, F.: Martingale Hardy spaces, BMO and VMO spaces with nonlinearly ordered stochastic basis. Anal. Math. 16, 227-239 (1990)
[197] Weisz, F.: Martingale Hardy spaces for $0 < p \leq 1$. Probab. Th. Rel. Fields 84, 361-376 (1990)
[198] Weisz, F.: Martingale Hardy spaces with continuous time. Probability Theory and Applications, Essays to the Memory of József Mogyoródi. Eds.: J. Galambos, I. Kátai. Kluwer Academic Publishers, Dordrecht, Boston, London, 1992, 47-75
[199] Weisz, F.: On duality problems of two-parameter martingale Hardy spaces. Bull. Sc. math. 114, 395-410 (1990)
[200] Weisz, F.: On the equivalence of some rearrangements of two-parameter Haar system. Analysis Mathematica. 18, 153-166 (1992)
[201] Weisz, F.: Two-parameter martingale Hardy spaces. Coll. Math. Soc. J. Bolyai, 58 Approximation Theory, Kecskemét (Hungary), 1990, 735-748
[202] Wong, E., Zakai, M.: Weak martingales and stochastic integrals in the plane. Ann. of Prob. 4, 570-586 (1976)
[203] Young, Wo-Sang.: Mean convergence of generalized Walsh-Fourier series. Trans. Amer. Math. Soc. 218, 311-320 (1976)
[204] Yosida, K.: Functional analysis. Berlin, Heidelberg, New York: Springer 1980
[205] Zygmund, A.: Trigonometric series I. Cambridge Press, London, 1959
[206] Zygmund, A.: Trigonometric series II. Cambridge Press, London, 1959

SUBJECT INDEX

LIST OF NOTATIONS

Printing: Weihert-Druck GmbH, Darmstadt
Binding: Buchbinderei Schäffer, Grünstadt

Vol. 1478: M.-P. Malliavin (Ed.), Topics in Invariant Theory. Seminar 1989-1990. VI, 272 pages. 1991.

Vol. 1479: S. Bloch, I. Dolgachev, W. Fulton (Eds.), Algebraic Geometry. Proceedings, 1989. VII, 300 pages. 1991.

Vol. 1480: F. Dumortier, R. Roussarie, J. Sotomayor, H. Żoładek, Bifurcations of Planar Vector Fields: Nilpotent Singularities and Abelian Integrals. VIII, 226 pages. 1991.

Vol. 1481: D. Ferus, U. Pinkall, U. Simon, B. Wegner (Eds.), Global Differential Geometry and Global Analysis. Proceedings, 1991. VIII, 283 pages. 1991.

Vol. 1482: J. Chabrowski, The Dirichlet Problem with L^2-Boundary Data for Elliptic Linear Equations. VI, 173 pages. 1991.

Vol. 1483: E. Reithmeier, Periodic Solutions of Nonlinear Dynamical Systems. VI, 171 pages. 1991.

Vol. 1484: H. Delfs, Homology of Locally Semialgebraic Spaces. IX, 136 pages. 1991.

Vol. 1485: J. Azéma, P. A. Meyer, M. Yor (Eds.), Séminaire de Probabilités XXV. VIII, 440 pages. 1991.

Vol. 1486: L. Arnold, H. Crauel, J.-P. Eckmann (Eds.), Lyapunov Exponents. Proceedings, 1990. VIII, 365 pages. 1991.

Vol. 1487: E. Freitag, Singular Modular Forms and Theta Relations. VI, 172 pages. 1991.

Vol. 1488: A. Carboni, M. C. Pedicchio, G. Rosolini (Eds.), Category Theory. Proceedings, 1990. VII, 494 pages. 1991.

Vol. 1489: A. Mielke, Hamiltonian and Lagrangian Flows on Center Manifolds. X, 140 pages. 1991.

Vol. 1490: K. Metsch, Linear Spaces with Few Lines. XIII, 196 pages. 1991.

Vol. 1491: E. Lluis-Puebla, J.-L. Loday, H. Gillet, C. Soulé, V. Snaith, Higher Algebraic K-Theory: an overview. IX, 164 pages. 1992.

Vol. 1492: K. R. Wicks, Fractals and Hyperspaces. VIII, 168 pages. 1991.

Vol. 1493: E. Benoît (Ed.), Dynamic Bifurcations. Proceedings, Luminy 1990. VII, 219 pages. 1991.

Vol. 1494: M.-T. Cheng, X.-W. Zhou, D.-G. Deng (Eds.), Harmonic Analysis. Proceedings, 1988. IX, 226 pages. 1991.

Vol. 1495: J. M. Bony, G. Grubb, L. Hörmander, H. Komatsu, J. Sjöstrand, Microlocal Analysis and Applications. Montecatini Terme, 1989. Editors: L. Cattabriga, L. Rodino. VII, 349 pages. 1991.

Vol. 1496: C. Foias, B. Francis, J. W. Helton, H. Kwakernaak, J. B. Pearson, H_∞-Control Theory. Como, 1990. Editors: E. Mosca, L. Pandolfi. VII, 336 pages. 1991.

Vol. 1497: G. T. Herman, A. K. Louis, F. Natterer (Eds.), Mathematical Methods in Tomography. Proceedings 1990. X, 268 pages. 1991.

Vol. 1498: R. Lang, Spectral Theory of Random Schrödinger Operators. X, 125 pages. 1991.

Vol. 1499: K. Taira, Boundary Value Problems and Markov Processes. IX, 132 pages. 1991.

Vol. 1500: J.-P. Serre, Lie Algebras and Lie Groups. VII, 168 pages. 1992.

Vol. 1501: A. De Masi, E. Presutti, Mathematical Methods for Hydrodynamic Limits. IX, 196 pages. 1991.

Vol. 1502: C. Simpson, Asymptotic Behavior of Monodromy. V, 139 pages. 1991.

Vol. 1503: S. Shokranian, The Selberg-Arthur Trace Formula (Lectures by J. Arthur). VII, 97 pages. 1991.

Vol. 1504: J. Cheeger, M. Gromov, C. Okonek, P. Pansu, Geometric Topology: Recent Developments. Editors: P. de Bartolomeis, F. Tricerri. VII, 197 pages. 1991.

Vol. 1505: K. Kajitani, T. Nishitani, The Hyperbolic Cauchy Problem. VII, 168 pages. 1991.

Vol. 1506: A. Buium, Differential Algebraic Groups of Finite Dimension. XV, 145 pages. 1992.

Vol. 1507: K. Hulek, T. Peternell, M. Schneider, F.-O. Schreyer (Eds.), Complex Algebraic Varieties. Proceedings, 1990. VII, 179 pages. 1992.

Vol. 1508: M. Vuorinen (Ed.), Quasiconformal Space Mappings. A Collection of Surveys 1960-1990. IX, 148 pages. 1992.

Vol. 1509: J. Aguadé, M. Castellet, F. R. Cohen (Eds.), Algebraic Topology - Homotopy and Group Cohomology. Proceedings, 1990. X, 330 pages. 1992.

Vol. 1510: P. P. Kulish (Ed.), Quantum Groups. Proceedings, 1990. XII, 398 pages. 1992.

Vol. 1511: B. S. Yadav, D. Singh (Eds.), Functional Analysis and Operator Theory. Proceedings, 1990. VIII, 223 pages. 1992.

Vol. 1512: L. M. Adleman, M.-D. A. Huang, Primality Testing and Abelian Varieties Over Finite Fields. VII, 142 pages. 1992.

Vol. 1513: L. S. Block, W. A. Coppel, Dynamics in One Dimension. VIII, 249 pages. 1992.

Vol. 1514: U. Krengel, K. Richter, V. Warstat (Eds.), Ergodic Theory and Related Topics III, Proceedings, 1990. VIII, 236 pages. 1992.

Vol. 1515: E. Ballico, F. Catanese, C. Ciliberto (Eds.), Classification of Irregular Varieties. Proceedings, 1990. VII, 149 pages. 1992.

Vol. 1516: R. A. Lorentz, Multivariate Birkhoff Interpolation. IX, 192 pages. 1992.

Vol. 1517: K. Keimel, W. Roth, Ordered Cones and Approximation. VI, 134 pages. 1992.

Vol. 1518: H. Stichtenoth, M. A. Tsfasman (Eds.), Coding Theory and Algebraic Geometry. Proceedings, 1991. VIII, 223 pages. 1992.

Vol. 1519: M. W. Short, The Primitive Soluble Permutation Groups of Degree less than 256. IX, 145 pages. 1992.

Vol. 1520: Yu. G. Borisovich, Yu. E. Gliklikh (Eds.), Global Analysis – Studies and Applications V. VII, 284 pages. 1992.

Vol. 1521: S. Busenberg, B. Forte, H. K. Kuiken, Mathematical Modelling of Industrial Process. Bari, 1990. Editors: V. Capasso, A. Fasano. VII, 162 pages. 1992.

Vol. 1522: J.-M. Delort, F. B. I. Transformation. VII, 101 pages. 1992.

Vol. 1523: W. Xue, Rings with Morita Duality. X, 168 pages. 1992.

Vol. 1524: M. Coste, L. Mahé, M.-F. Roy (Eds.), Real Algebraic Geometry. Proceedings, 1991. VIII, 418 pages. 1992.

Vol. 1525: C. Casacuberta, M. Castellet (Eds.), Mathematical Research Today and Tomorrow. VII, 112 pages. 1992.

Vol. 1526: J. Azéma, P. A. Meyer, M. Yor (Eds.), Séminaire de Probabilités XXVI. X, 633 pages. 1992.

Vol. 1527: M. I. Freidlin, J.-F. Le Gall, Ecole d'Eté de Probabilités de Saint-Flour XX – 1990. Editor: P. L. Hennequin. VIII, 244 pages. 1992.

Vol. 1528: G. Isac, Complementarity Problems. VI, 297 pages. 1992.

Vol. 1529: J. van Neerven, The Adjoint of a Semigroup of Linear Operators. X, 195 pages. 1992.

Vol. 1530: J. G. Heywood, K. Masuda, R. Rautmann, S. A. Solonnikov (Eds.), The Navier-Stokes Equations II – Theory and Numerical Methods. IX, 322 pages. 1992.

Vol. 1531: M. Stoer, Design of Survivable Networks. IV, 206 pages. 1992.

Vol. 1532: J. F. Colombeau, Multiplication of Distributions. X, 184 pages. 1992.

Vol. 1533: P. Jipsen, H. Rose, Varieties of Lattices. X, 162 pages. 1992.

Vol. 1534: C. Greither, Cyclic Galois Extensions of Commutative Rings. X, 145 pages. 1992.

Vol. 1535: A. B. Evans, Orthomorphism Graphs of Groups. VIII, 114 pages. 1992.

Vol. 1536: M. K. Kwong, A. Zettl, Norm Inequalities for Derivatives and Differences. VII, 150 pages. 1992.

Vol. 1537: P. Fitzpatrick, M. Martelli, J. Mawhin, R. Nussbaum, Topological Methods for Ordinary Differential Equations. Montecatini Terme, 1991. Editors: M. Furi, P. Zecca. VII, 218 pages. 1993.

Vol. 1538: P.-A. Meyer, Quantum Probability for Probabilists. X, 287 pages. 1993.

Vol. 1539: M. Coornaert, A. Papadopoulos, Symbolic Dynamics and Hyperbolic Groups. VIII, 138 pages. 1993.

Vol. 1540: H. Komatsu (Ed.), Functional Analysis and Related Topics, 1991. Proceedings. XXI, 413 pages. 1993.

Vol. 1541: D. A. Dawson, B. Maisonneuve, J. Spencer, Ecole d' Eté de Probabilités de Saint-Flour XXI - 1991. Editor: P. L. Hennequin. VIII, 356 pages. 1993.

Vol. 1542: J.Fröhlich, Th.Kerler, Quantum Groups, Quantum Categories and Quantum Field Theory. VII, 431 pages. 1993.

Vol. 1543: A. L. Dontchev, T. Zolezzi, Well-Posed Optimization Problems. XII, 421 pages. 1993.

Vol. 1544: M.Schürmann, White Noise on Bialgebras. VII, 146 pages. 1993.

Vol. 1545: J. Morgan, K. O'Grady, Differential Topology of Complex Surfaces. VIII, 224 pages. 1993.

Vol. 1546: V. V. Kalashnikov, V. M. Zolotarev (Eds.), Stability Problems for Stochastic Models. Proceedings, 1991. VIII, 229 pages. 1993.

Vol. 1547: P. Harmand, D. Werner, W. Werner, M-ideals in Banach Spaces and Banach Algebras. VIII, 387 pages. 1993.

Vol. 1548: T. Urabe, Dynkin Graphs and Quadrilateral Singularities. VI, 233 pages. 1993.

Vol. 1549: G. Vainikko, Multidimensional Weakly Singular Integral Equations. XI, 159 pages. 1993.

Vol. 1550: A. A. Gonchar, E. B. Saff (Eds.), Methods of Approximation Theory in Complex Analysis and Mathematical Physics IV, 222 pages, 1993.

Vol. 1551: L. Arkeryd, P. L. Lions, P.A. Markowich, S.R. S. Varadhan. Nonequilibrium Problems in Many-Particle Systems. Montecatini, 1992. Editors: C. Cercignani, M. Pulvirenti. VII, 158 pages 1993.

Vol. 1552: J. Hilgert, K.-H. Neeb, Lie Semigroups and their Applications. XII, 315 pages. 1993.

Vol. 1553: J.-L- Colliot-Thélène, J. Kato, P. Vojta. Arithmetic Algebraic Geometry. Trento, 1991. Editor: E. Ballico. VII, 223 pages. 1993.

Vol. 1554: A. K. Lenstra, H. W. Lenstra, Jr. (Eds.), The Development of the Number Field Sieve. VIII, 131 pages. 1993.

Vol. 1555: O. Liess, Conical Refraction and Higher Microlocalization. X, 389 pages. 1993.

Vol. 1556: S. B. Kuksin, Nearly Integrable Infinite-Dimensional Hamiltonian Systems. XXVII, 101 pages. 1993.

Vol. 1557: J. Azéma, P. A. Meyer, M. Yor (Eds.), Séminaire de Probabilités XXVII. VI, 327 pages. 1993.

Vol. 1558: T. J. Bridges, J. E. Furter, Singularity Theory and Equivariant Symplectic Maps. VI, 226 pages. 1993.

Vol. 1559: V. G. Sprindžuk, Classical Diophantine Equations. XII, 228 pages. 1993.

Vol. 1560: T. Bartsch, Topological Methods for Variational Problems with Symmetries. X, 152 pages. 1993.

Vol. 1561: I. S. Molchanov, Limit Theorems for Unions of Random Closed Sets. X, 157 pages. 1993.

Vol. 1562: G. Harder, Eisensteinkohomologie und die Konstruktion gemischter Motive. XX, 184 pages. 1993.

Vol. 1563: E. Fabes, M. Fukushima, L. Gross, C. Kenig, M. Röckner, D. W. Stroock, Dirichlet Forms. Varenna, 1992. Editors: G. Dell'Antonio, U. Mosco. VII, 245 pages. 1993.

Vol. 1564: J. Jorgenson, S. Lang, Basic Analysis of Regularized Series and Products. IX, 122 pages. 1993.

Vol. 1565: L. Boutet de Monvel, C. De Concini, C. Procesi, P. Schapira, M. Vergne. D-modules, Representation Theory, and Quantum Groups. Venezia, 1992. Editors: G. Zampieri, A. D'Agnolo. VII, 217 pages. 1993.

Vol. 1566: B. Edixhoven, J.-H. Evertse (Eds.), Diophantine Approximation and Abelian Varieties. XIII, 127 pages. 1993.

Vol. 1567: R. L. Dobrushin, S. Kusuoka, Statistical Mechanics and Fractals. VII, 98 pages. 1993.

Vol. 1568: F. Weisz, Martingale Hardy Spaces and their Application in Fourier Analysis. VIII, 217 pages. 1994.